Probabilistic Mechanics of Quasibrittle Structures

Strength, Lifetime, and Size Effect

ZDENĚK P. BAŽANT
Northwestern University

JIA-LIANG LE
University of Minnesota

CAMBRIDGE
UNIVERSITY PRESS

CAMBRIDGE
UNIVERSITY PRESS

University Printing House, Cambridge CB2 8BS, United Kingdom

One Liberty Plaza, 20th Floor, New York, NY 10006, USA

477 Williamstown Road, Port Melbourne, VIC 3207, Australia

4843/24, 2nd Floor, Ansari Road, Daryaganj, Delhi - 110002, India

79 Anson Road, #06-04/06, Singapore 079906

Cambridge University Press is part of the University of Cambridge.

It furthers the University's mission by disseminating knowledge in the pursuit of education, learning, and research at the highest international levels of excellence.

www.cambridge.org
Information on this title: www.cambridge.org/9781107151703

First published 2017

A catalogue record for this publication is available from the British Library.

Library of Congress Cataloging-in-Publication Data
Names: Bažant, Z. P. | Le, Jia-Liang, 1980–
Title: Probabilistic mechanics of quasibrittle structures : strength, lifetime, and size effect / Zdeněk P. Bažant, Northwestern University, Jia-Liang Le, University of Minnesota.
Other titles: Probabilistic mechanics of quasi brittle structures
Description: Cambridge : Cambridge University Press, [2017] | Includes bibliographical references and author index.
Identifiers: LCCN 2016041636 | ISBN 9781107151703 (hardback)
Subjects: LCSH: Fracture mechanics. | Brittleness. | Elastic analysis (Engineering) | Structural analysis (Engineering)
Classification: LCC TA409 .B39 2017 | DDC 624.1/76–dc23
LC record available at https://lccn.loc.gov/2016041636

ISBN 978-1-107-15170-3 Hardback

Dedicated to the memory of Alfred Martin Freudenthal (1906–1977), widely regarded as the founder of the field of structural safety and reliability, encompassing both the theories of probability and statistics and the theories of mechanics of materials and structures.

Contents

Foreword

Failure of materials was recognized centuries ago as a critical component of structural mechanics and consequently of structural design. Efficient (minimum amount of materials) and economical (minimum cost) designs are based on the basic principle of taking full advantage of the strength of the materials used, while at the same time carefully avoiding any type of material failure.

Whether or not they were able to quantify it, adequately, engineers realized very soon that material failure involved a high level of uncertainty. Later on, based on advances in the mathematical theory of probability, researchers in the field of mechanics identified two basic types of material behavior as far as failure is concerned – ductile and brittle failures – and were able to establish rigorous probabilistic models for both. The Gaussian and Weibull probability distributions have been the standard models for these two types of material failure, respectively.

However, there has always been a transition area between ductile and brittle failures with a behavior that was orders of magnitude more complex and challenging to model: it has been named "quasibrittle behavior." The two standard models for ductile and brittle failure were clearly not adequate in this transition area, and on top of that, there is a very wide range of materials falling in this category at the scale of laboratory testing and normal structures: concrete, various composites, toughened ceramics, many rocks, coal, ice, rigid foams, biological shells, bone, cartilage, dental ceramics, and many others. Furthermore, at the nano- and micrometer scales, virtually all materials become quasibrittle.

The first author of this book – one of the giants in the field of mechanics – has been instrumental over the years in developing a rigorous theoretical framework modeling the failures of structures made of quasibrittle materials. In a long series of seminal scientific papers, he has identified the challenges involved and has introduced a number of groundbreaking theories and models to address them. He and his coauthor, who, despite being much younger, has already impacted the field by his own seminal contributions, have provided in this volume the definitive treatment of this formidably challenging field, and in the process have established the complete theory of any type of material failure, ranging between the two limiting cases of ductile and brittle behaviors.

In particular, I would like to highlight the equal emphasis and importance given by the authors to the two disciplines of mechanics and probability, and their ingenious and highly successful blending of the two in a fully integrated theoretical framework. The volume is a true pleasure to read and will become immediately an indispensable tool for every scientist, scholar, and engineer interested in this critically important field.

George Deodatis
Santiago and Robertina Calatrava Family Endowed Chair and Chair,
Department of Civil Engineering and Engineering Mechanics, Columbia University
President, International Association for Structural Safety and Reliability, 2009–2013

Preface

Although some would vehemently deny it, many specialists would agree that, since the 1977 death of Freudenthal,[1] the research field of structural safety and reliability has been in a schism.

Alfred Freudenthal, the founder of this field of research in the 1960s, perceived the fields of (1) structural safety and (2) mechanics and physics of materials and structures as inseparable. He mastered both, and treated both to the depth of knowledge in his time. Since that time, unfortunately, most researchers have immersed themselves in one of these two fields in great detail and with high sophistication, while treating the other aspect simplistically and superficially. The connection has been weak.

On one side, there have been probabilists who develop and successfully market complex computer programs to assess safety, reliability, and lifetime of concrete structures without recognizing that failure probability of concrete structures cannot be predicted with simplistic or obsolete material models that eschew fracture mechanics and energetic size effect. Or there have been statistically minded experimenters who conduct extensive histogram testing of the strength of ceramics but ignore the scale effects, micromechanics, and microscale physics of failure.

On the other side, there have been mechanicians who construct highly refined constitutive and computational models for the mechanics of failure of concrete, geomaterials,

[1] Born in Poland in 1906, Alfred Martin Freudenthal received his engineering and doctoral degrees in Prague, in 1929 and 1930, respectively. His dissertation dealt with the theory of plasticity. He worked in Prague as a structural engineer in a well-known engineering design firm, collaborated with Prof. J. Melan, a leading bridge designer at that time, and simultaneously, in 1934, collected another engineering degree in Lwow. From 1936 to 1946 he was one the chief engineers of the new port of Tel Aviv and, after 1938, also a professor of bridge engineering at the Hebrew University of Technology in Haifa. In 1948 he arrived in the United States as a visiting professor at the University of Illinois, and between 1949 and 1969 he was a professor of civil engineering at Columbia University, New York, and then, until his death in 1977, a professor at George Washington University, Washington, DC.

A selection of his papers, published in 1981 by the American Society of Civil Engineers, shows a remarkably evenhanded attention to the mechanical behavior of materials and structures on one side, and the structural safety, reliability and fatigue on the other side, with both aspects intertwined as if they were one. His seminal works deal with viscoelasticity and nonlinear creep, plastic shells, orthotropic sandwich plates and shells, shrinkage stresses, concrete creep, consolidating media, strength of airframes, shear dilatancy in rock, seismic waves, work-hardening law for metals, relaxation spectra, second-order strain effects in metals, physical and statistical aspects of metal fatigue and residual stresses, fundamental theory of structural safety, safety of prestressed concrete, cumulative damage, lifetime estimation, random failure of structures with multiple load paths, reliability of reactor components, extreme value risk analysis, reliability of aircraft and of offshore platforms in seismic regions, structural optimization, and risk control.

and composites without recognizing that far greater prediction errors stem from simplistic or nonexistent treatment of the randomness of the material as well as the loads.

The present book attempts a step to rectify this schism. In a unified theoretical framework, it deals with the quasibrittle structures, which are those consisting of quasibrittle (or brittle heterogeneous) materials. These are commonplace materials, used more and more widely and increasingly important for modern technology, including much of high-tech. They encompass concretes (as the archetypical case), rocks, fiber composites, tough ceramics, sea ice, bone, wood, stiff soils, rigid foams, and so forth, as well as all brittle materials on the micrometer scale. They are characterized by a fracture process zone that is not negligible compared to the typical structural dimensions. This feature causes an intricate energetic size effect, which is intertwined with the classical statistical size effect, the only kind of size effect known in classical fracture mechanics of brittle materials. Compared to metals and ceramics, a probabilistic theory of strength, lifetime, and size effect of quasibrittle structures has been developed much more recently. Its comprehensive presentation, based mostly on previous studies at Northwestern University and the University of Minnesota, is the objective of this book.

Although the main purpose of this book is a comprehensive mathematical exposition of the subject, the book is also suitable as a text for an advanced course, as all the results are mathematically derived and the focus is on understanding rather than just description. Moreover, parts of the book can be covered in graduate courses dealing with the modeling of failure of various materials and structures, as featured in the curricula in civil, mechanical, aerospace, nuclear, offshore, geotechnical, and ocean engineering, as well as materials science and geophysics. Chapters 1–3, Sections 6.1, 6.4, 6.6, 12.1, 12.2, 13.1, 13.3, and Appendices A–C can also be covered in a broader graduate course on Quasibrittle (or Cohesive) Fracture and Scaling, which the first author has been teaching at Northwestern since the 1980s, and which the second author later introduced at the University of Minnesota.

Working on the journal articles that underlie our book, we benefited from outstanding collaborators, particularly Sze-Dai Pang, Marco Salviato, Miroslav Vořechovský, Jan Eliáš, Drahoš Novák, Augusto Cannoe Falchetto, Mihai Marasteanu, Joseph Labuz, Roberto Ballarini, Johnathan Manning, Bing Xue, and Mathieu Pieuchot. They deserve our deep thanks. The results presented here could not have been achieved without generous funding from the US National Science Foundation, the US Department of Energy, the Army Research Office, the US Department of Transportation, the Boeing Co., the Minnesota Department of Transportation, and the Center for Transportation Studies at the University of Minnesota, for all of which we are very grateful.

We also want to express our deep appreciation of the stimulating research environments provided by Northwestern University and the University of Minnesota. Last but not least, we wish to express our wholehearted thanks to our wives, Iva M. Bažant and Miao Pan, for their loving support of our research endeavor.

Zdeněk P. Bažant and Jia-Liang Le
Evanston and Minneapolis, July 18, 2016

1 Introduction

Without realistic failure mechanics, probabilistic analysis of structural safety is a fiction.

1.1 The Problem of Tail of Probability Distribution

Like most things in life, we must accept that the occurrence probability of any future event cannot be exactly zero. We must be contented with a structural failure probability that is negligible compared to other risks that people willingly take, such as car accidents. It is generally agreed that adequate safety of engineering structures is achieved by specifying a failure probability of 10^{-6} per lifetime as the maximum admissible in design [Nordic Committee for Building Structures (NKB) 1978; Melchers 1987; Duckett 2005; Ellingwood 2006]. This probability limit is generally accepted for engineering structures, whether bridges or aircraft (Duckett 2005; Department of National Defense of Canada 2007), although for some nuclear plant structures an even smaller limit is required.

The smallness of this probability limit is a source of great difficulty. To check the design merely by an experimental histogram, at least 10^8 tests of identical structures or specimens would be required. Even a direct computational verification would necessitate about 10^8 repetitions of Monte Carlo simulations with a fully realistic material model. Therefore, estimations of loads of such a small failure probability must rely on a model that is justified by a sound theory and is validated by experiments other than histogram testing.

For many years, realistic theoretical models have been available for the probability of ductile and brittle failures. The Gaussian and Weibull distributions, respectively, fit this purpose. Failures of structures made of quasibrittle materials are more difficult to predict and have been researched only recently. The difficulty is that the quasibrittle failures are transitional in nature between ductile and plastic failures.

Quasibrittle materials are heterogeneous materials with brittle constituents. At the scale of normal laboratory testing, they include concretes as the archetypical case, fiber-polymer composites, fiber-reinforced concretes, toughened ceramics, many rocks, coal, sea ice, wood, consolidated snow, particle board, rigid foams, particulate nanocomposites, biological shells, mortar, masonry, fiber-reinforced concrete, asphalt concrete (at low temperatures), stiff clays, silts, cemented sands, grouted soils, particle board, various refractories, bone, cartilage, dentine, dental ceramics, paper, carton, and cast iron.

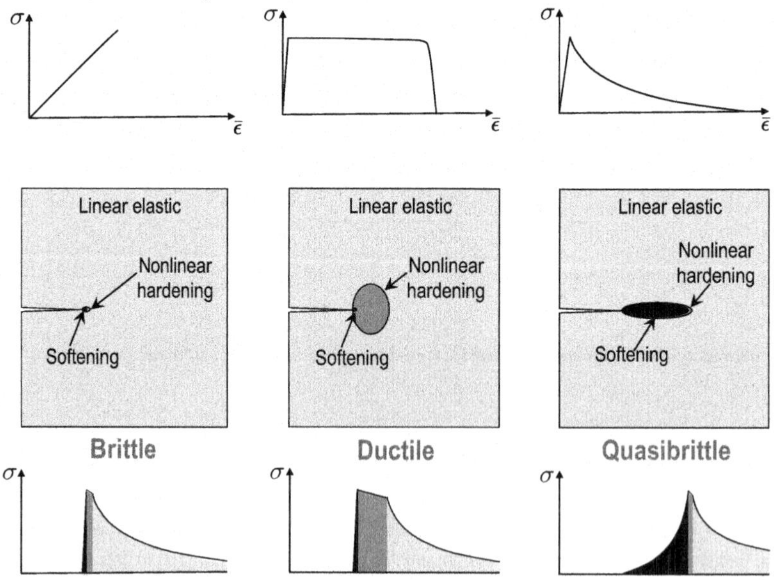

Figure 1.1. Types of fracture process. The diagrams at the top show the relationships between the cohesive stress and the averaged nominal strain across the crack extension line. The diagrams at the bottom show the stress profiles along the crack ligament.

On the nano- and micrometer scales, virtually all materials become quasibrittle, including silicon or thin metallic films.

In fracture, the degree of brittleness (or ductility, which is the opposite of brittleness) is manifested in the size of the fracture process zone (FPZ) formed at the tip of a propagating crack. In this regard, three cases, illustrated in Figure 1.1, may be distinguished:

1. In brittle failures, the FPZ is so small that it can be treated as a point (Fig. 1.1, left). Consequently, all the volume of the structure behaves elastically and the classical theory of linear elastic fracture mechanics (LEFM) is applicable.
2. In ductile (or plastic) failures, which are observed in elastoplastic materials (mainly metals) and are characterized by a stress-strain diagram with a horizontal yield plateau, there is a large nonlinear plastic (or yielding) zone. But the FPZ is still very small, micrometer size, which is almost pointwise for most applications (see Fig. 1.1, middle). Unlike the brittle case, the profile of the so-called cohesive stress transmitted in a quasibrittle material across the crack extension line is almost horizontal, with a steep stress drop at the crack tip.
3. The quintessential feature of quasibrittle failure is that the FPZ at the front of a crack is not negligible compared to the cross-sectional dimensions, and can sometimes occupy even the entire cross section of the structure. The profile of stress along the crack extension line has neither a long horizontal segment nor steep stress drop. Rather, it varies along the FPZ gradually, except for superposed statistical scatter due to heterogeneity (Fig. 1.1, right). The diagram of normal stress σ acting across the crack extension line versus the average normal strain $\bar{\epsilon}$ across the FPZ exhibits a gradual post-peak decline.

Figure 1.2. Portrait of Leonardo da Vinci (1452–1519). *Source:* Wikipedia

The FPZ length can vary enormously; it is typically about 50 cm in normal concretes, 5 cm in high-strength concretes, 10 μm to 1 mm in fine-grained ceramics, 10 nm in a silicon wafer, and 1 to 10 m in an Arctic sea ice floe. If the FPZ is negligible compared to the structure size, a quasibrittle structure becomes perfectly brittle, that is, follows LEFM. Thus concrete is quasibrittle on the scale of normal beams and columns, but perfectly brittle on the scale of a large dam. Arctic sea ice floe, fine-grained ceramic, and nanocomposites are quasibrittle on the scales of 10 m, 0.1 mm, and 0.1 μm, but brittle on the scales of 1 km, 1 cm, and 10 μm, respectively.

It is clear that the ratio between the overall structural size and the FPZ size determines the failure behavior of quasibrittle structures, transitioning from ductile to brittle as the ratio increases. Therefore, the main problem is the scaling of the failure behavior of quasibrittle structures, which has attracted significant research efforts over the last three decades. While previous research focused mainly on the scaling of the mean failure behavior, recent research has been directed toward the probabilistic aspect of this problem, in particular how such a size-dependent transitional failure behavior influences the strength and lifetime statistics of quasibrittle structures, and its consequences for the reliability-based structural design. This book presents a recently developed theory that addresses this topic, which is crucial for reliability-based analysis and design of quasibrittle structures.

1.2 History in Brief

1.2.1 Classical History

Let us start with a brief sketch of the history of the topic of size effect and scaling. Leonardo da Vinci was the pioneer in investigation of scaling law of material strength (Fig. 1.2). He speculated that "Among cords of equal thickness the longest is the least strong" (da Vinci 1500s). He also wrote that a cord "is so much stronger . . . as it is shorter". This scaling rule implied the nominal strength of a cord to be inversely proportional to its length, which is of course a strong exaggeration of the actual size effect.

Figure 1.3. Portrait of G. Galilei (1564–1642) and the title page of *Discorsi e Dimostrazioni Matematiche Intorno a Due Nuove Scienze. Source:* Wikipedia

More than a century later, Leonardo's scaling rule was rejected by Galileo Galilei (1638) in his famous book *Two New Sciences* (Fig. 1.3). Galileo argued that cutting a long cord at various points should not make the remaining part stronger (which is now known not to be true). Nevertheless, he proposed the famous "square-cube" scaling law to describe the effect of the object size on the ratio between the volume and surface area of the object. Based on this law, he explained the fact that large animals have relatively bulkier bones than small ones, which he called the "weakness of giants."

About fifty years later, Edmé Mariotte (1686) made a major advance by reexamining da Vinci's scaling rule (Fig. 1.4). He experimented with ropes, paper, and tin, and observed that "a long rope and a short one always support the same weight unless that in a long rope there may happen to be some faulty place in which it will break sooner than in a shorter. The same thing will happen in small slips of tin; for in a long one there may be perhaps some defect that may not be in a short one; and if you should take that part of it which did not break, it would sustain a greater weight. . . . " Mariotte later expressed this observation more generally in terms of the variability of the material strength, which is essentially the principle of "the Inequality of the Matter whose absolute Resistance is less in one Place than another." This statement can be considered as the first qualitative description of the statistical theory of size effect. However, at that time the probability theory was at its inception and was not yet ready to handle the problem within a formal mathematical framework. As discussed later, the mathematical formulation of statistical size effect for brittle solids was completed by Weibull (1939) almost three centuries later.

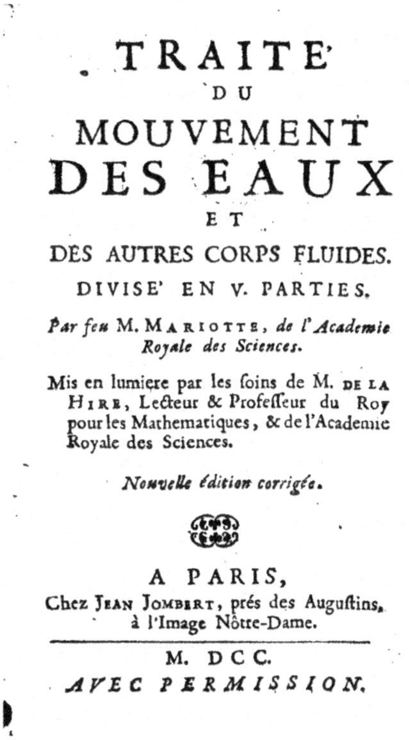

TRAITE'
DU
MOUVEMENT
DES EAUX
ET
DES AUTRES CORPS FLUIDES.
DIVISE' EN V. PARTIES.
Par feu M. MARIOTTE, *de l'Academie Royale des Sciences.*
Mis en lumiere par les ſoins de M. DE LA HIRE, Lecteur & Profeſſeur du Roy pour les Mathematiques, & de l'Academie Royale des Sciences.
Nouvelle édition corrigée.
A PARIS,
Chez JEAN JOMBERT, prés des Auguſtins, à l'Image Nôtre-Dame.
M. DCC.
AVEC PERMISSION.

Figure 1.4. Edmé Mariotte (1620–1684) and the title page of *Traité du mouvement des eaux*, posthumously edited by M. de la Hire. ("Principals découvertes de l'église" means "Principal discoveries of the chruch".) *Source:* Grzybowski, A, and Aydin, P. (2007) Edmé Mariotte (1620–1684) "Pioneer of Neurophysiology" Survey of Ophthalmology, 52(4), 443–451

Mariotte's conclusions were later rejected by Thomas Young (1807), who took a deterministic approach and stated that the length of a solid has no effect on its strength. This was a step backwards from Mariotte's idea on the statistical size effect. Nevertheless, Young did make a remark that the cohesive strength of a wire or bar is not always proportional to its diameter owing to some "accidental irregularities."

The second major advance was the seminal work of Griffith (1921), which laid the foundation of the theory of fracture mechanics. Meanwhile, in his paper he also pioneered the use of fracture mechanics to study the size effect. Griffith concluded that "the weakness of isotropic solids . . . is due to the presence of discontinuities or flaws . . . The effective strength of technical materials could be increased 10 or 20 times at least if these flaws could be eliminated" (Griffith 1921, page 179). He demonstrated this conclusion by an experimental investigation on the strength of glass fibers showing that the breaking stress was raised from 42,300 psi for a diameter of 0.0042 in. to 491,000 psi for a diameter of 0.00013 in. It is clear that Griffith's analysis of microscopic flaws or cracks provided a physical basis for Mariotte's statistical concept of size effect.

The mathematical description of the statistical size effect was made possible by the significant advances in probabilistic theories as well as extensive experimental investigations in the early twentieth century. Tippett (1925), Fréchet (1927), Fisher and Tippett (1928), and Peirce (1926) pioneered the mathematical formulation of extreme value statistics, which was later refined by von Mises (1936), Gnedenko (1943), and others (see also Gumbel 1958 and Freudenthal 1968). Weibull (1939) independently discovered a probability distribution function of structural strength using the weakest-link model (Weibull 1939), which is one of the three extreme value distribution functions proposed by Fisher, Tippett, and Fréchet.

The essential point of Weibull's analysis is that the structure can be statistically modeled by a chain of a large number of elements, and the tail distribution of strength of each element follows a power law. The resulting probability distribution of structural strength is now known as the Weibull distribution. The Weibull distribution of structural strength was later supported theoretically by probabilistic modeling of the distribution of microscopic flaws (see, e.g., Freudenthal 1968, 1981). It is noted that the Weibull distribution has also been widely applied to various other physical phenomena (Rinne 2009).

Another important aspect of Weibull's work is that he also derived a size effect equation on the mean structural strength based on the Weibull distribution. This is the first mathematical description of statistical size effect. Most subsequent studies until the 1980s dealt basically with refinements, justifications, and applications of Weibull's theory (Zaitsev & Wittmann 1974; Mihashi & Izumi 1977; Zech & Wittmann 1977; Mihashi & Zaitsev 1981; Mihashi 1983; Carpinteri 1986; Kittl & Diaz 1988; Carpinteri 1989; Kittl & Diaz 1989, 1990; Danzer, Supancic, Pascual, & Lube 2007; Danzer, Lube, Supancic, & Damani 2008). The essential feature of the Weibull statistical size effect is that it follows a power-law form, which implies the absence of a characteristic length (see detailed discussion in Section 1.5).

For a long time, it was generally assumed that, if a size effect was observed, it had to be of the Weibull type. Today we know this is not the case because the Weibull statistical size effect is limited to structures that (1) fail (or must be considered as failing) right at the initiation of the macroscopic fracture and (2) have at failure only a small FPZ compared to the structure size. Condition (1) is necessary for allowing the structure to be statistically modeled as a chain of elements, and condition (2) essentially implies that the number of elements in this chain is so large that it can be considered infinite. This is certainly the case for various fine-grain ceramics and for metal structures embrittled by fatigue. But this is not the case for *quasibrittle* materials, the main subject of this book.

1.2.2 Recent Developments

The most widely used quasibrittle material is concrete. The study of its fracture mechanics, initiated by Kaplan (1961), led to the discovery of a new type of size effect in quasibrittle fracture, which is fundamentally different from the statistical size effect.

Long ago it has been concluded that the classical LEFM is not applicable to concrete (Leicester 1969; Kesler, Naus, & Lott 1972; Walsh 1972). Leicester (1969) tested geometrically similar notched beams of different sizes and used a power law to fit the

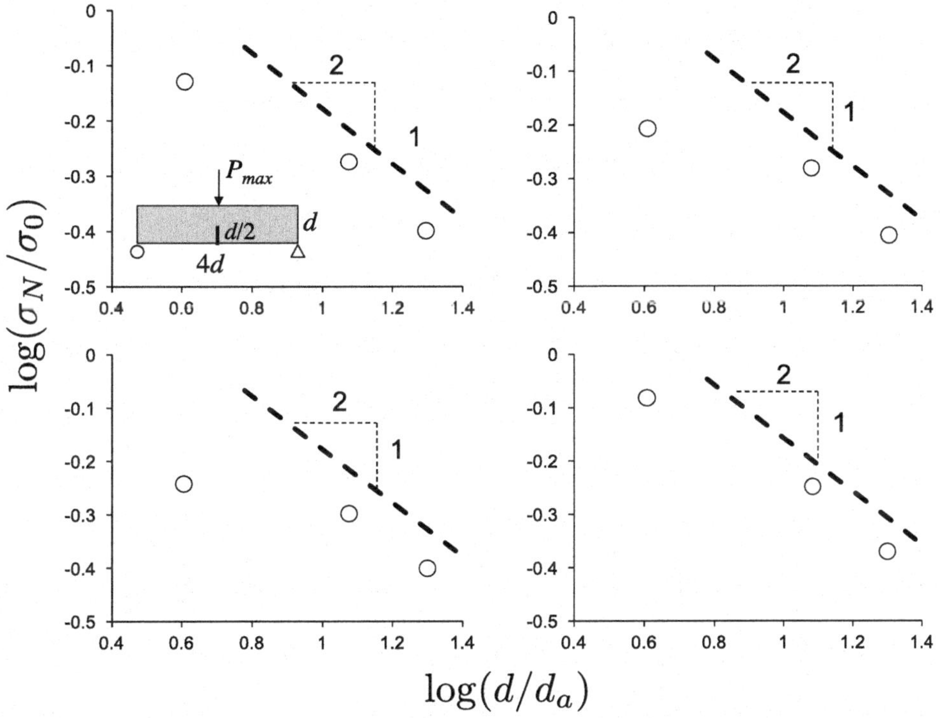

Figure 1.5. Measured size effect on the nominal strength of concrete beams.

measured size effect on the nominal strength: $\sigma_N \propto D^{-n} (n > 0)$. Based on the optimum fitting, the value of n was less than $1/2$. Such an n-value was inferred by assuming failure to be caused by a finite-angle notch, because its stress singularity exponent is weaker than that for a sharp crack. However, the optimum value of n was found to be less than the magnitude of the dominant stress singularity at the notch tip, as required by the LEFM. Furthermore, like the Weibull size effect, Leicester's power-law scaling also implied the nonexistence of a characteristic length, which cannot be the case for concrete owing to the large size of its FPZ.

The inapplicability of LEFM was further evidenced by Walsh's size effect experiments on geometrically similar notched beams (Walsh 1972, 1976). The nominal strength was plotted against the beam size in a double logarithmic diagram (Fig. 1.5), in which σ_0 is a reference strength and d_a is the average aggregate size. Without attempting a mathematical description, he made the point that this diagram deviates from a straight line of slope $-1/2$, and that this deviation must be regarded as a departure from LEFM.

The major milestone in application of fracture mechanics to concrete was the development of the fictitious crack model by Hillerborg et al. (1976). The concept of the fictitious crack model is analogous to the softening cohesive crack model of Barenblatt (1959, 1962), and is similar to the plastic fracture process zone model of Dugdale (1960) (later extended by Knauss 1973, 1974; Wnuk 1974; and Kfouri and Rice 1977). Hillerborg et al. used the fictitious crack model to simulate the failure of unnotched plane concrete beams in bending and further demonstrated a deterministic size effect, different from the Weibull statistical size effect, on the flexural strength of beams. The softening cohesive crack model by Palmer and Rice (1973) did the same for shear.

Around the same time, it was discovered (Bažant 1976) that the damage localization of strain-softening material would lead to a size effect on the post-peak deflections and energy dissipation of structures. The essential idea of crack band model was proposed as a remedy for realistic and objective finite element simulation of quasibrittle fracture (Bažant 1976, 1982; Bažant & Oh 1983). It was shown that the crack band model could accurately capture the size effect observed, by that time, on concrete structures. A more general nonlocal integral approach that can handle strain-softening damage in a more fundamental and, in some respect, more realistic manner followed soon (Bažant 1984*b*; Bažant, Belytschko, & Chang 1984; Pijaudier-Cabot & Bažant 1987; Bažant & Pijaudier-Cabot 1988; Bažant & Lin 1988*a*, 1988*b*). The nonlocal approach also led to the later development of the nonlocal implicit gradient models for quasibrittle materials (Peerlings, de Borst, Brekelmans, & de Vree 1996; Geers, Peerlings, Brekelmans, & de Borst 2000; Peerlings, Geers, de Borst, & Brekelmans 2001).

In early 1980s, Bažant (1984*b*) used an approximate energy analysis to derive a simple size effect law for the nominal strength of quasibrittle structures containing notches or traction-free (fatigued) large cracks formed prior to the peak load. This size effect law was later derived by using the equivalent LEFM with Taylor series expansion of the energy release rate function, by which the size effect curve was directly related to the fracture properties, such as the fracture energy, the softening law, and the R-curve, of quasibrittle materials (Bažant & Kazemi 1990*a*, 1991). Beginning with the mid-1980s, the interest in the size effect in quasibrittle fracture surged enormously. Besides an intensive focus on concrete (Petersson 1981; Carpinteri 1986; Planas & Elices 1988; Planas & Elices 1989; Bažant 1992*b*; Wittmann 1995), significant attention has also been paid to various other engineering materials such as ice (Bažant 1992*a*; Bažant & Li 1994; Li & Bažant 1994; Dempsey, Adamson, & Mulmule 1995; Mulmule, Dempsey, & Adamson 1995; Bažant & Kim 1998*a*, 1998*b*; Dempsey, Adamson, & Mulmule 1999); ceramics (McKinney & Rice 1981), rocks (Bažant, Gettu, & Kazemi 1991; Bažant, Lin, & Lippmann 1993; Le, Manning, & Labuz 2014), foam (Bažant, Zhou, Zi, & Daniel 2003), fiber composites (Bažant, Daniel, & Li 1996; Bažant et al. 1999, 2006; Bažant, Zhou, Novák, & Daniel 2004), braided composites (Caner et al. 2011), and bones (Bažant, Kim, & Yu 2013).

Recent research has focused back on the weakest-link model of strength statistics of quasibrittle structures that fails under controlled load at macrocrack initiation (Bažant & Pang 2006; Bažant & Pang 2007; Bažant, Le, & Bazant 2009; Le, Bažant, & Bazant 2011; Salviato, Kirane, & Bažant 2014). Different from Weibull's analysis, which is based on an infinite weakest-link model, recent studies proposed a finite weakest-link model to account for the fact that, for quasibrittle structures, the FPZ size is not negligible compared to the structure size. This model was further extended to lifetime statistics under both static and cyclic fatigue (Bažant et al. 2009; Le & Bažant 2011, 2012, 2014; Salviato et al. 2014). The finite weakest-link model predicts the size dependence of the probability distributions of structural strength and lifetime, which has been shown to have an important implication for reliability analysis of quasibrittle structures.

1.3 Safety Specifications in Concrete Design Codes and Embedded Obstacles to Probabilistic Analysis

The concept of structural reliability has penetrated into the design of many engineering structures. However, most existing design codes are not yet based on a calculation of the actual structural failure probability. Instead, they prescribe empirical safety factors of various kinds. The safety factor is understood as the ratio of the mean failure load measured by experiments to the failure load calculated deterministically. This kind of safety factor is used for some types of structures (e.g., for airframes, where it is generally considered to be 1.5). The design codes for structural concrete (as well as steel) were, in the 1970s, converted to the load factor and resistance design (LFRD), in which the safety factor is split into two types of partial factors:

1. The load factors, which account for the random variability of various kinds of load (live and dead loads, wind, earthquake, earth pressure, water pressure)
2. The resistance factors, aka strength reduction factors (popularly, "understrength" factors), applied to the calculated structural strength. These factors reflect the differences in risks in different types of failure; e.g., they are 0.9 for flexure and 0.75 for shear of concrete beams.

From the statistical viewpoint, three problems with these factors have recently been identified [for details, see Bažant & Frangopol (2002) and Bažant & Yu (2006)]:

1. While the safety factor of 1.6 for live loads seems reasonable, the safety factor (American Concrete Institute Committee 318 2011) for the self-weight acting alone, which is specified as 1.4, is excessive, by far. The value of 1.4 implies a 40% error in the self-weight, which is impossible to occur in practice. The errors in the mass density of concrete and in structural dimensions cannot justify errors of more than a few percent. Since, in a small-span bridge, the self-weight can represent 5% of the total load and in a large-span one 95%, such an excessive load factor unjustly penalizes large structures. So this penalty is equivalent to introducing in design a certain hidden size effect. But the trend of this hidden size effect is incorrect, with respect to not only the structure size, but also the structure type. For example, compared to the normal strength and non-prestressed concretes, the prestressed concrete structures and high-strength ones are lighter and thus receive less protection from the hidden size effect, but actually would need a higher protection because they are more brittle and exhibit a stronger size effect [for detailed discussion see Bažant & Yu (2006)].
2. Another problem is that various concrete design formulas have been set not to pass through the mean of experimental data but at the margin of a highly scattered data cloud, and that one must go to the original reports to find out. For example, in the case of shear failure of beams, the design formula passes about 65% below the mean

strength, which is near the bottom margin of the data cloud in which beams of different sizes, reinforcement, shear spans, aggregate sizes, etc. are commingled. This approach is necessary to make the deterministic design method adequately safe, but it implies covert understrength factors which, in many cases, lead to excessive safety, and in some cases not. The fact that the mean fit of the experimental data and the coefficient of variation (CoV) of its error (as well as the type of probability distribution) are unknown makes meaningful probabilistic analysis of structural strength impossible.

3. A similar problem occurs for the concrete strength. The design according to the code is based on the so-called required compression strength, f_c', which is set by the ACI code at about $\bar{f}_c - 1.34\delta_f$ where $\bar{f}_c$ = mean strength from cylinder tests and δ_f = their standard deviation (which is normally not known to the designer). Because the random scatter in concrete is high, f_c' can be 30% lower than $\bar{f}_c$. Again this approach is necessary to make the deterministic design approach safe, but it implies another covert understrength factor. The fact that the mean strength and its CoV are normally not reported and are unknown to the designer renders meaningful probabilistic design impossible [Model code 2010 sets $f_c' = \bar{f}_c - 8$ (units are in MPa) but the ratio of 8 MPa to δ_f is not known and varies].

Relatively sophisticated methods have been developed for the reliability analysis of civil engineering structures. They include the first-order and second-order reliability methods (FORM and SORM), where the failure risk of the structure can be estimated by using reliability indices, such as the Cornell and Hasofer–Lind indices (Benjamin & Cornell 1970; Hasofer & Lind 1974; Rackwitz & Fiessler 1976; Rackwitz & Fiessler 1978; Ang & Tang 1984; Haldar & Mahadevan 2000*a*). However, these methods are usually applied without regard to, or even in ignorance of, the aforementioned obstacles. Such an attitude limits their applicability to quasibrittle structures. Besides, these methods need restructuring to take into account the size effect on the strength distribution and its deviations from Gaussian and Weibull distributions established in this book.

Although the design code must be satisfied, it is nevertheless always prudent to also check the design of large and daring structures probabilistically, using the data from concrete strength tests conducted on the concrete chosen for the structure, or at least a reasonably estimated mean and standard deviation of such tests. This is how the probabilistic theory in this book is applicable to concrete structures, in spite of the aforementioned obstacles to probabilistic analysis. Another application should, of course, be the improvement of concrete design codes. And, of course, in the engineering practice of many other quasibrittle materials, e.g., fiber composites, tough ceramics, and rocks, the theory expounded here does not face similar obstacles. In aeronautical engineering there are no design codes, only performance requirements.

1.4 Importance of Size Effect for Strength Statistics

If the scaling of a theory is not understood, the theory itself is not understood. This motto has guided the development of fluid mechanics for more than 100 years, and recently

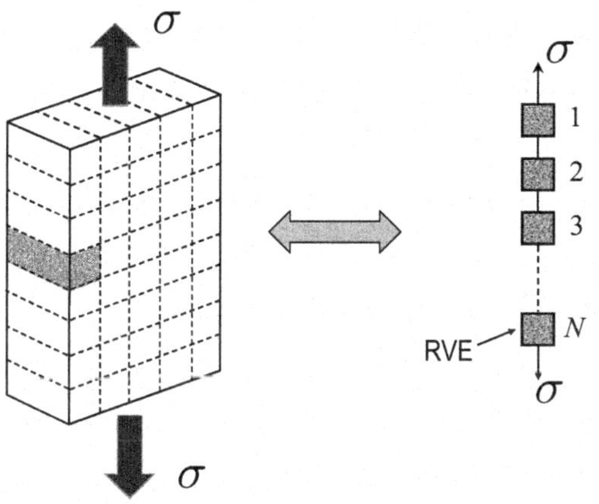

Figure 1.6. A prism under uniaxial tension (left) and the weakest-link model (right).

became particularly important for quasibrittle structures. For elasticity with a strength limit and for plasticity, which are the classical theories of solid mechanics, the scaling properties are so simple and automatic that they have required no particular attention in testing as well as structural design. The point to note is that in these classical theories there is no size effect, i.e., geometrically similar structures of different sizes fail at the same maximum stress, equal to the strength limit.

On the other hand, in quasibrittle failures there is a strong and intricate size effect, whose source is partly deterministic (due to energy release caused by stress redistribution during crack growth) and partly statistical (due to material strength randomness). To validate and calibrate a probabilistic theory of quasibrittle failure, it is necessary to take into account the scaling of structural strength with respect to the structure size, briefly the size effect. This is the major theme of this book.

The importance of size effect for statistics can be explained by considering a prism under homogeneous tension (Fig. 1.6, left). If the material is quasibrittle, the specimen failure is triggered by the failure of one representative volume element (RVE) of the material. For the purpose of quasibrittle strength statistics, the RVE cannot be defined in the classical way, i.e., as the smallest volume for which its material properties do not change significantly when this volume is moved through the material (Hill 1963). Rather, the RVE must be defined as the smallest volume whose failure causes the whole structure to fail (Bažant & Pang 2007) [provided the geometry is positive (Bažant & Planas 1998), which is a term to be explained in Section 2.1].

The failures triggered by one element are characterized by the weakest-link model, pictured as the failure of a chain with many links (Fig. 1.6, right). In a very small specimen, only few RVEs come into play. However, if the length and width of a thin specimen consist of 1000 RVEs, the specimen contains 10^6 RVEs and triggering the failure in one particular RVE has a probability on the order of 10^{-6}. Or, if in a square prism with a cross section of 30×30 RVEs and length of 1111 RVEs the relevant failure probability of one RVE is again 10^{-6}. Obviously, testing such a large specimen actually samples the far-out tail of failure probability distribution.

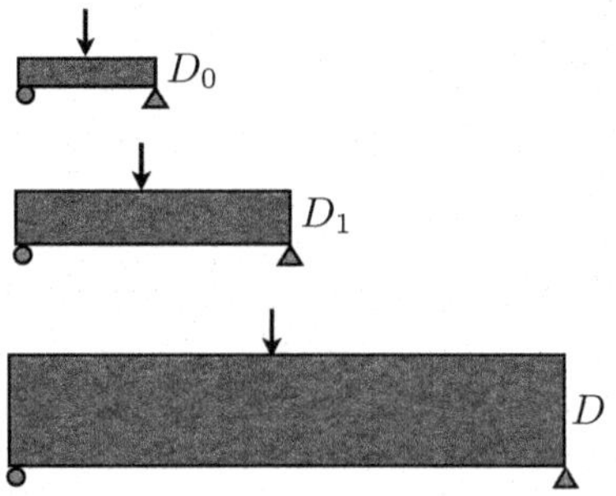

Figure 1.7. Geometrically similar structures of different sizes.

1.5 Power-Law Scaling in the Absence of Characteristic Length

The basic and simplest type of spatial scaling occurs in physical systems that lack a characteristic length, l (or $l = 0$). The scaling is the ratio of responses of systems of different sizes that are geometrically similar. Geometric similarity occurs when one structure can be obtained from another by polar affinity transformation (after proper positioning). This means that all the homologous dimensions of the structures are scaled in the same ratio. In the case of two-dimensional (2D) scaling of a three-dimensional (3D) structure, only the in-plane dimensions are scaled while the structure width is kept the same (see, e.g., the beams shown in Fig. 1.7).

We seek to obtain the scaling law for any structure response Y (e.g., the maximum stress in the structure) as a function of the characteristic size (or dimension) D of the structure. The scaling law relates response Y for size D to the corresponding response Y_0 of a structure of some chosen reference size D_0, i.e.,

$$\frac{Y}{Y_0} = f(\lambda), \quad \text{where } \lambda = \frac{D}{D_0} \tag{1.1}$$

λ is the scaling ratio and $f(\lambda)$ is a dimensionless function.

To determine function $f(\lambda)$ when a characteristic length is absent (or zero, or infinite), we imagine geometrically similar structures of three different sizes, D_0, D_1, and D, with the corresponding responses Y_0, Y_1, and Y. The ratio of the responses for sizes D and D_1 is $Y/Y_1 = f(D/D_0)\,/\,f(D_1/D_0)$. Now, if and only if there is no characteristic length, size D_1 may be taken as the reference size instead of D_0, which means that $Y/Y_1 = f(D/D_1)$. This yields the condition:

$$\frac{f(D/D_0)}{f(D_1/D_0)} = f\left(\frac{D}{D_1}\right) \tag{1.2}$$

[for solids, see Bažant (1993) and Bažant & Chen (1997); for fluids and in general see Sedov (1959) and Barenblatt (1979)]. This is a functional equation for the unknown scaling law $f(\lambda)$. To solve it, we note that if the scaling law is valid for both D and $D_1 = D + \delta D$ at any D, then, by induction, it must be valid for all D. Therefore, we differentiate this equation with respect to D

$$\frac{f'(D/D_0)}{D_0 f(D_1/D_0)} = \frac{1}{D_1} f'\left(\frac{D}{D_1}\right) \tag{1.3}$$

where $f'(x) = \mathrm{d}f(x)/\mathrm{d}x$. Substitution of $D_1 = D$ into Eq. 1.3 yields the differential equation $f'(\lambda)/f(\lambda) = f'(1)/\lambda$, or

$$\frac{\mathrm{d}f(\lambda)}{f(\lambda)} = s\,\frac{\mathrm{d}\lambda}{\lambda}, \quad \text{with } s = \left.\frac{\mathrm{d}f(\lambda)}{\mathrm{d}\lambda}\right|_{\lambda=1} = \text{constant} \tag{1.4}$$

which can easily be solved by separation of variables. Because $Y = Y_0$ for $D = D_0$, the initial condition is $f(1) = 1$. Thus it is found that Eq. 1.2 has one and only one solution, namely the power law with an unknown constant exponent, s:

$$f(\lambda) = \lambda^s \tag{1.5}$$

That is why power-law scaling plays such a prominent and fundamental role. It is a necessary indicator and manifestation of self-similarity (Barenblatt 1979).

In conclusion, power-law scaling must apply for every physical theory in which there is no characteristic length. In deterministic solid mechanics, such failure theories include the theories of elasticity (with a strength limit), perfect plasticity, and hardening elasto-plasticity. Also included is the LEFM, because the FPZ, whose size is an indicator of the material characteristic length, is in LEFM considered to be shrunken into a point. As shown in Section 3.6 and Appendix A, $s = 0$ for the scaling of stress in elasticity and plasticity (which is by definition the case of no size effect), and $s = -1/2$ for the scaling of stress in LEFM.

Vice versa, if the scaling law is a power law, there can be no characteristic length. As will be seen, this simplest class of scaling laws includes the Weibull statistical strength theory.

1.5.1 Nominal Strength of Structure and Size Effect

It has become standard to characterize the scaling of structural strength in terms of the so-called nominal strength σ_N, which represents a maximum load parameter with the dimension of stress. Depending on whether the geometric similarity is 2D or 3D, the nominal strength is defined as

$$\sigma_N = c_n P_m / bD \ \text{ for 2D}, \qquad \sigma_N = c_n P_m / D^2 \ \text{ for 3D} \tag{1.6}$$

Here P_m is the maximum load the structure can carry, and b is the structure thickness in the third dimension, in the case of 2D similarity. The coefficient c_n is a chosen dimensionless constant, which does not depend on the structure size and is chosen for convenience. It may be chosen so as to make σ_N coincide, for example, with the maximum stress, or with the average stress in a cross section, or with the stress at any homologous points of similar structures of different sizes (for examples, see Bažant 1984). Often σ_N is chosen to represent no particular stress and is simply set as $c_n = 1$.

When the nominal strength is independent of D, we say that there is no size effect. Thus the *size effect* came to be understood as the dependence of nominal strength on the structure size, D. Since only the relative changes of the structure size are considered, D can be defined as any characteristic dimension, e.g., in the case of a beam as its depth or its span, or in the case of a circular plate as its diameter or thickness, or in the case of a hollow tube as its diameter, or wall thickness or length. To determine exponent s of Eq. 1.5 for the nominal strength or for any other mechanical variable such as the maximum displacement, the material failure criterion must be taken into consideration. This can be most simply done by dimensional analysis.

In this book, the term "strength" without any modifier is understood in the usual sense of a material property. In the sense of a structural failure characteristic with the dimension of stress, the term "structural strength" or "strength of structure" is used. The terms "load capacity," "ultimate load," or "maximum load" have the dimension of force.

1.6 Statistical and Deterministic Size Effects

Since the 1980s (Bažant & Oh 1983; Bažant 1984*b*; Bažant & Planas 1998; Bažant 2005), it has been known that Weibull's is not the only size effect and that, in quasibrittle structures, there are two basic kinds of size effect caused by fracture:

1. Statistical, which is of Weibull type.
2. Deterministic (or energetic), which is caused by energy release engendered by stress redistribution associated with
   a. either the formation of a large enough fracture process zone at a smooth surface – *Type 1* size effect,
   b. or the stable growth of a large crack prior to the maximum load – *Type 2* size effect.

Types 1 and 2 follow different laws. The distinction between them was identified later, in the 1990s (Bažant & Li 1995*b*; Bažant & Planas 1998; Bažant 2005; Bažant & Yu 2009). The transition between types 1 and 2 is a more difficult problem and was experimentally explored only recently (Le 2011; Hoover & Bažant 2013; Hoover et al. 2013; Le & Xue 2013; Hoover & Bažant 2014*c*; Le, Pieuchot, & Ballarini 2014).

1.7 Simple Models for Deterministic Size Effects

There are two simple types of deterministic size effect. They are analyzed in detail in Chapter 3, but their elementary explanation is simple, and is as follows.

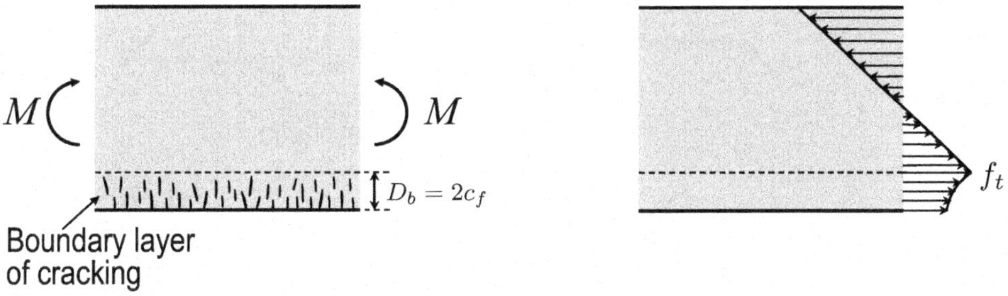

Figure 1.8. Stress redistribution in the distributed cracking layer.

1.7.1 Type 1 Size Effect for Failures at Crack Initiation

A simple example of this size effect is found in the modulus of rupture test, which consists in the bending of a simply supported beam. Let L = its span, D = its depth and b its width, and P_m = maximum of the concentrated load applied at mid-span. Because of stress redistribution in the evolving FPZ at the bottom face (Fig. 1.8), the maximum load is not decided by the maximum stress at the bottom face according to the theory of bending, which is $\sigma_1 = 3P_mL/2bD^2 = (3L/2D)\sigma_N$, where $\sigma_N = P_m/bD$. Rather it is decided by the stress value $\bar{\sigma}$ roughly in the middle of the damage layer at the bottom face, representing the evolving FPZ. Approximately, $\bar{\sigma} = \sigma_1 - \sigma' c_f$, where $\sigma' = 2\sigma_1/D$ = stress gradient and c_f = half of the thickness of the boundary layer of cracking. Setting $\bar{\sigma} = f_t$ = tensile strength of the material, we have $(3L/2D)\sigma_N(1 - 2c_f/D) = f_t$. This gives

$$\sigma_N = \frac{\sigma_0}{1 - D_b/D} \tag{1.7}$$

in which $\sigma_0 = (2D/3L)f_t$ and $D_b = 2c_f$; σ_0 and D_b are constants because the ratio D/L remains constant during geometrical scaling.

Equation 1.7 for σ_N, however, is valid only when $D \gg D_b$ (i.e., when the cracking zone is a small fraction of the beam depth). But since the derivation is valid only for small enough c_f/D (i.e., up to the first-order term of the asymptotic expansion of σ_N in terms of $1/D$), one may replace Eq. 1.7 by the following asymptotically equivalent size effect formula (Fig. 1.9a):

$$\sigma_N = \sigma_0 \left(1 + \frac{rD_b}{D}\right)^{1/r} \tag{1.8}$$

which has (for any r) the same first two terms of the Taylor series expansion, but has the advantage of being (for positive r) acceptable for the entire range of D; r is a positive constant. The value of r is empirical and was taken between 1 and 2 for concrete (Bažant & Novák 2000*a*), while $r = 1.47$ appeared optimum in Bažant and Novák's analysis of test data.

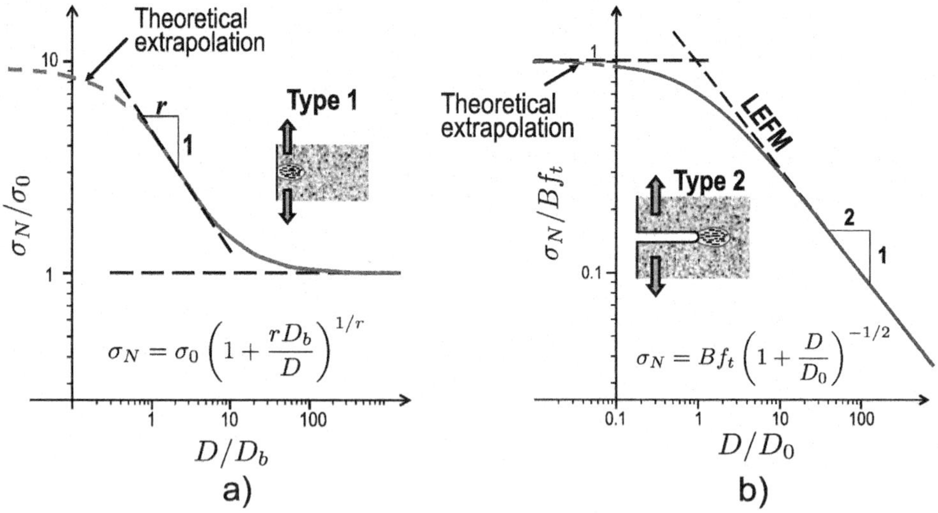

Figure 1.9. Size effect curves of nominal strength of quasibrittle structures. (a) Type 1 failure and (2) type 2 failure (with left-side extrapolation by cohesive crack model).

Alternatively, the foregoing type 1 size effect formula can be derived from the energetic formula for the type 2 size effect (Bažant & Novák 2000*a*; Bažant 2005) as the limiting case for structures with a zero notch depth, which is explained next.

1.7.2 Type 2 Size Effect for Structures with Deep Cracks or Notches

This size effect occurs in notched fracture specimens or in structures that fail only after stable growth of large cracks. Such failures exhibit a stronger energetic size effect than type 1 (Iguro, Shioya, Nojiri, & Akiyama 1985; Marti 1989; Gettu, Bažant, & Karr 1990; Bažant & Kazemi 1991; Shioya & Akiyama 1994; Walraven 1995; Bažant & Planas 1998). They are typical of reinforced concrete structures, as well as some cases of fiber composites (Wisnom 1992; Bažant et al. 1996), and are also exhibited by some unreinforced concrete structures; e.g., the dipping fracture in a gravity dam stabilized by vertical compression (Bažant & Planas 1998), or by floating sea ice plates in the Arctic (Bažant & Li 1994, 1995*a*; Li & Bažant 1994; Bažant & Kim 1998*a*, 1998*b*; Bažant 2002*a*, 2002*b*).

Consider the rectangular panel in Figure 1.10, which is initially under a uniform stress equal to σ_N. Introduction of a crack of length a and an FPZ of width h may be imagined to cause a full relief of stress from the shaded triangles on the flanks of the crack band shown in Figure 1.10. The original strain energy density in these triangles is $\sigma_N^2/2E$, where $E =$ Young's modulus. The slope k of the effective triangular boundary of the stress relief zone need not be determined, since the only important fact is that k is independent of size D (slope k giving the correct energy release magnitude depends, of course, on the specimen geometry; it can be determined by comparison with the exact solution according to the theory of elasticity).

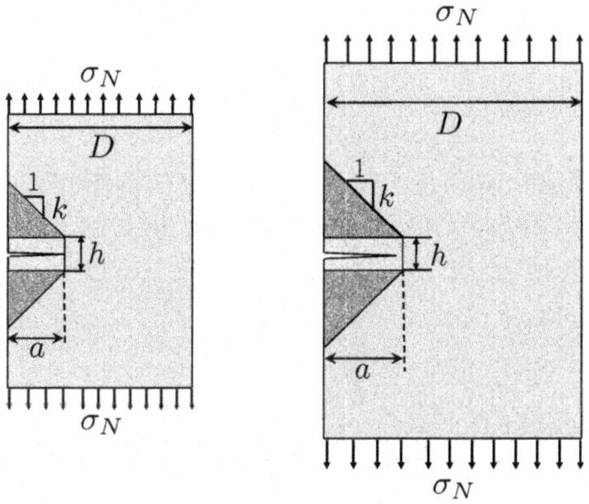

Figure 1.10. Stress relief zones in small and large structures.

For the normal range of interest, the shapes of the cracks at maximum load for various sizes D are usually similar and their lengths at maximum load are usually approximately proportional to the structure size D, while the width h of the FPZ is essentially a constant, related to the size of the largest inhomogeneities in the material (such as aggregate pieces in concrete). Except for h, the geometric similarity is usually confirmed by experiments [for diagonal shear of beams, see, e.g., Fig. 1.11 adapted from Syroka-Korol & Tejchman (2014)], as well as by finite element solutions based on the crack band, cohesive, or nonlocal models.

The stress reduction in the triangular zones, having areas $ka^2/2$ (Fig. 1.10), causes (for width $b = 1$) the energy release

$$U_a = 2\,\frac{ka^2}{2}\,\frac{\sigma_N^2}{2E} \tag{1.9}$$

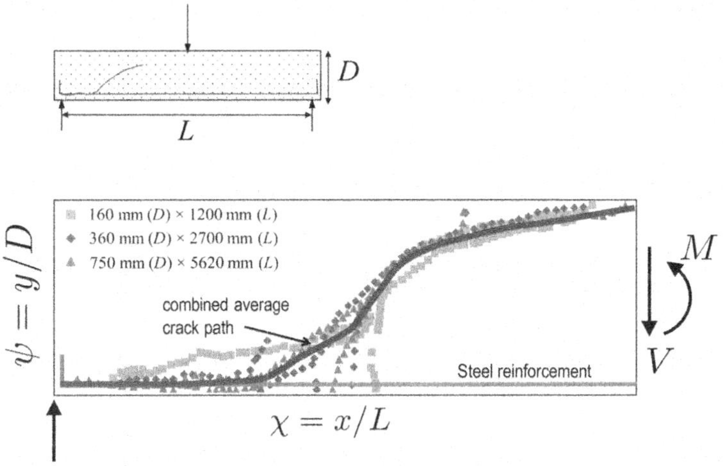

Figure 1.11. Geometrically similar diagonal shear cracking in reinforced concrete beams.

The stress drop within the crack band of width h causes a further energy release,

$$U_b = ha\,\frac{\sigma_N^2}{2E} \tag{1.10}$$

The total energy dissipated by the fracture is $W = aG_f$, where G_f is the fracture energy, a material property representing the energy dissipated per unit area of the fracture surface. Energy balance during the initial static increment of crack length a requires that

$$\frac{\partial(U_a + U_b)}{\partial a} = \frac{\mathrm{d}W}{\mathrm{d}a} \tag{1.11}$$

Setting $a = D(a/D)$, where a/D is approximately a constant in the case of full geometric similarity, the solution of the last equation for σ_N yields Bažant's approximate size effect law (Bažant 1984*b*) (Fig. 1.9b):

$$\sigma_N = \frac{Bf_t}{\sqrt{1+\beta}}, \quad \beta = \frac{D}{D_0} \tag{1.12}$$

$$\text{where} \quad D_0 = \frac{h}{2k(a/D)}, \quad Bf_t = \sqrt{\frac{2}{h}EG_f} \tag{1.13}$$

Here D_0 is called the transitional size; B is a dimensionless structural geometry parameter, independent of D; and f_t is the material tensile strength introduced solely for the convenience of dimensionality.

A rigorous derivation of this law, applicable to any structure geometry, is given in Chapter 3 on the basis of the asymptotic properties of the cohesive crack model. This law was also derived in full generality on the basis of the integral equation of the smeared-tip method (Bažant 2005), and also on the basis of the equivalent LEFM (Bažant 1997) or of Rice's path-independent J-integral (Bažant & Planas 1998). Further verifications were provided by nonlocal finite element analysis, and by random particle (or discrete element) models. The size effect law in Eq. 1.12 applies to many geometries. It was experimentally verified and theoretically justified for a broad range of different materials and structures (Bažant 1984*b*, 1997, 2002*b*, 2004*b*, Bažant & Kazemi 1990*a*; Gettu et al. 1990; Bažant & Chen 1997; Bažant & Planas 1998; Dempsey et al. 1999; Bažant, Zi, & McClung 2003; Bažant & Yavari 2005; RILEM TC QFS 2005). Yu et al. (2016) gave a general derivation of Eq. 1.12 based on dimensional analysis and energy conservation, eschewing fracture mechanics concepts. The most complete experimental verification of this law has been provided in the recent comprehensive tests of Hoover et al. (Hoover & Bažant 2013, 2014*c*; Hoover et al. 2013).

Note that, for a very large size ($D \gg D_0$), the size effect law, Eq. 1.12, asymptotically approaches the power law $\sigma_N \propto D^{-1/2}$, which represents the size effect of LEFM for geometrically similar structures and corresponds to the inclined asymptote of slope $-1/2$ in the doubly logarithmic plot. For a very small size, this law asymptotically approaches the case of no size effect, σ_N = constant, which represents a horizontal asymptote and gives the same scaling as in strength theory or plastic limit analysis.

The ratio $\beta = D/D_0$ is called the brittleness number of a structure. For $\beta \to \infty$, quasibrittle structures become perfectly brittle (i.e., follow LEFM), in which case the size effect is the strongest possible, while for $\beta \to 0$, the structure is non-brittle (or ductile,

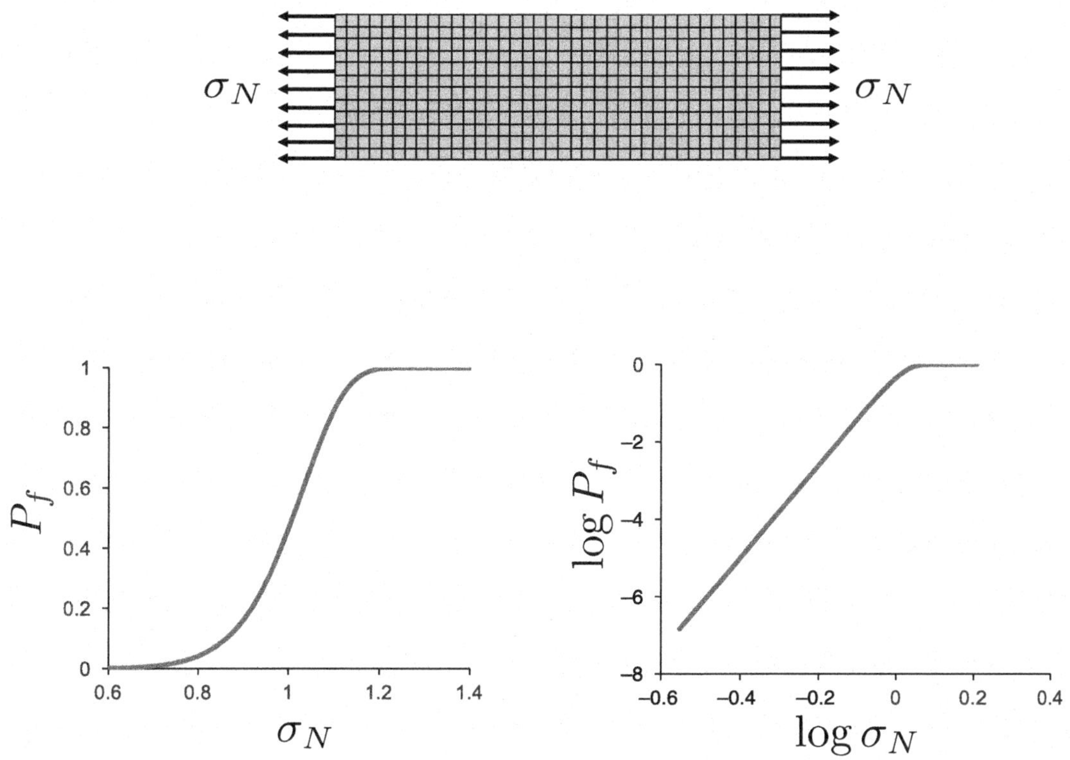

Figure 1.12. Weibull strength distribution of brittle structures.

plastic), in which case there is no size effect. Regardless of geometry, quasibrittle structures are approximately those for which $0.03 \leq \beta \leq 30$, and through this range the size effect gives a smooth transition, bridging the power laws of size effects for the two asymptotic cases. Thus the size effect law has the character of asymptotic matching.

The transitional size D_0, delineating the predominantly brittle behavior from predominantly non-brittle (or ductile) behavior, has been shown to be proportional to the effective size of the FPZ, with the proportionality constant depending on the structure geometry.

1.8 Probability Distributions of Strength of Ductile and Brittle Structures

As we discussed in previous sections, the limiting cases of quasibrittle failure are the brittle failure and the ductile failure. The brittle failure is progressive and becomes dynamic and unstoppable as soon as one structural element (the RVE) initiates the propagation of failure across the structure. By contrast, the ductile failure is non-progressive and occurs simultaneously over the whole cross section of the structure, as assumed in the plastic limit analysis. These two distinct failure characteristics lead to very different probability distributions of strength of brittle and ductile structures.

Consider brittle failure of a structure with many RVEs (Fig. 1.12, top), which are much smaller than the overall structure size. Since the brittle failure is irreversibly

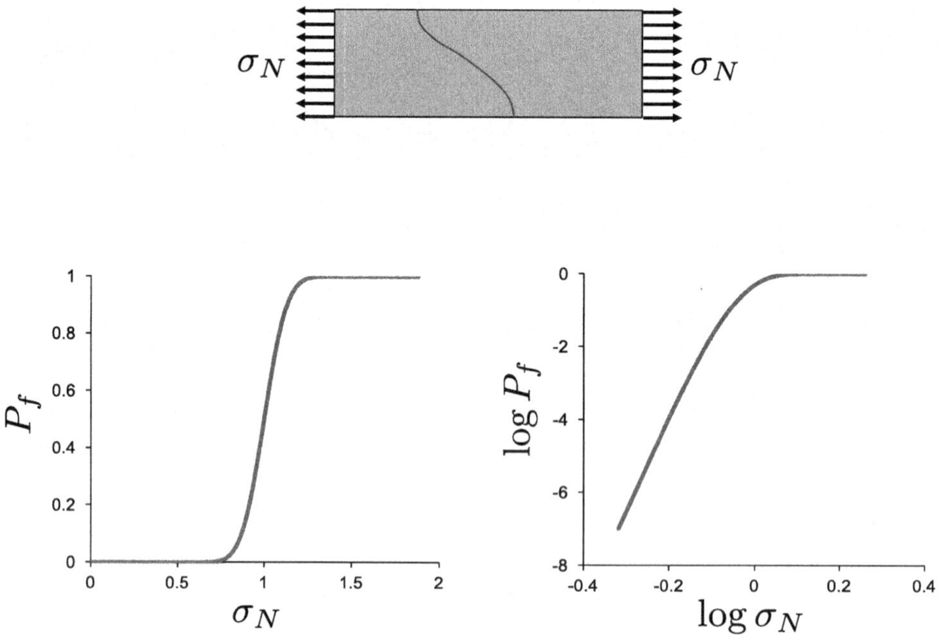

Figure 1.13. Gaussian strength distribution of ductile structures.

initiated by a single RVE, it is statistically equivalent to the failure of the weakest link in a very long chain, which is called the infinite weakest-link model. As is well known, and will be explained in Section 2.1, the probability distribution of structural strength (or nominal strength) is in this case the Weibull distribution, pictured in Figure 1.12 bottom.

In ductile failure, by contrast, all the material elements (RVEs) along the failure surface are mobilized to their strength limit simultaneously (Fig. 1.13, top). So the maximum load is a sum of the contributions (weighted by the ratios of applied load to various local stresses) of all the material elements (or RVEs) along the failure surface. Therefore, the probability distribution of the structural strength is the distribution of a weighted sum of randomly distributed variables, the local strengths of material elements. It is well known, and will be explained in Section 2.7, that if there are sufficiently many elements in the sum, the distribution of structural strength must be the Gaussian (or normal) distribution, regardless of the strength distribution of the material elements (Fig. 1.13 bottom) (although, as will be shown later, those distributions must also be Gaussian except for the far-left tail).

Often it is argued that the strength distribution in ductile failure should be taken as lognormal because, unlike the Gaussian, it does not extend to the negative strength values. But this argument is faulty, because the Central Limit Theorem implies that the Gaussian distribution is not valid in the tail reaching into the negative strength values. Besides, from the mechanics viewpoint, assuming the lognormal distribution would imply the failure load to be a product, rather than a sum, of the contributions

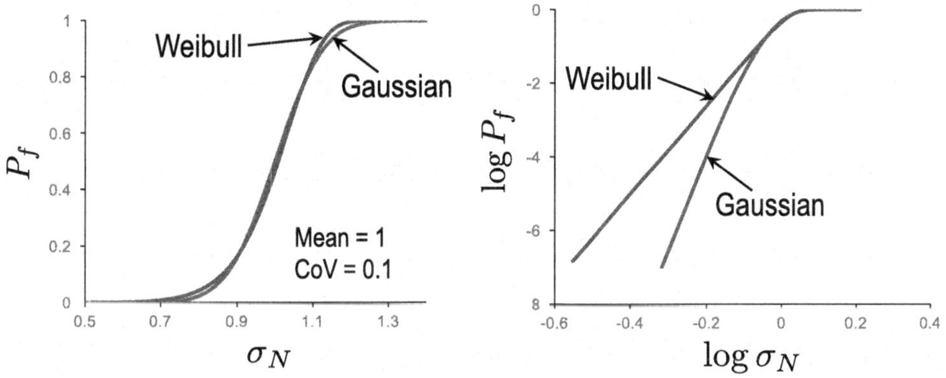

Figure 1.14. Comparison of Gaussian and Weibull distributions.

from all the structural elements, which is evidently impossible. Similarly, the Weibull distribution cannot be used for ductile failures, but this opposite fact is generally accepted.

In Figure 1.14, the Gaussian and Weibull distributions are plotted for a mean equal to 1 and for the same CoV with the value of 10% typical of real concrete structures. With respect to structural safety, the locations of the tail points with failure probability 10^{-6} ought to be particularly noticed. For the Weibull distribution, this tail point is almost twice as far from the mean than it is for the Gaussian distribution. This confirms the generally perceived wisdom that the brittle failures are far more dangerous. In any case, if it is known that the distribution is Gaussian or Weibull, the location of the 10^{-6} tail point can be determined from the CoV and structural safety realistically assessed.

This observation signals a difficult problem for quasibrittle failures. Such failures are transitional between the ductile limit and the brittle limit, which is an aspect not considered in design practice and structural reliability programs. Thus the correct structural strength distribution is some combination of Gaussian and Weibull distributions. The combination varies strongly with the structure size, and less strongly also with the structure shape. In current practice, and also in the structural safety literature and at conferences, in which only elementary mechanics is used, the distance of the 10^{-6} tail point from the mean is not known, and the question is not even posed, rendering dubious the current assessments of the safety of engineering structures that are made of concrete, composites, and ceramics. Several recent theoretical studies (Bažant & Pang 2006, 2007; Bažant et al. 2009; Le & Bažant 2011, 2012; Le et al. 2011) have proposed an answer to this problem and the present book deals with it in detail.

2 Review of Classical Statistical Theory of Structural Strength and Structural Safety, and of Statistics Fundamentals

2.1 Weakest-Link Model

The simplest statistical theory for structural strength is Weibull's theory. It is based on the infinite weakest-link model, imagined as an infinite chain (Fig. 2.1) in which each link corresponds to one representative volume element (RVE) of the material. The Weibull theory is based on the following two hypotheses:

1. The structure fails if one RVE fails, as implied by the weakest-link model.
2. The number of RVEs that could trigger the structural failure is extremely large and can be considered as infinite.

The structural geometry for which the former hypothesis is valid is often called the *positive geometry*. The positive geometry is defined as the structural geometry for which the stress intensity factor (or the energy release rate) at constant load increases as the crack grows. For structures under load control, the failure for such geometry occurs right at the beginning of crack propagation, as soon as the full fracture process zone (FPZ) forms. In the opposite case of negative geometry, the structure does not fail when one RVE softens. Rather, the crack extends under an increasing load in a stable manner, as typically seen in reinforced concrete structures and also achieved in some fracture specimens such as a large panel with a small center crack, loaded on the crack, or in a reverse-taper double-cantilever specimen.

Hypothesis 2 means that, in comparison to structure size D, the FPZ is so small that it can be treated as a point. For geometrically similar structures of various sizes, the stress distribution as a function of relative coordinate vector $\boldsymbol{\xi} = \boldsymbol{x}/D$ of material points is then independent of D ($\boldsymbol{x}$ = actual coordinate vector).

Hypothesis 2 has often gone unmentioned and has usually been considered automatically applicable. But recently it transpired that, for structures made of concretes, fiber composites; tough ceramics; and other quasibrittle materials, the number of potentially failing RVEs is often far too small to justify the application of Weibull statistics. (As will be seen, the number of RVEs must exceed 10^4, in the sense of a weighted equivalent number; see Section 2.4.)

Denote P_k = failure probability of the kth RVE ($k = 1, 2, \ldots N$) of structure, and P_f = failure probability of the structure as a whole. Since positive geometry is assumed, the structure must fail as soon as one RVE fails. In other words, the structure behaves

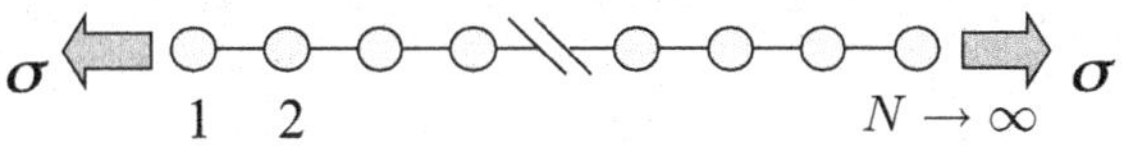

Figure 2.1. Weakest-link model of strength statistics.

statistically as a chain, which fails as soon as one link fails (Fig. 2.1). The way to determine P_f is to switch attention from failure probabilities P_k to survival probabilities, which are $1 - P_k$ for each RVE. Obviously, the structure will survive if all its RVEs survive. So, the survival probability of the structure is the joint probability of survival of all the RVEs. Thus, if we assume that all P_k are statistically independent (which is acceptable when the autocorrelation length of the random strength field in the structure is not appreciably larger than the spacing of the RVEs), we have, according to the joint probability theorem,

$$1 - P_f = (1 - P_1)(1 - P_2)\cdots(1 - P_N) \tag{2.1}$$

$$\text{or} \quad \ln(1 - P_f) = \sum_{k=1}^{N} \ln(1 - P_k) \approx -\sum_{k=1}^{N} P_k \tag{2.2}$$

Here we introduced the small P_k approximation

$$\ln(1 - P_k) \approx -P_k \tag{2.3}$$

because, in a very long chain, $P_k \ll 1$. The reason is that a very long chain is likely to fail at stress that is in the range of very low probability.

2.2 Weibull Theory

Based on his experiments of unprecedented scope (not yet surpassed), Weibull (1939, 1951) realized that, to fit his test data, the left (low-probability) tail of the cumulative distribution function (cdf) of RVE strength (i.e., failure probability P_k of one RVE) must be a power law,[1] i.e.,

$$P_k = [\sigma(\boldsymbol{x}_k)/s_0]^m \qquad \text{for small } \sigma(\boldsymbol{x}_k) \tag{2.4}$$

Here s_0 (of the dimension of stress) and m (dimensionless) are material constants; s_0 is called the scale parameter and m the Weibull modulus (or shape parameter); and $\sigma(\boldsymbol{x}_k)$ is the positive part of the maximum principal stress at a point of coordinate vector $\boldsymbol{x}_k$ (we take the positive part because negative normal stresses do not cause tensile fracture). Substituting Eq. 2.4 into Eq. 2.2 and making the limit transition from a discrete sum to an integral over structure volume V (which is justified if the structure consists of many

[1] In 1928, Fisher and Tippett (1928), working at Cambridge University, evidently unknown to Weibull, came to the same conclusion by mathematical arguments based on the stability postulate of extreme value statistics (as explained later in this chapter).

RVEs each of which is much smaller than D), we get the well-known Weibull probability integral;

$$-\ln(1-P_f)=\sum_k\left(\frac{\sigma(\boldsymbol{x}_k)}{s_0}\right)^m\approx\int_V\left(\frac{\sigma(\boldsymbol{x})}{s_0}\right)^m\frac{\mathrm{d}V(\boldsymbol{x})}{{l_0}^{n_d}} \tag{2.5}$$

where n_d = number of spatial dimensions in which the structure is scaled ($n_d = 1, 2$ or 3). The integrand

$$c_f(\boldsymbol{x})=\frac{[\sigma(\boldsymbol{x})/s_0]^m}{{l_0}^{n_d}} \tag{2.6}$$

is called the spatial concentration of failure probability and is the continuum equivalent of P_k of one RVE, whose volume is ${l_0}^{n_d}$. Because the structural strength depends on the minimum strength value in the structure, which is always in the low-probability range if the structure is large, the validity of Eq. 2.5 for large enough structures is unlimited.

Equation 2.5 is contingent upon the assumption that the brittle failure of material occurs in tension (rather than shear or a shear–tension combination), and that the random material strength is the same for each spatial direction, i.e., that the strengths in the three principal stress directions are perfectly correlated. Then it is justified to interpret σ in Eq. 2.5 as the positive part of the maximum principal stress at each continuum point (this stress must be considered as nonlocal when the finite elements are smaller than the RVEs). However, if the random strengths in the principal directions at the same continuum point were statistically independent, then $\sigma^m(\boldsymbol{x})$ in Eq. 2.5 would have to be replaced by $\sum_{I=1}^{3}\bar{\sigma}_I^m(\boldsymbol{x})$ where $\bar{\sigma}_I(\boldsymbol{x})$ are the positive parts of the principal nonlocal stresses at that point (Freudenthal 1968; Bažant & Planas 1998). Nevertheless, assuming this kind of statistical independence seems unrealistic.

2.3 Scaling of Weibull Theory and Pure Statistical Size Effect

Consider now geometrically similar structures of different sizes D. In such structures, the dimensionless stress fields $\bar{\sigma}(\boldsymbol{\xi})$ are the same functions of dimensionless coordinate vector $\boldsymbol{\xi}=\boldsymbol{x}/D$, i.e., they depend only on structure geometry but not on structure size D. In Eq. 2.5, we may then substitute

$$\sigma(\boldsymbol{x})=\sigma_N\bar{\sigma}(\boldsymbol{\xi})\quad(\boldsymbol{\xi}=\boldsymbol{x}/D) \tag{2.7}$$

where σ_N = nominal stress = P/bD, P is the applied load or a conveniently defined load parameter, and b = structure width (which may but need not be scaled with D). Further we may set

$$\mathrm{d}V(\boldsymbol{x})=D^{n_d}\,\mathrm{d}V(\boldsymbol{\xi}) \tag{2.8}$$

After rearrangements, Eq. 2.5 yields $-\ln(1-P_f) = (\sigma_N/S_0)^m$ or[2]

$$P_f = 1 - \mathrm{e}^{-(\sigma_N/s_0)^m \Psi (D/l_0)^{n_d}} = 1 - \mathrm{e}^{-(\sigma_N/S_0)^m} \tag{2.9}$$

where $$S_0 = s_0 (l_0/D)^{n_d/m} \Psi^{-1/m}, \quad \Psi = \int_V [\bar{\sigma}(\boldsymbol{\xi})]^m \, \mathrm{d}V(\boldsymbol{\xi}) \tag{2.10}$$

According to Eq. 2.9, the tail probability of structural failure is a power law:

$$P_f \approx (\sigma_N/S_0)^m \quad (\text{for } \sigma_N \to 0) \tag{2.11}$$

For $P_f \leq 0.02$ (or 0.2), its deviation from Eq. 2.9 is <1% (or < 10%) of P_f.

The effect of structure geometry is embedded in integral Ψ, which is independent of the structure size. Because exponent m in this integral is typically > 20 (and $0.8^{20} = 0.012$), the regions of structure in which the stress is less than about 80% of material strength have a negligible effect.

Note that P_f depends only on the parameter

$$s_0^* = s_0 {l_0}^{n_d/m} \tag{2.12}$$

and not on s_0 and l_0 separately. So, the material characteristic length l_0 is used here only for convenience, to serve as a chosen unit of measurement. The Weibull statistical theory of strength, per se, has no characteristic length [which is manifested by the fact that the scaling law for the mean strength is a power law (Bažant 2005)]. However, in generalization to the probabilistic-energetic theory of failure and size effect (which is expounded in Chapter 7), the use of a material characteristic length is essential. This is why introducing here l_0 is convenient.

The last expression in Eq. 2.9 is the Weibull distribution in standard form, with scale parameter S_0. From Eq. 2.10 one finds that

$$\sigma_N = C_0 (l_0/D)^{n_d/m} \tag{2.13}$$

where $$C_0 = C_f \Psi^{-1/m}, \quad C_f = s_0 [-\ln(1-P_f)]^{1/m} \tag{2.14}$$

This equation, in which C_0 and s_0 are independent of D, describes the scaling of nominal strength of structure when the failure probability P_f is specified. The mean nominal strength is calculated as (Fig. 2.2):

$$\bar{\sigma}_N = \int_0^\infty \sigma_N p_f(\sigma_N) \, \mathrm{d}\sigma_N = \int_0^1 \sigma_N \, \mathrm{d}P_f = \int_0^\infty (1-P_f) \, \mathrm{d}\sigma_N \tag{2.15}$$

[2] There is a more direct way to derive Eq. 2.9 if we consider all the N links in the chain to have the same failure probability P_1. Then, for an infinite chain,

$$1 - P_f = \lim_{N\to\infty} (1-P_1)^N = \lim_{N\to\infty} \left(1 - \frac{x}{N}\right)^N, \quad x = NP_1$$

According to the Euler relation, the last limit is equal to e^{-x}. So,

$$P_f = 1 - e^{-x} = 1 - e^{-NP_1} = 1 - e^{-N(\sigma/s_0)^m}$$

where we substituted the tail approximation from Eq. 2.4. Note that, unlike Eq. 2.2, this derivation does not require explicitly assuming P_1 to be small (although, for $N \to \infty$, P_1 of the weakest link is inevitably small).

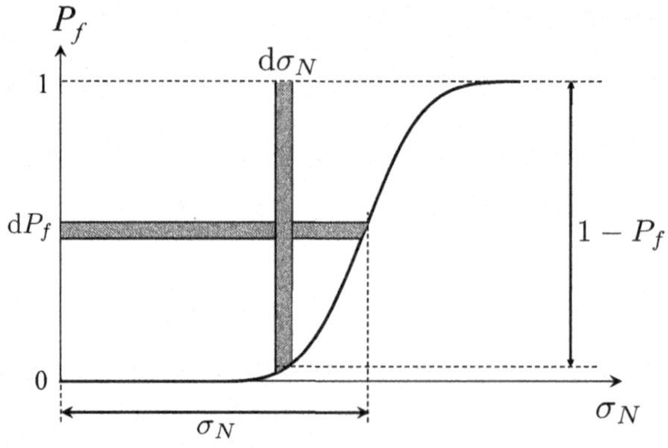

Figure 2.2. Calculation of mean structural strength.

where $p_f(\sigma_N) = \mathrm{d}P_f(\sigma_N)/\mathrm{d}\sigma_N$ = probability density function (pdf) of structural strength, while $P_f(\sigma_N)$ = strength cdf. Substituting Eq. 2.9 and noting that $\int_0^\infty t^{z-1}e^{-t}\,\mathrm{d}t = \Gamma(z)$ = gamma function, one gets, after rearrangements, the well-known Weibull scaling law for the mean nominal strength as a function of structure size D and geometry parameter Ψ:

$$\bar{\sigma}_N(D, \Psi) = S_0\Gamma(1 + 1/m) = C_s(\Psi)\, D^{-n_d/m} \tag{2.16}$$

where $$C_s(\Psi) = \Gamma(1 + 1/m)\, l_0{}^{n_d/m} s_0/\Psi^{1/m} \tag{2.17}$$

For the gamma function one may use the approximation $\Gamma(1 + 1/m) \approx 0.6366^{1/m}$, which is accurate within the range $5 \le m \le 50$ [Eq. 12.1.22 in Bažant and Planas (1998)].

The coefficient of variation (CoV) of σ_N is calculated as

$$\omega_N{}^2 = \bar{\sigma}_N{}^{-2} \int_0^\infty (\sigma_N - \bar{\sigma}_N)^2 p_f \mathrm{d}\sigma_N = \bar{\sigma}_N{}^{-2} \int_0^1 \sigma_N{}^2\, \mathrm{d}P_f(\sigma_N) - 1 \tag{2.18}$$

where $\mathrm{d}P_f(\sigma_N)/\mathrm{d}\sigma_N = p_f(\sigma_N)$ = pdf of σ_N. Substitution of Eq. 2.9 gives, after rearrangements, the following well-known expression for the strength CoV:

$$\omega_N = \sqrt{\frac{\Gamma(1 + 2/m)}{\Gamma^2(1 + 1/m)} - 1} \tag{2.19}$$

which is independent of structure size as well as geometry. Approximately, $\omega_N \approx (0.462 + 0.783m)^{-1}$ for $5 \le m \le 50$ [Eq. 12.1.28 in Bažant and Planas (1998)].

2.4 Equivalent Number of Elements

It is conceptually useful to introduce the equivalent number, N_{eq}, of RVEs for which a chain with N_{eq} links gives the same cdf. For a chain under the same tensile stress $\sigma = \sigma_N$ in each element, we have

$$P_f = 1 - \mathrm{e}^{-N_{\mathrm{eq}}(\sigma_N/s_0)^m} \tag{2.20}$$

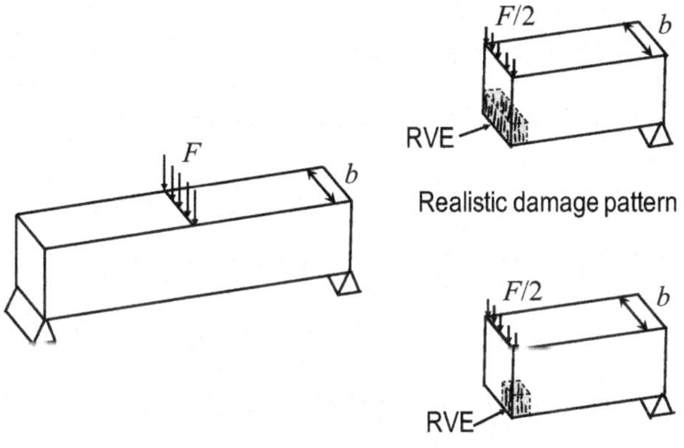

Figure 2.3. Damage pattern in a three-dimensional structure and the corresponding RVE.

Setting this equal to Eq. 2.9 and solving for N_{eq}, we obtain

$$N_{\text{eq}} = (s_0/S_0)^m = (D/l_0)^{n_d}\Psi \tag{2.21}$$

The equivalent number of RVEs, N_{eq}, is here a more convenient alternative to what is called the Weibull stress, σ_W, in the Beremin model (Beremin 1983) for large crack-tip plastic zone. That model is defined by setting

$$P_f = 1 - \mathrm{e}^{-(\sigma_W/s_0)^m} = 1 - \mathrm{e}^{-(V_{\text{eff}}/V_0)(\sigma_N/s_0)^m} \tag{2.22}$$

where $V_{\text{eff}}/V_0 = (D/l_0)^{n_d}\Psi$. Equating this to Eq. 2.9, we see that

$$\sigma_W = \sigma_N\ \Psi^{1/m}(D/l_0)^{n_d/m} \tag{2.23}$$

i.e., the Weibull stress is the nominal stress corrected for the size and geometry factors.

Equation 2.22 is the reason why the statistical size effect has often been called the "volume effect." But this term may be misleading. In Eq. 2.21 the integration over volume V of a three-dimensional structure may have to be made two-dimensionally (i.e., with $n_d = 2$) if the structure must fail, for reasons of mechanics, simultaneously through its entire width b (Fig. 2.3). This happens when the fracture must propagate simultaneously along the whole width of the fracture front, since mechanics prevents the fracture from propagating forward within only a small portion of the width. According to the "volume effect," the widening of a narrow beam, which increases the structure volume V, would be predicted to reduce the σ_N of the beam, while experience shows virtually no effect of the widening on σ_N. For this reason, calling the statistical size effect the volume effect is not quite accurate and can be misleading.

2.5 Stability Postulate of Extreme Value Statistics

The Weibull distribution is an asymptotic distribution of survival of a structure (or chain) in which the number n of elements or RVEs (or links) tends to infinity. It is called the extreme value distribution because the strength of the structure (or chain) is decided by the weakest element (or weakest link), and the extreme is here meant to be the minimum (there are other applications in which the maximum is of interest). Since a generalization to elements subjected to different stresses (or different cross sectional areas) is easy, we consider all the elements (or links) to be subjected to the same stress σ. Furthermore, we consider the strengths of all the elements (or links) to be independent identically distributed random variables.

The trick to derive the extreme value distribution purely mathematically is to subdivide the set of all elements (or links), N in number, into n identical subsets, each of which contains ν elements. To obtain the asymptotic distribution, we must consider each subset to be infinite, i.e., $\nu \to \infty$, in which case the total set (or chain) of elements (or links) is also infinite, i.e., $N = n\nu \to \infty$. Let $f(\sigma)$ and $F(\sigma)$ be the asymptotic cumulative probability of survival of each infinite subset and of the whole infinite set, respectively, subjected to stress σ. Evidently, all the subsets must survive for the whole set to survive, and so $F(\sigma)$ represents the joint probability of survival of all n infinitely many identical subsets. So, according to the joint probability theorem, we may write

$$F(\sigma) = f^n(\sigma) \tag{2.24}$$

Furthermore, since the whole set as well as all the subsets are infinite, their survival distributions must be of the same type, though not identical since the smallest random strength value among n subsets must be equal or smaller than in one subset. Although the distribution functions $F(\sigma)$ and $f(\sigma)$ must be of the same type, they can have different means and different CoVs. So they must be related by a linear transformation, i.e.,

$$F(\sigma) = f(a_n\sigma + b_n) \tag{2.25}$$

where a_n and b_n are coefficients depending on number n. Combining the last two equations, we obtain the following functional equation for the unknown asymptotic cumulative survival distribution $f(\sigma)$:

$$f^n(\sigma) = f(a_n\sigma + b_n) \tag{2.26}$$

This functional equation is called the *stability postulate of extreme value statistics* – "stability," because, in passing from the subset to the whole set, the distribution must remain of the same type, i.e., stable.

2.6 Distributions Ensuing from Stability Postulate

In arguably the most famous probability paper of the last century, Fisher and Tippett (1928) derived three asymptotic distributions that satisfy the stability postulate. These distributions are called the extreme value distributions (or stable extreme values distributions). In terms of failure probability under stress σ, $P_f(\sigma) = 1 - f(\sigma)$, they are[3]

$$1)\text{ Weibull distribution:}\quad P_f(\sigma) = 1 - e^{-\sigma^m} \tag{2.27}$$

$$2)\text{ Gumbel distribution:}\quad P_f(\sigma) = 1 - e^{-e^{\sigma}} \tag{2.28}$$

$$3)\text{ Fréchet distribution:}\quad P_f(\sigma) = 1 - e^{|\sigma|^{-m}} \tag{2.29}$$

Which of the extreme value distributions actually occurs is characterized by the so-called domain of attraction (Gnedenko 1943; Gumbel 1958; Ang & Tang 1984; Bouchaud & Potters 2000). Gnedenko (1943) derived the conditions for the elementary distribution function to converge to one of these extreme value distributions. With Fisher and Tippett's results, this leads to the Fisher–Tippett–Gnedenko theorem, which states that if the distribution of the normalized maximum converges to a non-degenerate distribution, then the limiting distribution $f(\sigma)$ would belong to one of aforementioned extreme value distributions. It should be noted that not all elementary distribution functions would lead to one of these three extreme value distributions (Vanmarcke 2010). However, the distributions whose extreme value distribution are not Weibull, Gumbel, or Fréchet are rather unusual and do not occur in applications to strength.

For Weibull distribution, the domain of attraction includes all the elementary distributions with a left tail in the form of a power law of non-negative numbers; for Gumbel, all those with an exponential tail decaying to minus infinity; and for Fréchet, all those with an inverse power law infinite tail decaying to plus infinity. Among these three, only the first (Weibull distribution) is suitable for strength of structures following the weakest-link model.

Although Fisher and Tippett's mathematical proof that no other distribution can satisfy the stability postulate is difficult,[4] it is easy to check that the Weibull, Gumbel, and Frěchet distributions satisfy it. For example, if we substitute the Weibull distribution, we subsequently get

$$(e^{-\sigma^m})^n = e^{-(a_n\sigma + b_n)^m} \tag{2.30}$$

$$n\sigma^m = (a_n\sigma + b_n)^m \tag{2.31}$$

$$(n^{1/m} - a_n)\sigma = b_n \tag{2.32}$$

Since the last equation must be valid for any σ, we conclude that

$$a_n = n^{1/m}, \quad b_n = 0 \tag{2.33}$$

[3] Since both Eqs. (2.27) and (2.28) were derived by Fisher and Tippett (1928), they should, in fairness, be called the Fisher–Tippett–Weibull and Fisher–Tippett–Gumbel distributions; Fisher and Tippett, however, did not demonstrate any applications nor experimental comparisons. Weibull and Gumbel did both, extensively so. Fréchet (1927) preceded Fisher and Tippett.

[4] It has been pointed out that these three extreme value distributions may not be exhaustive (Ang & Tang 1984; Vanmarcke 2010). However, no one presented a counterexample of another pdf satisfying the stability postulate, whose validity is generally accepted.

which are the conditions for satisfying the stability postulate. Note that here n is proportional to structure size D. Thus we have shown another way to derive the Weibull size effect, previously obtained in Eq. 2.9 (see Sec. 9.6 for more detail).

2.7 Central Limit Theorem and Strength Distribution of Ductile Structures

As we discussed in Section 1.8, the main feature of ductile (or plastic) failure is that, at the failure point, all material elements (RVEs) would reach their strength capacities simultaneously. Therefore, the overall structural strength is the sum of contributions of all the RVEs along the failure surface. The strength distribution of ductile structures must follow the Gaussian distribution except for its extreme far-left tail, which is normally beyond the range of practical interest. This can be understood from the Central Limit Theorem (CLT), which is the classical pillar of the theory of probability originated by Gauss (1809).

For convenience of readers, the original version of CLT is now presented and its simplified proof given. Let $X_1, X_2, \dots X_n$ be a set of independent random variables with arbitrary but identical pdfs $f_X(x)$. The mean and variance of each are $\mu = \int_{-\infty}^{\infty} x f_X(x) dx = \mathrm{E}(X_k)$ and $\delta^2 = \int_{-\infty}^{\infty} (x-\mu)^2 f_X(x) \mathrm{d}x = \mathrm{E}[(X_k - \mu)^2]$ where $\mathrm{E}(x)$ denotes the expectation of x. The CLT states that

$$\lim_{n\to\infty} \frac{\sum_{k=1}^{n} X_k - n\mu}{\delta\sqrt{n}} = \lim_{n\to\infty} \frac{\mu_n - \mu}{\delta/\sqrt{n}} \to N(0,1) \tag{2.34}$$

$$\text{where } \mu_n = \frac{1}{n}\sum_{k=1}^{n} X_k \tag{2.35}$$

$N(0,1)$ = standard normal (or Gaussian) distribution, i.e., the normal distribution with mean 0 and standard deviation 1. It is clear that, as n becomes large, the average value of a set of n independent identically distributed random variables will approach the Gaussian distribution, where the mean value equals the mean of the individual random variables and the standard deviation decreases as $1/\sqrt{n}$.

Prior to the proof, consider first the Fourier transform:

$$\Phi(\omega) = \int_{-\infty}^{\infty} e^{i\omega x} f_X(x)\,\mathrm{d}x = \mathrm{E}(e^{i\omega x}) \tag{2.36}$$

which is called the characteristic function ($i^2 = -1$). This function is unique, since inverse transformation exists and is $f_X(x) = \frac{1}{2\pi}\int_{-\infty}^{\infty} e^{-i\omega x}\Phi(x)\mathrm{d}\omega$. Introducing the expansion $e^{i\omega x} = 1 + \frac{i^1}{1!}\omega x + \frac{i^2}{2!}\omega^2 x^2 + \cdots$ and integrating term by term, one gets

$$\Phi(\omega) = 1 + \frac{i^1}{1!}\omega m_1 + \frac{i^2}{2!}\omega^2 m_2 + \cdots, \quad m_k = \int_{-\infty}^{\infty} x^k f_X(x)\,\mathrm{d}x \tag{2.37}$$

where $m_1, m_2, \ldots$ are the central moments of the cdfs of the random variables. Replacing X with the sum $X_1 + X_2 + \cdots + X_n$, we get the characteristic function of the sum:

$$\Phi(\omega) = \mathrm{E}(e^{i\omega(X_1+X_2+\cdots+X_n)}) = \mathrm{E}(e^{i\omega(X_1)})\mathrm{E}(e^{i\omega(X_2)})\cdots\mathrm{E}(e^{i\omega(X_n)})$$
$$= \Phi_{X_1}(\omega)\Phi_{X_2}(\omega)\cdots\Phi_{X_n}(\omega) \tag{2.38}$$

Now, to prove the CLT [e.g., Feller (1957)], let us switch to standardized random variables:

$$Y_k = \frac{X_k - \mu}{\delta} \quad (k - 1, 2, \ldots n) \tag{2.39}$$

$$\text{for which} \quad \mathrm{E}(Y_k) = 0, \quad \mathrm{E}(Y_k^2) = 1 \tag{2.40}$$

i.e., the mean is 0 and the standard deviation is 1. According to Eq. 2.37,

$$\Phi_{Y_k}(z) = 1 - \frac{z^2}{2} + \mathrm{o}(z^2) \quad (k = 1, 2, \ldots n) \tag{2.41}$$

where $\mathrm{o}(x^2)$ refers to any function that goes to 0 faster than x^2 as $x \to 0$. Note that the linear term and all odd powers of z are missing, because of symmetry, which is why the standardized variables are used. Next consider the standardized sum:

$$R_n = \sum_{k=1}^{n} \frac{X_i - \mu}{\delta\sqrt{n}} = \frac{1}{\sqrt{n}} \sum_{k=1}^{n} Y_k \tag{2.42}$$

Its characteristic function is

$$\Phi_{R_n}(z) = \mathrm{E}(e^{izR_n}) = \mathrm{E}\left[\exp\left(\frac{iz}{\sqrt{n}}\sum_{k=1}^{n} Y_k\right)\right] = \prod_{k=1}^{n} \mathrm{E}\left[\exp\left(\frac{izY_k}{\sqrt{n}}\right)\right]$$
$$= \prod_{k=1}^{n} \Phi_{Y_k}\left(\frac{z}{\sqrt{n}}\right) = \left[\Phi_{Y_k}\left(\frac{z}{\sqrt{n}}\right)\right]^n \tag{2.43}$$

(These transformations would not be true if Y_k were not identical and independent random variables.) So, according to Eq. 2.41 and based on the binomial theorem,

$$\Phi_{R_n}(z) = \left(1 - \frac{z^2}{2n} + \mathrm{o}(z^2)\right)^n \tag{2.44}$$

Finally, according to the Euler relation $\lim_{n\to\infty}(1 + a/n)^n = e^a$,

$$\lim_{n\to\infty} \Phi_{R_n}(z) = \lim_{n\to\infty} \left(1 - \frac{(z^2/2)}{n} + \mathrm{o}(z^2)\right)^n = e^{-z^2/2} \tag{2.45}$$

Which distribution has this characteristic function? – Standard normal (Gaussian) distribution. Indeed,

$$\Phi_N(\omega) = \int_{-\infty}^{\infty} e^{i\omega x}\left(\frac{1}{\sqrt{2\pi}}e^{-x^2/2}\right)\mathrm{d}x = \frac{1}{\sqrt{2\pi}}\int_{-\infty}^{\infty} e^{-x^2/2}\cos\omega x\,\mathrm{d}x = e^{-\omega^2/2} \tag{2.46}$$

Q.E.D.[5] The CLT can be traced back to Gauss (1809), and, as a limiting distribution of binomial distribution, to de Moivre (1733).

Example: Consider many independent statistical variables with rectangular distribution between a and b, for which

$$f_X(x) = \frac{1}{b-a}\left[H(x-a) - H(x-b)\right] \tag{2.47}$$

where H denotes the Heaviside step function (Fig. 2.4a). As an example, we can calculate the histograms of the average of n random variables (i.e., $y = n^{-1}\sum_{i=1}^{n} x_i$, $n = 1, 2, 6, 20$), which are sampled from a rectangular distribution between 0 and 1. Denote the cdf of y as $\Psi_n(y)$. Figures 2.4b–d present the relationship between the inverse of the standard Gaussian cdf of $\Psi_n(y)$ and the random average value y. If this relationship is linear, then $\Psi_n(y)$ must be Gaussian. Furthermore, according to CLT, the slope of this linear relationship is equal to $\sqrt{n}/\delta$, where $\delta =$ the standard deviation of the individual random variable x. For the present example, $\sqrt{n}/\delta \approx 8.49$ for $n = 6$ and $= \sqrt{240} \approx 15.49$ for $n = 20$. As shown in Figures 2.4c and d, for $n = 6$ the distribution of the averaged sum is already very close to Gaussian and follows the prediction of CLT. In fact, a sum of six randomly generated values of a rectangularly distributed variable is often used in huge computer sampling based on the Gaussian distribution, because it is much faster to compute than the exact formula.

2.8 Failure Probability When Both the Strength and Load Are Random, and Freudenthal Integral

One problem of interest to designers is the maximum load that the structure can sustain without exceeding the specified failure probability, such as 10^{-6}, which is what we considered so far. More realistic is the risk-based design problem in which the randomness of the load is also taken into account. The load randomness is characterized by a certain pdf, $g(\lambda_N)$, where λ_N is a load parameter, here chosen as the nominal stress, $\lambda_N = c_n P/dD$, based on load P rather than on the structure strength. To calculate the failure probability P_{FL} when both the structural strength and the load are random, one must sum the contributions from all the candidate catastrophic combinations of λ_N with nominal strength σ_N based on structural strength, which are those for which $\sigma_N < \lambda_N$.

For structural strength in the interval $d\sigma_N$ and load in the interval $d\lambda_N$, the joint probability theorem indicates that the contribution is $[f(\sigma_N)\,d\sigma_N][g(\lambda_N)\,d\lambda_N]$, where $f(\sigma_N)$ is the pdf of the structural strength, which is what has been discussed so far, and $g(\lambda_N)$ is the pdf of the load.[6] Therefore the failure probability when both the strength and the

[5] For a proof in terms of moment generating functions, see Bulmer (1979, p. 115).

[6] $g(\lambda_N)$ typically has a high positive skewness, i.e., a long tail on the right, which has often been ignored in calculations.

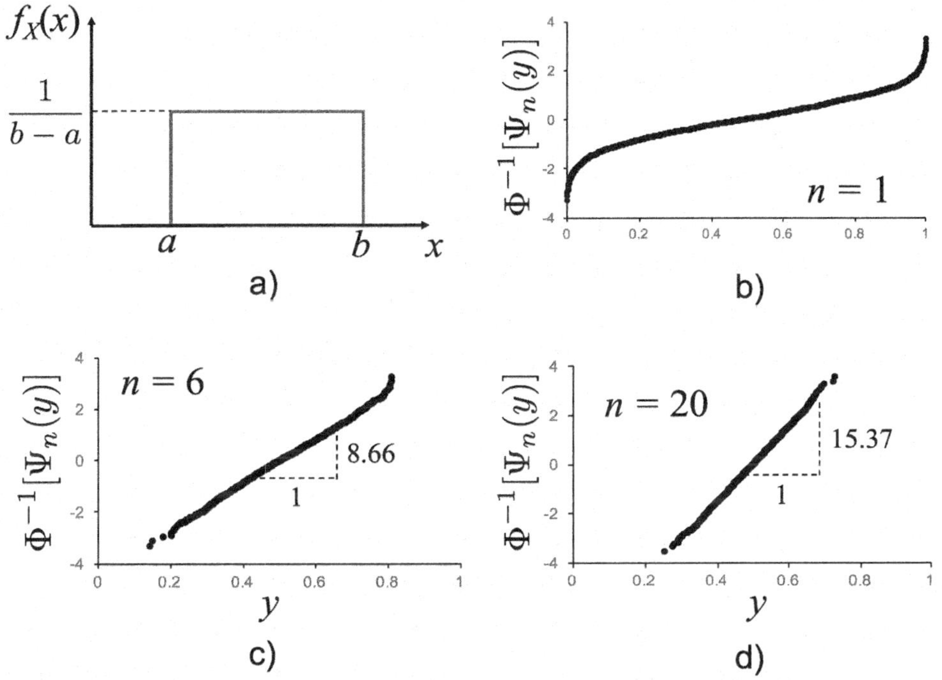

Figure 2.4. Demonstration of the Central Limit Theorem. (a) Sampling function and (b)–(d) relationship of the inverse of the standard Gaussian cdf of $\Psi_n(y)$ and the random average value y for $n = 1$, $n = 6$, and $n = 20$.

load are random is

$$P_{FL} = \text{Prob}(\sigma_N < \lambda_N) = \iint\limits_{\sigma_N < \lambda_N} f(\sigma_N) g(\lambda_N)\, d\sigma_N\, d\lambda_N \tag{2.48}$$

where the integration must be carried out only over the domain in which $\sigma_N < \lambda_N$. In the case of several nonproportional random loads of different distributions, this equation needs to be expanded into multiple semi-infinite integrals with several cdfs $g_i(\lambda_i)$ ($i = 1, 2, \ldots n$) over a domain where σ_N is less than λ_N for all possible load combinations. The evaluation of this integral becomes very complicated, and special techniques such as importance sampling are required.

For a single random load parameter λ_N, on the other hand, the evaluation is easy and, as pioneered in civil engineering by Freudenthal (1956, 1966), can even be converted to a single convolution integral (sometimes called Freudenthal's reliability integral), as follows:

$$P_{FL} = \int_0^\infty \left(\int_0^{\lambda_N} f(\sigma_N)\, d\sigma_N \right) g(\lambda_N)\, d\lambda_N = \int_0^\infty F(\lambda_N) g(\lambda_N)\, d\lambda_N \tag{2.49}$$

The upper limit of the second (parenthesized) integral ensures that the condition $\sigma_N < \lambda_N$ is never violated and that all the cases $\sigma_N \geq \lambda_N$ be excluded. Function $F(\lambda_N) = \int_0^{\lambda_N} f(\sigma_N)\, d\sigma_N$ is the cdf of the structural strength [see also Haldar and Mahadevan

(2000a), Section 7.3]. Using integration by parts, one can also convert the last equation to

$$P_{FL} = 1 - \int_0^{\infty} f(\sigma_N) G(\sigma_N)\, \mathrm{d}\sigma_N \tag{2.50}$$

where $G(\sigma_N)$ is the cdf of the load parameter.

In practice, the structural failure probability is often computed efficiently by the Cornell or Hasofer–Lind reliability indices within the framework of the first-order (FORM) and second-order (SORM) reliability methods (Haldar & Mahadevan 2000a). For quasibrittle structures, though, they need improvement owing to the size-dependence of the probability distributions of structural resistance; see Chapter 11. Except for that chapter, the rest of the book is focused on structural and material uncertainty.

3 Review of Fracture Mechanics and Deterministic Size Effect in Quasibrittle Structures

The statistical and deterministic aspects of failure and size effect are usually intertwined. Therefore, before focusing on the statistical aspects, we briefly review in this chapter the main results on the deterministic (or energetic) aspects of quasibrittle failure and the associated size effect.

3.1 Linear Elastic Fracture Mechanics

Linear elastic fracture mechanics (LEFM) is the simplest model of fracture mechanics and the basis for all its ramifications. In LEFM, the fracture process is assumed to occur within a point, i.e., the tip of a perfectly sharp crack (Fig. 3.1). Thus all the volume of the structure is elastic and there are no nonlinear boundary conditions on both the external surface and the crack faces. So, linear elasticity with the principle of superposition applies.

The LEFM criterion of crack propagation can, for example, be derived from the expression for the Helmholtz free energy $\mathcal{F}$ of the structure with N sharp cracks of length a_i ($i = 1, 2, \ldots N$):

$$\mathcal{F} = U(a_1, a_2, \ldots, a_N; \lambda) + \sum_{i=1}^{N} \int G_F \, \mathrm{d}a_i \tag{3.1}$$

where U is the strain energy of the structure; λ is a loading parameter, which controls the boundary displacements (or related input such as temperature changes); and $G_F =$ fracture energy. A necessary condition for the crack propagation of crack i at equilibrium is that $\partial \mathcal{F}/\partial a_i = 0$ at constant λ, which gives the fracture propagation criterion:

$$\mathcal{G} = -\frac{\partial U}{\partial a_i} = G_F \tag{3.2}$$

where $\mathcal{G} =$ energy release rate. If the Gibbs free energy $\mathcal{F}^*$, or complementary energy U^*, is used, we have

$$\mathcal{F}^* = -U^*(a_1, a_2, \ldots, a_N; \lambda^*) + \sum_{i=1}^{N} \int G_F \, \mathrm{d}a_i \tag{3.3}$$

where λ^* is parameter of applied forces (or loads). The necessary condition of propagation of crack i at equililbrium is that $\partial \mathcal{F}^*/\partial a_i = 0$ at constant λ^*, which yields an

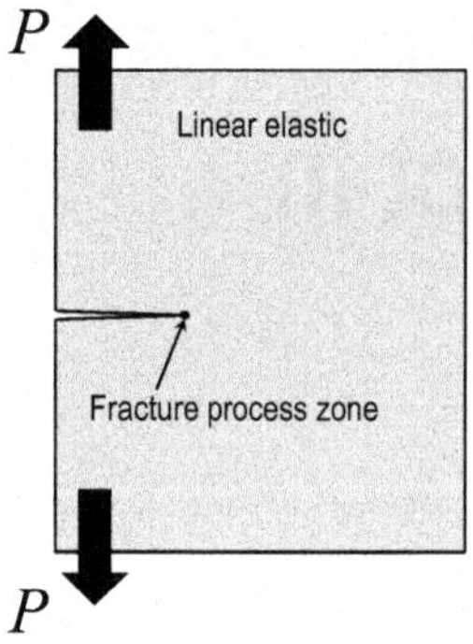

Figure 3.1. A brittle structure subjected to mode I loading.

alternative form of the fracture propagation criterion:

$$\mathcal{G} = \frac{\partial U^*}{\partial a_i} = G_F \tag{3.4}$$

$\mathcal{G}$ is related to the stress intensity factor K_I by Irwin's relation (e.g., Bažant & Planas 1998, p. 41):

$$\mathcal{G} = \frac{K_I^2}{E'} \tag{3.5}$$

where $E' = E =$ Young's modulus for plane stress and $E' = 1/(1 - \nu^2)$ for plane strain ($\nu =$ Poisson ratio); in K_I subscript I refers to mode I, which is the opening mode of fracture. Using K_I, one may write the criterion of crack propagation as $K_I = K_c$ where K_c is called the fracture toughness, and according to Eq. 3.5, one has

$$K_c = \sqrt{E'G_f} \tag{3.6}$$

Relations similar to Eq. 3.5 hold for mode II, the in-plane shear mode, and mode III, the out-of-plane shear mode (Fig. 3.2). Each of K_I (or K_{II}, K_{III}) is a parameter of its associated asymptotic near-tip stress, strain, and displacement fields. These fields have the general forms

$$\sigma_{ij} = r^{-1/2} f_{ij}(\theta) \tag{3.7}$$

$$\epsilon_{ij} = r^{-1/2} g_{ij}(\theta) \tag{3.8}$$

$$u_i = r^{1/2} F_i(\theta) \tag{3.9}$$

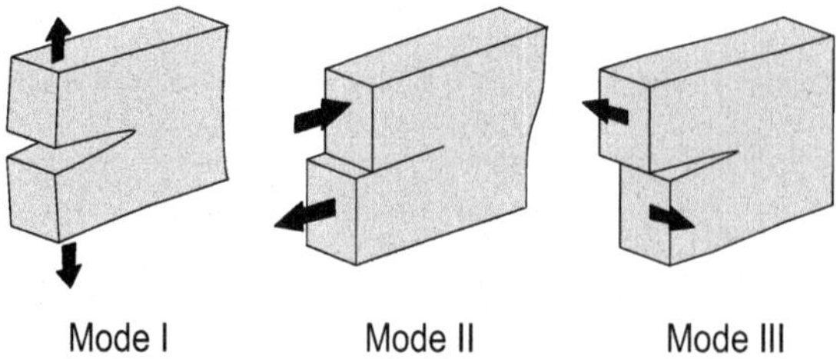

Figure 3.2. Three basic fracture modes: mode I (opening mode), mode II (in-plane shear mode), and mode III (out-of-plane shear mode).

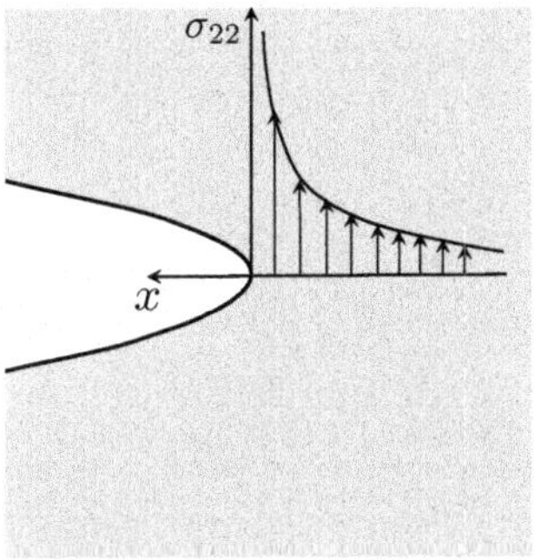

Figure 3.3. Near-tip crack mouth opening profile.

where (r, θ) are the polar coordinates centered at the crack tip, with the angle normally measured counterclockwise from the crack extension line, and $f_{ij}(\theta), g_{ij}(\theta), F_i(\theta) =$ certain functions of θ, which do not depend on the structure geometry (e.g., Bažant & Planas 1998). The equation for u_2 $(i = 2)$ means that the near-tip opening of a LEFM crack has a parabolic profile $w = 8K_{\mathrm{I}}\sqrt{x}/\sqrt{2\pi}E'$ (Fig. 3.3), and the equation for σ_{22} in mode I gives, just ahead of the tip, the stress profile $\sigma_{yy} = K_{\mathrm{I}}/\sqrt{2\pi x}$.

The stress intensity factor is proportional to the applied load. For example, for an infinite plane loaded at infinity by uniaxial tensile stress σ, a crack of length $2a$, normal to σ, has the stress intensity factor

$$K_{\mathrm{I}} = \sigma\sqrt{\pi a} \tag{3.10}$$

For multiple loads, the entire singular fields are additive and if these loads produce the same mode of fracture, the stress intensity factors are also additive.

The energy release rate $\mathcal{G}$, defined by Eq. 3.2, is in LEFM always equal to the energy flux into the propagating crack tip. This flux may be calculated by Rice's J-integral (Rice 1968*b*):

$$J = \int_{\Gamma}\left(\frac{1}{2}C_{ijkl}\sigma_{ij}\sigma_{kl}n_1 - n_j\sigma_{ij}\frac{\partial u_i}{\partial x_1}\right)\mathrm{d}s \tag{3.11}$$

which is path independent; here $C_{ijkl} =$ the compliance tensor, $\Gamma =$ any closed integration path running counterclockwise around the crack tip, $n_i =$ unit normal to this path, and $x_1 =$ the direction of propagation (Fig. 3.4).

3.2 Cohesive Crack Model

In quasibrittle materials, fracture is a nonlinear problem because the fracture process zone (FPZ) is non-negligible in size. The FPZ undergoes progressive fracturing, which represents softening damage due to random microcracking and frictional-dilatant macroslips. Under the hypothesis that the fracture localizes into a single crack, the cohesive crack model lumps the inelastic strains $\epsilon''(x, y)$ over the width h of the FPZ into one line, the crack extension line (whose direction is denoted as x). In other

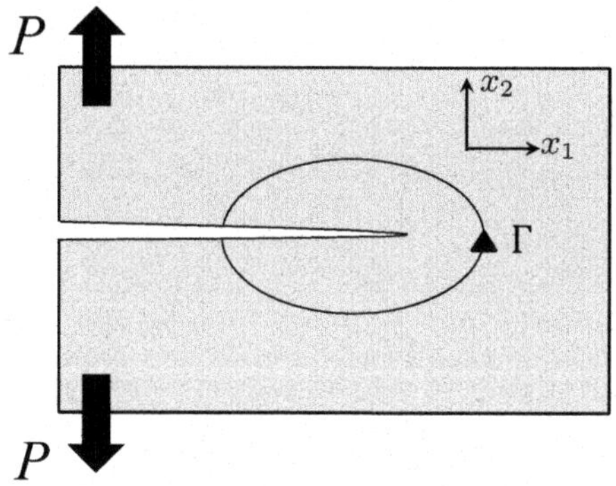

Figure 3.4. Schematic of the contour for computation of the J-integral.

words, $w(x) = \int_0^h \overline{\epsilon''}(x, y)\,\mathrm{d}y$ = opening width of the cohesive crack (Fig. 3.5), where $\overline{\epsilon''}(x, y)$ = average inelastic strain across the FPZ.

The cohesive crack model was conceived by Barenblatt (1959, 1962), while for ductile fracture of metals a similar model was conceived by Dugdale (1960). The foundation of cohesive crack model was completed by an essential contribution of Rice (1968*a*, 1968*b*) who proved that the flux of energy J into the moving fracture process zone is, in stationary propagation, equal to the area, G_f, under the cohesive softening curve (which is a point mostly overlooked in concrete literature); see Eq. 3.14.

The cohesive crack model was subsequently developed and applied by many researchers (Knauss 1973; Palmer & Rice 1973; Knauss 1974; Wnuk 1974; Kfouri & Rice 1977; Bao & Suo 1992; Li & Bažant 1994, 1997; Bažant & Li 1994, 1995*a*, 1995*b*, 1997; Broberg 1999; Ortiz & Pandolfi 1999; Freed & Banks-Sills 2008; Park, Paulino, & Roesler 2009; Barenblatt 2014). Hillerborg, with Modéer and Petersson (Hillerborg, Modéer, & Petersson 1976), deserve credit for popularizing the cohesive crack model (under the name of the "fictitious crack model") for concrete, and for developing an effective finite element algorithm.

The decline of the cohesive stress σ_{yy} is, in the cohesive crack model, defined by the softening cohesive law (Fig. 3.6):

$$\sigma_{yy} = f(w) \tag{3.12}$$

in which $f(w)$ is a monotonically decreasing function, $f'(w) < 0$. This function is assumed to characterize the material properties of a single localized crack. Such characterization is sufficient as long as the normal stresses σ_{xx} and σ_{zz} along the crack plane

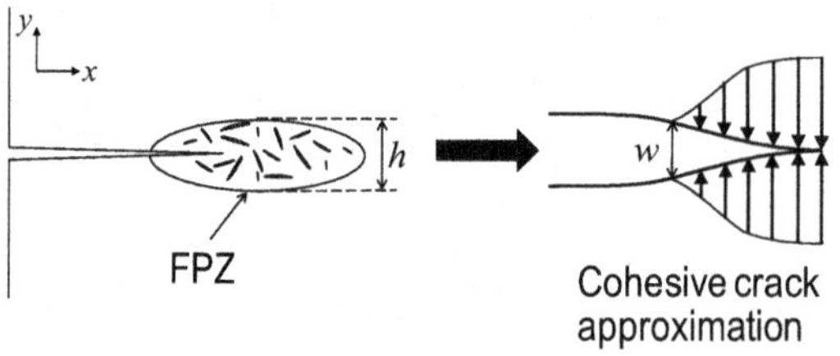

Figure 3.5. Approximation of the FPZ using the cohesive crack model.

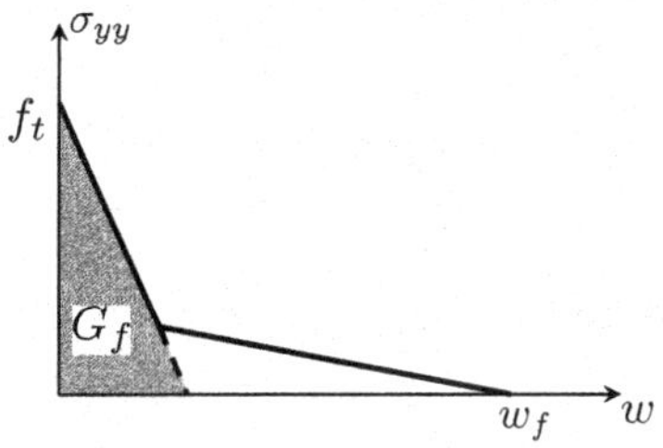

Figure 3.6. Schematic of the bilinear cohesive law.

are negligible compared to the compressive and tensile strength limits of the material (otherwise a crack-band or nonlocal model is required). At the start of crack opening, $f(0) = f_t$ = tensile strength of the material. The total fracture energy of the cohesive crack model is defined as

$$G_F = \int_0^{w_f} f(w)\,\mathrm{d}w \tag{3.13}$$

where w_f is the crack opening at which the cohesive stress is reduced to zero. The point at which $w = w_f$ is the point that delineates the real crack from the cohesive crack. Instead of G_F, one can equivalently characterize the fracture properties by the corresponding fracture toughness $K_c = \sqrt{EG_F}$.

For steady-state propagation, it was shown by Rice (1968*a*, 1968*b*) that the energy flux into the FPZ is equal to the fracture energy, i.e.,

$$J = G_F \tag{3.14}$$

where the path of the J-integral must lie outside the FPZ. This equality does not hold for non-steady states of crack propagation.

The cohesive crack model removes the LEFM stress singularity at the crack tip, and requires that the total stress intensity factor

$$K_{\mathrm{I}t} = K_{\mathrm{I}} + K_\sigma = 0 \tag{3.15}$$

where K_{I}, K_σ = stress intensity factors due to applied loads and to cohesive stresses. This ensures that the cohesive crack, unlike a LEFM crack, has a smooth closure at its tip. The smooth closure condition $K_{\mathrm{I}t} = 0$ must be imposed in analytical as well as numerical calculations.

The fact that both fracture energy G_F and material strength f_t are independent material properties, having different dimensions (J/m^2 and N/m^2), implies a material characteristic length,

$$l_0 = \frac{E\,G_F}{f_t^2} \tag{3.16}$$

called Irwin's length (Irwin 1958). Its physical meaning is that it characterizes the length of the FPZ (and is proportional to it).

The cohesive crack model has the advantage that the entire volume of the structure can be treated as elastic and all the nonlinearity of the FPZ is embedded in the boundary conditions at the crack face. This feature makes some analytical approaches possible. The crack opening $w(x)$ and load-point displacement u_P can be expressed with the help of Green's (or influence) functions as integrals over the cohesive stress distribution $\sigma_{yy}(x)$ and the applied load P. This leads to a set of two integral equations which, together with another integral representing the smooth closure condition, define the cohesive crack problem. It was shown that the calculation of the maximum load can be reduced to an eigenvalue problem, which allows very efficient direct calculation of the maximum load and its size dependence without solving the history of crack growth (Bažant & Li 1994, 1995*a*, 1995*b*, 1997; Li & Bažant 1994, 1997; Bažant & Planas 1998).

Recently it has been shown that, over a broader size range, the cohesive crack model is very sensitive to the shape of the cohesive curve. It was found that a realistic cohesive law for concrete (Cusatis & Schauffert 2009; Bažant & Yu 2011) as well as for bone (Bažant et al. 2013) is a bilinear curve with four independent parameters, having an abrupt change of slope (at the "knee point") and a long tail (Fig. 3.6). The area under the initial steep tangent is called the initial fracture energy, G_f, and typically $G_f \approx 0.3G_F$ to $0.5G_F$; G_f of course implies a second material characteristic length,

$$l_1 = \frac{E\,G_f}{f_t^2} \tag{3.17}$$

which is shorter than l_0 and is usually approximately equal to the double of the maximum size of material inhomogeneities such as the maximum aggregate size in concrete. In large structures, the stress values reach the tail segment already at maximum load. The transition from G_f to G_F as the specimen size increase explains why some experimenters using a fixed two-parameter softening curve (e.g., linear or exponential) erroneously concluded that the fracture energy increases with specimen size.

It turns out that the maximum loads of normal concrete structures are controlled by G_f rather than G_F because normally the stress values do not yet reach the tail segment of the post-peak softening curve. They reach into it only during post-peak softening, which is important for energy absorption under impact.

3.3 Crack Band Model

The *crack band model*, developed by Bažant and Oh (1983) (also Bažant 1982) as an extension of simpler models with sudden post-peak stress drop instead of gradual softening (Bažant & Cedolin 1979, 1980), generalizes the cohesive crack model in a manner that is both convenient for finite element analysis and enabling a more realistic representation of the distributed fracture.

The cohesive crack model, on the other hand, can be regarded as an approximation of the crack band model ignoring the effect of FPZ width and of the normal stresses parallel to the crack plane. While the cohesive crack model is, in finite element programs, represented as an interelement crack, the crack band model is modeled as a stack (or a band, strip) of finite elements undergoing damage. It has been checked (Bažant & Cedolin 1979) that, if a sudden post-peak stress drop is assumed and normal mesh size is used (about 20 elements over the cross section), the energy release rate obtained from the crack band model gives for LEFM K_I errors no greater than 1%.

Among various types of nonlinear fracture models, the crack band model is what now predominates for concrete in the structural and geotechnical engineering industry and for composites in the aircraft industry, and is also an embedded automatic feature of various commercial softwares (e.g., SBETA, ATENA, DIANA) and open source code OOFEM.

The inelastic damage strain $\epsilon''_{yy} = w/h = \overline{\epsilon''}_{yy}$ is assumed to be distributed uniformly over the width of the so-called crack band, equal to the finite element width. In contrast to the cohesive crack model, the damage in the crack band can be defined by a general nonlinear triaxial constitutive law:

$$\boldsymbol{\sigma} = \boldsymbol{F}(\boldsymbol{\epsilon}) \quad \text{or} \quad \sigma_{ij} = F_{ij}(\epsilon_{kl}) \tag{3.18}$$

where $\boldsymbol{\sigma}$, $\boldsymbol{\epsilon}$ are the stress and strain tensors; i, j, k, l are cartesian tensorial subscripts; and $\boldsymbol{F}$ is a function of a history functional.

A big advantage of the crack band model is that it uses a tensorial damage constitutive law (Eq. 3.18), which is much richer than the cohesive law (Eq. 3.12), this law being only uniaxial. A sophisticated, highly realistic, damage constitutive model such as microplane model M7 (Caner & Bažant 2013) can be used. Such a constitutive law is automatically sensitive to stress components other than σ_{yy}, i.e., to $\sigma_{xx}, \sigma_{zz}, \sigma_{xy}, \sigma_{xz}$, and σ_{yz}. Among these, the normal compressive stresses parallel to crack plane are the most important since they can promote fracture and can alone cause splitting fracture (even when $\sigma_{yy} = 0$). The shear components, on the other hand, cause a change in crack band growth direction.[1] The microplane model, without or with creep or rate effects, has been used as a powerful constitutive law for realistic crack band fracture analysis of concrete, gas shale, fiber–polymer composites, fiber-reinforced concrete, and so forth, modeling effects that cannot be captured with the cohesive crack model.

As established in a number of studies beginning in 1976 (Bažant 1976; Bažant & Cedolin 1979, 1980; Pietruszczak & Mróz 1981; Bažant et al. 1984), a softening constitutive model can give objective results only if it is combined with a material characteristic length, h (which is not the same as l_0 or l_1). Length h controls, and can be measured as, the minimum possible spacing of parallel cracks when the cracks do not localize. Systems of nonlocalized parallel cracks of constant spacing can occur under certain types of loading, e.g., when localization is prevented by embedded reinforcement, plates (e.g., laminates) bonded on the surface, or by a zone of compressed material in front of the parallel cracks (an example is the cracks in cooling a half-space). Bažant and Pijaudier-Cabot (1989) devised a test of the crack band width that is based

[1] The possibility of using a realistic damage constitutive law is also an advantage over the phase-field models (Borden et al. 2012, 2014) recently co-opted from physics. These models use a primitive damage constitutive law in which the degree of damage is characterized by a single damage parameter c varying smoothly between 1 and 0 across the field. For a realistic constitutive law fitting the existing variety of triaxial test data for concrete or shale, one would need to vary continuously across the phase field over a dozen parameters (such as the yield limits on the microplanes). This would be prohibitively impractical. So would be a transition of continuous damage field into a single localized phase-field band.

on preventing localization in a concrete layer by bonded steel plates. The fact that the crack band model (as well as the nonlocal model) can model nonlocalized distributed damage and its transition to localized damage (representing the localization instability) is a major advantage over the cohesive crack model (and also over the phase-field model).

Note that while the cohesive softening curve must not include elasticity, i.e., cannot have a rising segment (which is a condition violated in many papers), the damage law used in the crack band model must include elasticity, i.e., must begin with a rising elastic segment that represents the average elastic strain over the band width. If it is determined that the distributed damage will localize into a single isolated crack band, the crack band model allows increasing or decreasing the band width h to another width h^*, as suits the finite element calculations. But to do that, the steepness of the post-peak softening part of the damage law must be adjusted to maintain the same energy dissipation, which is a crucial feature of the crack band model.

It is instructive to consider the most primitive relation between the normal stress σ across the crack band and the corresponding average strain $\bar{\epsilon}$ – a bilinear curve of rising slope E and descending slope E_t ($E_t < 0$). The fracture energy is $G_F = (hf_t^2/2)(1/E - 1/E_t)$ where f_t = tensile strength. To preserve the same G_F when the crack band width is changed from h to h^*, the inverse descending slope must be adjusted to

$$\frac{1}{E_t^*} = \frac{1}{E} - \frac{2G_F}{h^* f_t^2} \tag{3.19}$$

Note that when

$$h^* > \frac{2EG_F}{f_t^2} = 2l_0 \tag{3.20}$$

one gets a negative E_t^*, which represents a snapback on the stress-strain relation and is difficult to handle in static analysis. So, $2l_0$ represents a limit on increasing the crack band width to reduce the number of elements needed for analysis.

For a realistic but complex constitutive law such as the microplane model, it is not a trivial task to identify which material parameters should be adjusted to regularize the energy dissipation in the softening regime. To resolve this issue for the case in which the element size is larger than the crack band size, it has been proposed to keep the material properties and parameters unchanged, while the crack band is coupled with fictitious elastic elements (springs) in the principal stress directions as shown in Figure 3.7 (Červenka, Bažant, & Wierer 2005). In the finite element implementation, the elastic property of the added spring is the same as the elastic property of the material and the effective length of the spring is much larger than the crack band size. Therefore, during the fracture process, the energy stored in the elastic spring can be transferred to the crack band and get dissipated. It is with this algorithm that the crack band model has been implemented (together with complex damage constitutive law) in commercial finite element softwares such as ATENA.

The crack band model gives excellent results when the mesh is laid out so that the mesh lines match the direction of crack propagation. If the direction of propagation

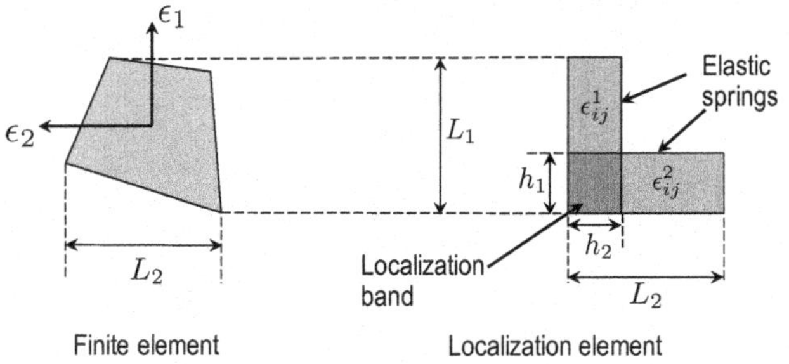

Figure 3.7. Simplified 2D representation of spring–crack band arrangement.

neither is known nor can be estimated by preliminary analysis, the crack band will propagate zig-zag across the mesh (Fig. 3.8a). In this case, the accuracy is considerably impaired because the development of shear stresses in the direction parallel to the overall zig-zag band would cause some degree of interlocking. Compared to propagation along the mesh line, errors of about $\pm 17\%$ of the applied force can occur in this case (Bažant 1982). Such errors can be significantly mitigated by using a triangular mesh (Fig. 3.8b), which consists of horizontal, vertical, and diagonal mesh lines (note that the widths of vertical and inclined crack bands in Fig. 3.8b are different, which must be reflected in a change of postpeak slope). The mesh in Fig. 3.8b allows a better approximation of arbitrary fracture propagation direction in the case in which the actual propagation direction is unknown. An alternative remedy, which has been adopted in commercial software such as ATENA, is to introduce an empirical correction factor depending on the angle between the mesh line and crack band direction (Červenka 1998).

The concept of the crack band model has recently been extended to continuum-based probabilistic modeling of quasibrittle fracture (Le & Eliáš 2016). The model is anchored by a probabilistic treatment of damage initiation, localization, and propagation. In addition to the regularization of fracture energy for localized damage, the model takes into account the probabilistic onset of damage localization inside the finite element. The random location of the crack band is related to the random material strength, whose statistics is described by a finite weakest-link model. It was shown that the model could effectively mitigate the issue of spurious mesh dependence of stochastic simulations of quasibrittle fracture (Le & Eliáš 2016).

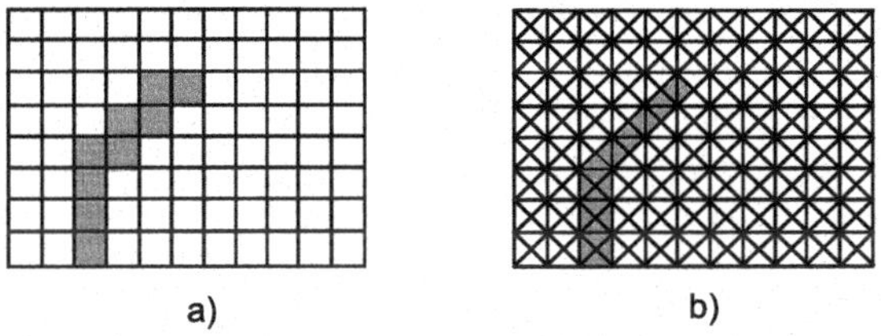

Figure 3.8. Modeling of crack propagation inclined to mesh lines. (a) Zig-zag propagation of a crack band in a square mesh, and (b) a better mesh representation of inclined fracture path.

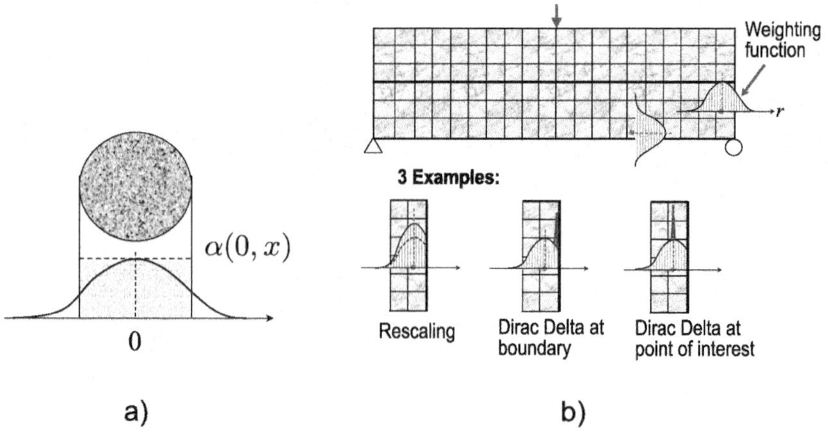

Figure 3.9. Nonlocal damage model of quasibrittle fracture. (a) Nonlocal weighting function, and (b) treatment of weighting function near the structural boundary.

3.4 Nonlocal Damage Models and Lattice-Particle Model

The nonlocal damage model is the most general model to handle distributed softening damage and control its localization instability. In the first and simplest model (Bažant et al. 1984),[2] the stress tensor at point $\boldsymbol{x}$ is calculated from the nonlocal continuum strain tensor $\bar{\boldsymbol{\epsilon}}(\boldsymbol{x})$ at point $\boldsymbol{x}$, which is defined as the weighted average of the strain tensors $\boldsymbol{\epsilon}(\boldsymbol{x}')$ at adjacent points $\boldsymbol{x}'$ within a certain neighborhood of point $\boldsymbol{x}$, with volume V_0, i.e.,

$$\bar{\boldsymbol{\epsilon}}(\boldsymbol{x}) = \int_{V_0} \alpha(\boldsymbol{x}, \boldsymbol{x}')\boldsymbol{\epsilon}(\boldsymbol{x}')\,\mathrm{d}V(\boldsymbol{x}') \tag{3.21}$$

where $\alpha(\boldsymbol{x}, \boldsymbol{x}')$ is an empirical bell-shaped weighting function (Fig. 3.9a). The size of the neighborhood is given by the material characteristic length l_1.

The original model (Eq. 3.21), however, had two problematic features: (1) wave propagation that was dispersive even for a homogeneous elastic continuum, and (2) the possibility of zero-energy periodic deformations for some weight functions $\alpha(\boldsymbol{x}, \boldsymbol{x}')$. These problems were overcome by the nonlocal continuum with local elastic strain (Pijaudier-Cabot & Bažant 1987; Bažant & Pijaudier-Cabot 1988; Bažant & Jirásek 2002), in which only the softening damage strain $\boldsymbol{\epsilon}''(\boldsymbol{x})$ or its parameter is subjected to nonlocal averaging of the type in Eq. 3.21. Besides the softening damage strain, the nonlocal averaging procedure can also be applied to other state variables of the constitutive model that characterize the damage process (Bažant & Jirásek 2002).

For all nonlocal models, a difficulty arises near the structure boundary where the averaging neighborhood tends to protrude outside the boundary (Fig. 3.9b). Various

[2] The use of nonlocal concept in continuum mechanics is much older, dating to the 1960s. The first nonlocal models, however, were introduced to capture, in a continuous manner, the sizes and spacing of material inhomogeneities in elastic and plastic-hardening materials (Eringen 1966, 1972; Kröner 1966, 1967; Eringen & Edelen 1972) and were not able to prevent spurious damage localization.

convenient ad hoc assumptions were introduced. For example, one may rescale the weighting function within the body (Bažant & Jirásek 2002) or place a Dirac delta function either along the structural boundary, or at the center point of the nonlocal integral (Borino, Failla, & Parrinello 2003). However, these approaches were not based on physical arguments. The problem with the boundary can be eliminated by the nonlocal boundary layer model. In this model, a boundary layer of thickness l_1 is excluded from the body in which the nonlocal averaging is carried out; see Section 8.1.

An alternative to the nonlocal integral formulation is the nonlocal gradient models. These are weakly nonlocal models in which the nonlocal quantities are calculated by using the spatial gradients. Early attempts focused on the development of explicit-gradient models, where the nonlocal variables can be calculated based on the explicit second-order derivative of the corresponding local variable (Aifantis 1984; Bažant 1984*a*), i.e.,

$$\bar{\epsilon} = \epsilon + c\nabla^2 \epsilon \tag{3.22}$$

where $c = l^2 \times$ constant and $l =$ material characteristic length. The inclusion of the second-order derivative imposes challenges in numerical implementation. A more robust approach is to use an implicit gradient model, in which the nonlocal variables are calculated by solving the Helmholtz differential equation (Peerlings et al. 1996, 2001), i.e.:

$$\bar{\epsilon} - c\nabla^2 \bar{\epsilon} = \epsilon \tag{3.23}$$

with the boundary condition $\mathbf{n}\nabla\epsilon = 0$, where $\mathbf{n}$ is the unit normal vector of the structure boundary. However, this Neumann boundary condition for the implicit gradient model has not been physically justified, which is somewhat analogous to the aforementioned issue of the treatment of the nonlocal weighting function near the structure boundary.

The most realistic approach to softening damage in brittle heterogeneous materials is the discrete models in the form of either lattice or particle models. The discrete lattice model was first proposed by Hrennikoff (1941) to solve elasticity problems, and was later developed for both 2D and 3D simulations of quasibrittle fracture (Zubelewicz & Bažant 1987; Bažant, Tabbara, Kazemi, & Pijaudier-Cabot 1990; Schlangen & van Mier 1992; Jirásek & Bažant 1995*a*, 1995*b*; Cusatis, Bažant, & Cedolin 2003*a*, 2003*b*). The discrete particle model was originated from the so-called distinct element model developed for simulation of the behavior of particulate materials such as cohesionless soils and rock masses (Cundall 1971; Cundall & Strack 1979). The model was further extended to concrete materials (Zubelewicz & Bažant 1987), in which the material domain is discretized by a set of rigid polyhedral elements.

In further extensions of this approach, these elements were connected by nonlinear springs of zero length (Kawai 1978; Bolander & Saito 1998; Bolander, Hong, & Yoshitake 2000). All these discrete models involve some characteristic length scale (e.g., particle size or lattice size), which acts as a localization limiter regularizing the energy dissipation in the fracture process. In discrete particle models, the particles could either represent the actual material inhomogenieties (e.g., aggregates in concrete materials) or simply be used as an efficient means to introduce a length scale to the model (Jirásek & Bažant 1995*a*, 1995*b*).

Over the last decade, significant advances have been made toward realistic discrete models for various concrete materials. A realistic model, called the lattice discrete particle model (LDPM), has been developed by Cusatis and coworkers (Cusatis et al. 2003*a*, 2003*b*; Cusatis, Pelessone, & Mencarelli 2011). This model has been shown to give very realistic predictions of fracture and cracking damage in many situations, in excellent agreement with experiments. The LDPM was extended to fiber-reinforced concrete (Schauffert & Cusatis 2011, Schauffert et al. 2011) and to rate-dependent failure of concrete (Smith et al. 2014). One attractive feature of discrete models is that they provide an explicit representation of the heterogeneities of the material, which makes them well suited to handle the random size distribution of the material heterogeneities. With further incorporation of the randomness of material properties, the discrete element model has been used successfully for stochastic modeling of quasibrittle fracture (Grassl & Bažant 2009; Eliáš, Vořechovský, Skočeк, & Bažant 2015).

Discrete element models usually require excessive computational powers, especially for analysis of large-scale structures. Therefore, recent efforts have been directed toward the development of robust multiscale computational frameworks, which aim to bridge the discrete element model and the continuum finite element model (Rezakhani & Cusatis 2016).

3.5 Overcoming Instability of Tests of Post-Peak Softening of Fiber–Polymer Composites

In addition to the challenges in computation, the post-peak softening behavior of quasibrittle structures has posed difficulties in experiments. Until 1963, it was believed that concrete and rock explode at maximum load and the load applied by the testing machine drops suddenly to zero. Beginning in 1963, it was discovered that a gradual post-peak decline of compressive or tensile load could be measured by using a stiff loading frame and fast hydraulic servo-control (Rüsch & Hilsdorf 1963; Hughes & Chapman 1966; Evans & Marathe 1968; Heilmann, Hilsdorf, & Finsterwalder 1969; Wawersik & Fairhurst 1970; Hudson, Brown, & Fairhurst 1971). The stabilizing effect of machine stiffness was mathematically demonstrated by static stability analysis by Bažant in 1976 [cf. Bažant & Cedolin (1991, Section 13.2)]. The controllability of post-peak response was further enhanced by electronic control with a gauge measuring the crack-mouth opening displacement (CMOD) or the crack-tip opening displacement (CTOD).

Unfortunately, the stiff machine frames and fast CMOD control did not work for composites. The specimens still failed dynamically upon reaching the peak load, and no stable post-peak could be measured. Consequently, for decades many experts doubted the applicability of fracture mechanics to composites, especially the cohesive crack model, and used stress-based failure criteria despite their deficiencies. In 2016, though, a breakthrough came.

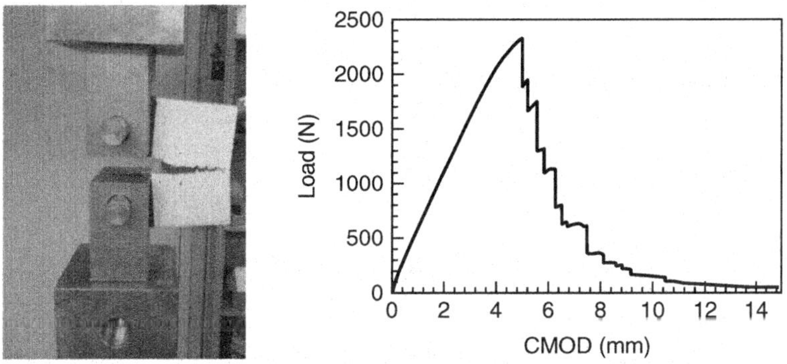

(a) (b)

Figure 3.10. Compact tension fracture test of woven carbon-epoxy laminate (Salviato et al. 2016). (a) Test set-up with new stiff and massive grips (patent pending, July 2016), and (b) example of a measured load-deflection curve with post-peak softening.

Based on mathematical analysis of dynamic stability and controllability of the test setup, it was found that, for specimens made of very strong and very light materials requiring loading grips (or fixtures), one needs far stiffer and more massive grips than supplied by manufacturers prior to 2016 (for composites, about 250 times stiffer and 10 times heavier). With the new grip design, stable measurement of post-peak softening was demonstrated for compact tension specimens of woven carbon-epoxy laminates, not only for the CMOD control but also for the load-point control (Fig. 3.10) (Salviato et al. 2016). The area under the complete load-deflection curve agreed with the fracture energy deduced by matching notched size effect tests with the size effect law. This proves that quasibrittle (or cohesive) fracture mechanics is perfectly applicable to carbon composites.

3.6 Dimensional Analysis of Asymptotic Size Effects

The asymptotic fracture behavior for structure size much smaller or much bigger than that of the FPZ size can often be determined by dimensional analysis. The number of governing nondimensional (or dimensionless) variables can be determined from Vashy–Buckingham's Π theorem of dimensional analysis (Vashy 1892; Buckingham 1914, 1915; Barenblatt 1979, 1987, 2014). This theorem states that the number of nondimensional variables (say, n) governing any physical problem is equal to the total number of variables (say, n_{tot}) minus the number of parameters with independent dimensions (say, n_{ind}); i.e.,

$$n = n_{\text{tot}} - n_{\text{ind}} \tag{3.24}$$

(a) Elastic or Elasto-Plastic Materials

The failure of elastic or elasto-plastic materials is governed by the strength limit or the yield stress, τ_0, whose metric dimension is N/m^2. The failure also depends on the nominal stress σ_N, whose metric dimension is also N/m^2. Further, it depends on the characteristic structure size D and other dimensions such as span L, notch length a and various other geometric characteristics, all of which have the metric dimension of m. So, according to the Vashy–Buckingham theorem, the equation governing failure must have the functional form

$$\Phi\left(\frac{\sigma_N}{\tau_0}, \frac{L}{D}, \frac{a}{D}, \ldots\right) = 0 \tag{3.25}$$

Since τ_0 is a constant, and since for geometrically similar structures $L/D, a/D, \ldots$ are also constants, it follows that the nominal stress at failure, σ_N, must be proportional to τ_0 and thus a constant when only the structure size D is varied.

So, when the failure criterion is defined in terms of only the material strength (or yield limit), we have in Eq. 1.5:

$$s = 0, \quad \sigma_N \propto D^0 = \text{constant} \tag{3.26}$$

The same scaling must apply to stresses and strains at any homologous points of structures of different sizes. Since the strains are the derivatives of displacement, the displacements u at homologous points scale as $u \propto D^1$.

(b) Linear Elastic Fracture Mechanics

Here the failure is determined by the value of the critical stress intensity factor K_c, the metric dimension of which is N m$^{-3/2}$. The other quantities determining failure are the same as before, including σ_N, D, L, a, and so forth. Again, the number of nondimensional variables on which the failure can depend follows from Vashy–Buckingham's Π theorem and it turns out that the failure condition must now have the form

$$\Phi\left(\frac{\sigma_N\sqrt{D}}{K_c}, \frac{L}{D}, \frac{a}{D}, \ldots\right) = 0 \tag{3.27}$$

Since K_c is a material constant, and since the ratios $L/D, a/D, \ldots$ are all constant (for geometrically similar structures), it follows that $\sigma_N\sqrt{D}$ must also be constant. Hence,

$$s = -1/2, \quad \sigma_N \propto D^{-1/2} \tag{3.28}$$

provided that not only the structure dimensions but also the crack length (and shape) are geometrically scaled.

(c) Scaling for Critical Crack Size Independent of Structure Size

Note first that there is a fundamental difference between the classical applications of fracture mechanics to metallic structures and the modern applications to quasibrittle structures such as concrete structures:

- In typical metallic structures, the maximum load often occurs (or must be assumed to occur) while the crack size is still negligible compared to the structural dimensions

and is determined by material characteristics such as the spacing of major defects, the grain size, or Irwin's material characteristic length.
- In concrete and other quasibrittle structures, there often is large stable crack growth (with distributed damage) before the maximum load is reached. The maximum load often occurs when either a large FPZ develops or the crack extends over a significant portion of the cross section (in reinforced concrete structures, typically 50% to 90%).

Consider geometrically similar metallic structures of different sizes which, at maximum load, often contain cracks that are small compared to the cross section dimension. Often the cracks at maximum load are roughly of the same size for different structure sizes, and are so small that the disturbance of the stress field caused by the crack is only local. So, if we, for instance, consider a crack of length $2a$ in two dimensions, the stress intensity factor (obtained from the energy release rate $\mathcal{G}$ as $K_{\mathrm{I}} = \sqrt{E\mathcal{G}}$, E being Young's modulus) is approximately calculated from Eq. 3.10, $K_{\mathrm{I}} = \sigma\sqrt{\pi a}$, which is exact for a crack in a homogeneously stressed infinite solid. The condition of no failure is written as $K_{\mathrm{I}} < K_c$ where K_c is the given fracture toughness of the material. Obviously, this condition of no failure is equivalent to

$$\sigma < f_t, \qquad \text{with} \qquad f_t = K_c(\pi a)^{-1/2} \tag{3.29}$$

where f_t may be treated as the tensile strength. So this is the usual strength criterion. In this instance, the only size effect is statistical, of Weibull type, as described in Chapter 2 (Weibull 1939; Freudenthal 1968; Bolotin 1969; Elishakoff 1983).

(d) Asymptotic Size Effects in Quasibrittle Structures

Here the problem of scaling is more complicated because the material possesses a characteristic length, l_0, which is what matters. It is nevertheless clear that, for a sufficiently large size, the scale of the material inhomogeneities, the FPZ size, and thus the material characteristic length, should become unimportant and are negligible compared to structural dimensions, same as in LEFM. So the power-law scaling of LEFM must apply asymptotically for sufficiently large sizes. If there is a large crack at failure, the exponent of this asymptotic power law must be $-1/2$, which is represented by the inclined asymptote of slope $-1/2$ in Figure 3.11.

The characteristic length l_0 (called also the material length) must also become unimportant for very small structure sizes. For example, it can happen for concrete that the specimen size is smaller than l_0. This means that, for very small sizes, the size effect should again asymptotically approach a power law. Because, for such small sizes, a discrete crack cannot be discerned (as the entire specimen is occupied by the fracture process zone), the exponent of the power law should be zero, corresponding to the strength criterion (see the horizontal dashed asymptote in Fig. 3.11).

The difficulty is that most applications of quasibrittle materials fall into the transitional range between these two asymptotes, which typically spans three order of magnitude. The scaling law must bridge these two power laws, following a certain transitional curve (Fig. 3.11). Determining its law requires more refined arguments, which are

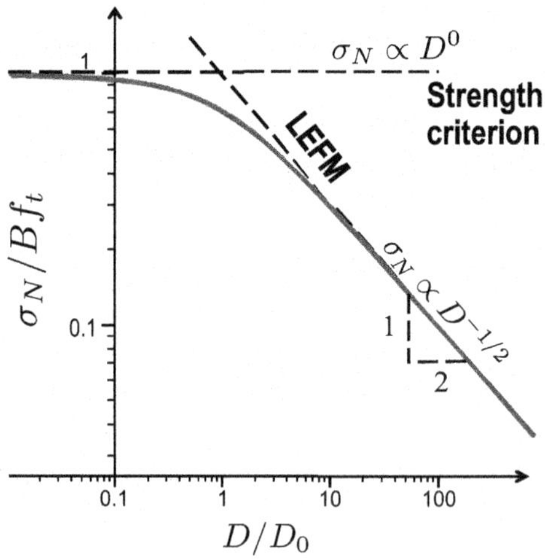

Figure 3.11. Size effect on nominal strength of quasibrittle structures with a preexisting crack.

presented in Section 3.9, and more fundamentally in Appendices B and C (for more detail see also pp. 189–204 in Bažant 2005, where type 1 and 2 size effects were originally called Cases 2 and 1, respectively). There also exists a type 3 size affect (Case 3 in Bažant 2005), which is so close to type 2 that it is hard distinguish in experiments.

For those who are not familiar with fracture mechanics, the type 2 size effect has been derived in Yu et al. (2016) without recourse to fracture mechanics, simply using dimensional analysis and energy conservation (or first law of thermodynamics).

3.7 Second-Order Asymptotic Properties of Cohesive Crack or Crack Band Models

For determining the transition of the deterministic size effect between the small-size and large-size asymptotes, it is useful to obtain the asymptotic expansions of nominal strength σ_N. Based on the first two terms of such expansions, one can distinguish three and only three types of the transitional deterministic (or energetic) size effect:

$$\text{For } D \to 0: \quad \sigma_N = b_0 - c_0 D^p + \cdots \qquad \text{(all types)} \tag{3.30}$$

$$\text{For } D \to \infty: \quad \sigma_N = b_1 + c_1 D^{-1} + \cdots \qquad \text{(type 1)} \tag{3.31}$$

$$\sigma_N = D^{-1/2}(b_2 - c_2 D^{-1} + \cdots) \quad \text{(type 2)} \tag{3.32}$$

$$\sigma_N = D^{-1/2}(b_3 - c_3 D^{-2} + \cdots) \quad \text{(type 3)} \tag{3.33}$$

where $p, b_0, c_0, \ldots c_3$ are positive constants determined by structure geometry. This was demonstrated by second-order dimensional analysis in Bažant (2004*b*) and by smeared-tip method in Bažant (2005), and also by asymptotic analysis of the boundary value problem in Bažant (2001); see Appendices B and C. The first terms in Eqs. 3.32 and 3.33

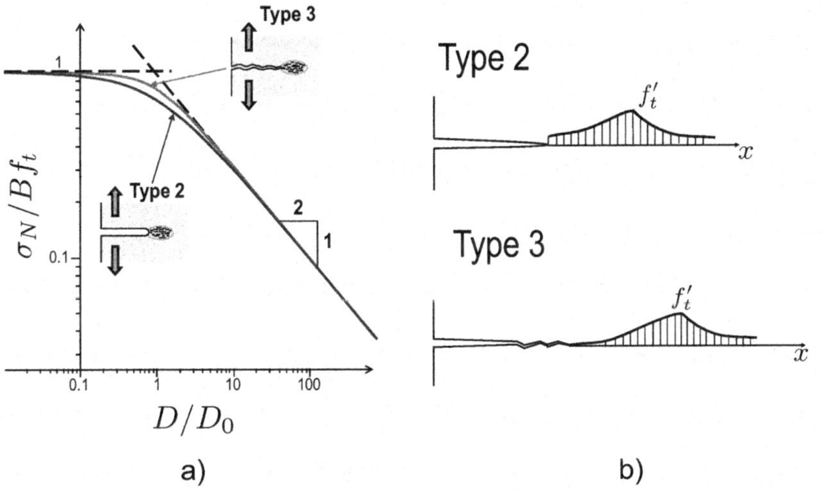

Figure 3.12. Comparison of type 2 and type 3 size effects. (a) Type 2 and type 3 size effect curves, and (b) typical cohesive stress profiles.

represent power-law scaling.[3] The second-order small-size asymptotics (Eq. 3.30) was derived based on the cohesive crack model, in which the exponent p describes the initial descent of the cohesive law (provided it begins after the peak with a finite slope); see Appendix C for detailed analysis. The second terms in Eqs. 3.31 to 3.33 distinguish various types of deterministic size effect which are discussed next.

3.8 Types of Size Effect Distinguished by Asymptotic Properties

Type 1 size effect occurs if the geometry is such that P_{max}, the maximum load, is reached while a crack initiates from one FPZ that is still attached to a smooth surface. Type 2 size effect (Fig. 3.11) occurs if there is a large notch or preexisting stress-free (fatigued) crack and if the geometry is positive, in which case that P_{max} occurs while the FPZ is still attached to the tip of the notch or fatigued crack.

One can also distinguish the type 3 size effect, which occurs if a large crack can grow stably prior to P_{max}, which is the case of initially negative geometry. This implies that the type 3 size effect can occur only in sufficiently large specimens such that the FPZ is smaller than the structural dimension. The size effect types 1 and 2 are very different, but types 2 and 3 are quite similar and hardly distinguishable in fracture testing (Fig. 3.12a).[4]

[3] The transition from one power law to another is in physics traditionally handled by renormalization group transformation (Barenblatt 2003). This transformation, however, merely identifies the relations between different power laws. It does not describe the gradual transition between them. In quasibrittle fracture mechanics, the transition can be very broad, typically spanning three or four orders of magnitude.

[4] For details, see Bažant (2005) and note that in this original study the type 1 size effect was called Case 2 and type 2 case 1.

However, these two types of size effect can be differentiated by their cohesive stress profiles. For type 2 size effect, the stress (at maximum load) at the tail end of the cohesive zone (or FPZ) is nonzero when the FPZ is still attached to the tip of a notch or to the tip of a previously formed fatigued crack. For type 3 size effect, the stress at the tail end of the FPZ is zero because the FPZ is attached to a traction-free crack that is formed as a result of the stable propagation of the real crack (Fig. 3.12b).

3.9 Derivation of Quasibrittle Deterministic Size Effect from Equivalent LEFM

Useful approximations of cohesive crack model can be obtained by LEFM if the tip of a sharp LEFM crack is placed roughly into the middle of the FPZ, i.e., the softening zone (for a precise definition of tip location, one may use the condition that the stiffness of the structure be the same as for the cohesive crack model).

The equivalent LEFM has been used to express the coefficients of the size effect law in terms of LEFM functions and material parameters (Bažant & Pfeiffer 1987; Bažant & Kazemi 1990*a*, 1991). In LEFM, the stress intensity factor is generally expressed as

$$K_I = \frac{P}{b\sqrt{D}}\,k(\alpha) \tag{3.34}$$

where P = applied load or loading parameter, b = structure width, and $k(\alpha)$ = dimensionless stress intensity factor, which depends on the relative crack or notch length $\alpha = a_0/D$ and, of course, structure geometry. Introducing the nominal strength $\sigma_N = P_{max}/bD$ (P_{max} = the maximum load capacity of the structure) and Irwin's relation $K_I^2 = E'\mathcal{G}$ where $\mathcal{G} = G_f$ = fracture energy of the material (a material constant), we get

$$\sigma_N = \sqrt{\frac{E'G_f}{Dg(\alpha)}} \tag{3.35}$$

where $E' = E$ = Young's modulus for plane stress and $E' = E/(1-\nu^2)$ for plane strain (ν = Poisson ratio), and $g(\alpha) = k^2(\alpha)$ = dimensionless energy release function.

In quasibrittle materials the FPZ has a finite size, which is about the same for structures of different sizes. The length of an equivalent LEFM crack may be written as $a = a_0 + c_f$, where length c_f is a constant representing about one half of the FPZ length; $c_f = \gamma_f l_1$, where l_1 = Irwin's material characteristic length corresponding to G_f and the precise value of γ_f can be calculated for a given specimen geometry from the softening law of the cohesive crack model; e.g., in Bažant and Yu (2011), $\gamma_f \approx 0.3$. In terms of the relative crack length $\alpha = a/D$, one has $\alpha = \alpha_0 + \theta$ where $\alpha_0 = a_0/D$ and $\theta = c_f/D$. The dimensionless energy release rate function can be expanded into the Taylor series:

$$g(\alpha) = g_0 + g_0'\theta + \tfrac{1}{2}g_0''\theta^2 + \tfrac{1}{3!}g_0'''\theta^3 + \cdots \tag{3.36}$$

where $g_0 = g(\alpha_0)$, $g_0' = \mathrm{d}g(\alpha)/\mathrm{d}\alpha|_{\alpha=\alpha_0}$, and so on.

3.9.1 Type 2 Size Effect

Type 2 size effect is simpler than type 1, so let us discuss it first. In this case, $a_0 \neq 0$ and, assuming the positive geometry, applicable in many situations, we thus have $g_0' > 0$ and we truncate the Taylor series after the second, linear, term. In fact, the higher-order terms cannot be used because the resulting size effect formula would violate the two-sided asymptotic matching condition that $\lim_{D\to 0} \sigma_N$ be finite, even though it would improve the size effect approximation for very large D.

Let us now assume α_0 to be independent of D, which is exactly true for notched geometrically similar fracture specimens of different sizes and is usually approximately true for stably grown cracks in structures [which is verified for diagonal shear cracks in beams in, e.g., in Syroka-Korol & Tejchman (2014), Fig. 1.11]. Substitution of the truncated Eq. 3.36 into Eq. 3.35 yields the size effect law of the form (Bažant & Kazemi 1990*a*, 1991; Bažant & Planas 1998; Bažant 2005):[5]

$$\sigma_N = \sqrt{\frac{E'G_f}{g_0' c_f + g_0 D}} = Bf_t \left(1 + \frac{D}{D_0}\right)^{-1/2} \tag{3.37}$$

$$\text{where} \quad D_0 = c_f \frac{g_0'}{g_0}, \quad Bf_t = \sqrt{\frac{E'G_f}{c_f g_0'}} \tag{3.38}$$

Note that the transitional size D_0, delineating the brittle behavior from nonbrittle behavior, is proportional to the effective size c_f of the FPZ.

Function $g(\alpha)$ introduces the effect of structure geometry. Thus, the size effect law in Eq. 3.37 expresses not only the effect of size but also the effect of structure geometry (or shape). It might better be called the "size-shape" effect law, and may be applied even to structures that are not geometrically similar.

It has been demonstrated that Eq. 3.37 agrees well with the experimentally observed size effect for structures made of many quasibrittle materials including concretes, composites, ceramics, sea ice, foam, bone, and limestone (Fig. 3.13).

To determine the material fracture characteristics from the measured maximum loads of notched fracture specimens of different sizes, one may rearrange Eq. 3.37 into a linear regression plot (Fig. 3.14):

$$Y = AX + C \quad \text{with} \quad X = D, \quad Y = \frac{E'}{g_0\, \sigma_N^2} \tag{3.39}$$

$$A = \frac{1}{G_f}, \quad C = c_f \frac{g_0'}{G_f g_0} \tag{3.40}$$

[5] This size effects may alternatively be expressed in terms of an apparent LEFM stress intensity factor $\bar{K}_{\mathrm{I}}$. Defining $\bar{K}_{\mathrm{I}} = \sigma_N \sqrt{D}$ and noting that $E'G_f = K_c^2$ where K_c = fracture toughness (a material constant), one has:

$$\bar{K}_{\mathrm{I}} = K_c \sqrt{\frac{D}{g_0' c_f + g_0 D}} = K_c \sqrt{\frac{D/\ell_f}{1 + D/D_0}} \tag{a}$$

where $\ell_f = g_0' c_f$ (Eq. 15 in Bažant & Kazemi 1990*aa*).

The use of K_c is preferred in the ceramics literature. The size dependence of $\bar{K}_{\mathrm{I}}$ can also be converted to size dependence of the R-curve (Bažant, Gettu, & Kazemi 1991)

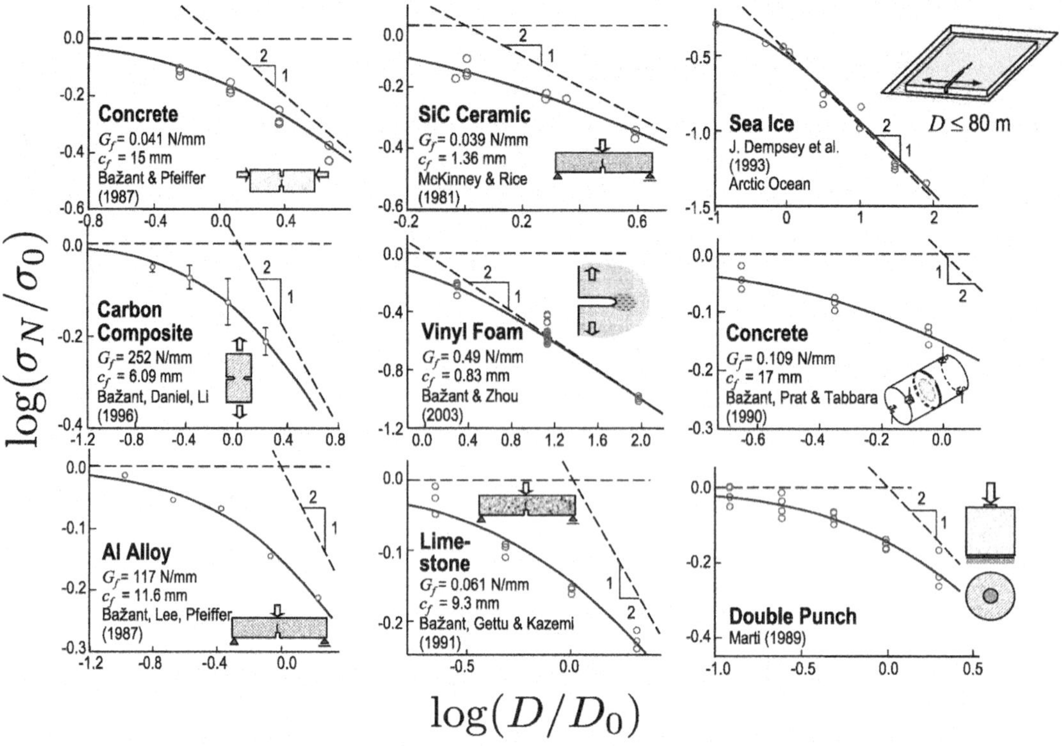

Figure 3.13. Experimental evidence of type 2 size effect.

After identifying A and C by linear regression of test data on the maximum loads of structures of different sizes, the fracture characteristics are then obtained as $G_f = 1/A$, $c_f = (C/A)(g_0/g_0')$. This size effect method of fracture testing became an international standard recommendation of a fracture test (RILEM TC 89-FMT 1990).

3.9.2 Type 1 Size Effect

When there is no notch or a preexisting stress-free crack, we have $g_0 = 0$. Then in Eq. 3.36 we must take the second and third terms in the series expansion, i.e.,

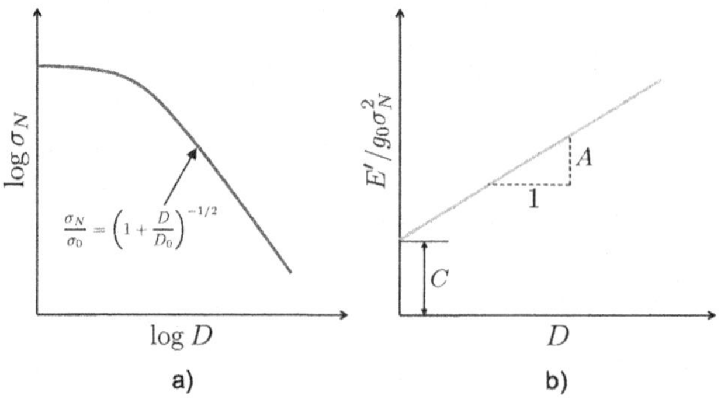

Figure 3.14. Plots of type 2 size effect curve. (a) Bilogarithmic and (b) linear regression.

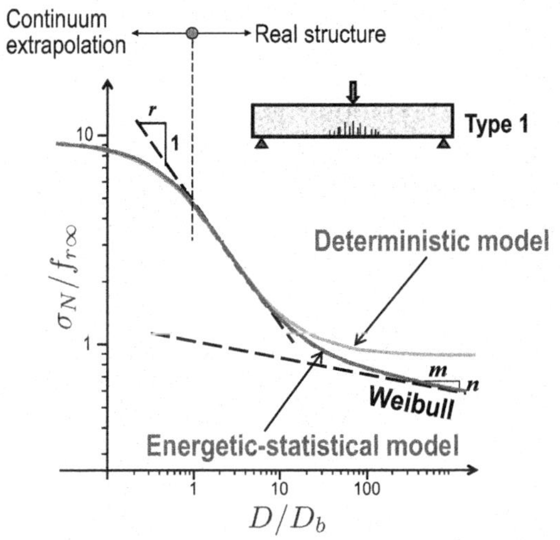

Figure 3.15. Type 1 size effect curve of nominal strength of quasibrittle structures.

$g(\alpha) = g_0'\theta + \frac{1}{2}g_0''\theta^2$. Substitution into Eq. 3.35 yields

$$\sigma_N = \sqrt{\frac{E'G_f}{g_0'c_f + \frac{1}{2}g_0''c_f^2/D}} \tag{3.41}$$

Normally, however, $g_0'' < 0$. This causes a problem; σ_N becomes imaginary for small enough D. Even though the approximation in Eq. 3.41 is valid only for large enough D, we want a two-sided asymptotic matching formula of general applicability. To this end, Eq. 3.41 for large D may be approximated as

$$\sigma_N = f_{r\infty}(1-x)^{-1/2} \tag{3.42}$$

$$\text{where} \quad f_{r\infty} = \sqrt{\frac{E'G_f}{g_0'c_f}}, \quad x = \frac{-g_0''}{g_0'}\frac{c_f}{2D} \tag{3.43}$$

Again, Eq. 3.42 is valid for only large enough D. It can be modified by a binomial expansion that does not change the first two terms of the expansion. Therefore, it is legitimate to replace this equation by

$$\sigma_N = f_{r\infty}(1+x)^{1/2} \quad \text{or} \quad = f_{r\infty}[1 + r(x/2)]^{1/r} \tag{3.44}$$

where r is an arbitrary positive number as needed to fit the small-size data. Setting $x/2 = D_b/D$ where

$$D_b = \left\langle \frac{-g_0''}{4g_0'} \right\rangle c_f \tag{3.45}$$

we arrive at the final form of the type 1 size effect law (Fig. 3.15):

$$\sigma_N = f_{r\infty}\left(1 + \frac{rD_b}{D}\right)^{1/r} \tag{3.46}$$

This expression not only has the correct first- and second-order asymptotic properties for $D \to \infty$, like Eq. 3.41, but unlike Eq. 3.41 it also has a realistic form for small D, such that $\lim_{D\to 0} \sigma_N$ is neither 0 nor imaginary.

The feature that $\sigma_N \to \infty$ for $D \to 0$ is shared by the famous, widely used, Hall–Petch formula for the strength of polycrystalline materials. One might prefer a finite limit for f_r or σ_N but this does not matter in practice because D cannot be less than about three maximum aggregate sizes (as the material could no longer be treated as a continuum). On the other hand, a finite limit of f_r is useful to introduce the small-size asymptotics anchoring of a continuum model, as discussed in Chapter 12.

Comparison with the derivation of Eq. 1.8 further shows that D_b has the meaning of the double of the thickness of the boundary layer of cracking. Note that, in unnotched three point bend tests, σ_N in Eq. 3.46 for $r = 2$ is the same as the modulus of rupture, usually denoted as f_r. Also note that, in Eq. 3.45, the signs $\langle .. \rangle$ represent the Macauley brackets, which mean the positive part of the argument; they were inserted because the ratio g_0''/g_0' can sometimes be negative, in which case there can be no size effect, i.e., D_b would vanish.[6]

3.10 Nonlocal Weibull Theory for Mean Response

If the material strength is random and the characteristic length, l_0, is not negligible compared to D, then the structure inevitably exhibits a combined statistical-energetic size effect, and if D is not much larger than l_0 then the energetic part will dominate. In the mean sense, this can generally be captured by the nonlocal Weibull theory (Bažant & Xi 1991), in which the material failure probability depends not on the continuum stress at a given point $\boldsymbol{x}$ but on the nonlocal strain $\hat{\epsilon}(\boldsymbol{x})$ at that point, defined by spatial averaging of the positive part of the maximum principal local strain $\epsilon(\boldsymbol{x})$ over a neighborhood of size ℓ surrounding point $\boldsymbol{x}$. Equation 2.9 is then generalized as

$$P_f = 1 - \exp\left[-\int_V \left(\frac{E\hat{\epsilon}(\boldsymbol{x})}{s_0} \right)^m \mathrm{d}V(\boldsymbol{x})/{l_0}^{n_d} \right] \tag{3.47}$$

When $\ell \leq l_0$, this formulation becomes essentially equivalent to the classical Weibull theory, Eq. 2.9, and becomes identical when $D \gg l_0$ and $D \gg \ell$.

The nonlocal averaging accounts for the spatial concentration of failure probability, and represents an autocorrelated random field with autocorrelation length l_a. Although l_a could, in theory, differ from ℓ, it seems reasonable to assume that $l_a \approx \ell$. Figure 3.16 shows the comparison between the numerical simulation using the nonlocal Weibull theory and the available experimental data on concrete beams from 10 different labs (Bažant & Novák 2000*a*, 2000*b*; Bažant et al. 2004).

[6] The special case of Eq. 3.46 for $r = 1$ was derived in Bažant and Li (1995*b*). The special case for $r = 2$ was proposed by Carpinteri et al. (1994, 1995*b*), who called it the "multifractal" scaling law (MFSL) and tried to justify it by fracture fractality of crack surface or cracking damage using, however, strictly geometric (nonmechanical) arguments. The term "multifractal," though, is unjustified because, on closer scrutiny (Bažant & Yavari 2005, 2007), the formula does not follow from the hypothesis of fractality.

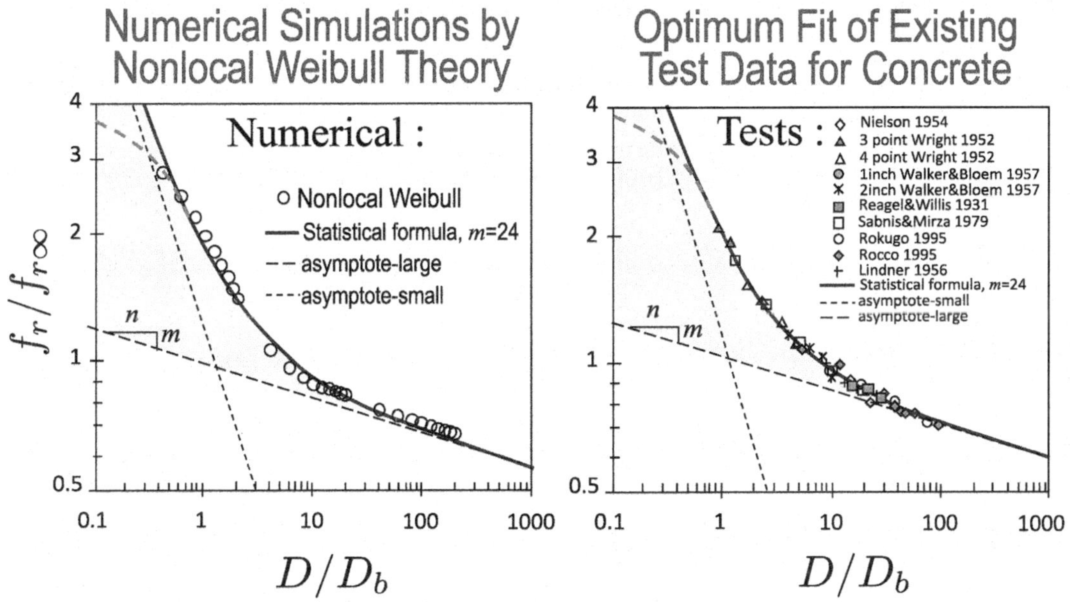

Figure 3.16. Measured type 1 size effect of concrete beams and simulations by the nonlocal Weibull theory.

3.11 Combined Energetic-Statistical Size Effect Law and Bridging of Type 1 and 2 Size Effects

While for very large sizes the type 1 size effect tends to be purely statistical (i.e., Weibullian), for small sizes, as we have seen, it is caused mainly by stress redistribution due to nonnegligible finite size of the FPZ. This redistribution is deterministic and is associated with energy release.

Although structure sizes D smaller than the inhomogeneity size make physically no sense, a theoretical extrapolation to size $D \to 0$ is helpful because it allows two-sided asymptotic matching. This possibility is offered by deriving the size effect law from the cohesive crack model. This is a continuum model extendable to zero size, for which the small-size asymptotic expansion of size effect is possible and was already presented in Eq. 3.30.

By asymptotic matching of the large-size asymptotic expansion of Eq. 3.47 and of the first two terms of the small-size expansion in terms of D according to Eq. 3.30, the following approximate energetic-statistical size effect on the mean σ_N in type 1 failures was derived (Bažant & Novák 2000*a*; Bažant 2004*a*, 2004*b*), as shown in Fig. 3.15:

$$\sigma_N = A\left(\theta^{rn_d/m} + r\kappa\theta\right)^{1/r}, \quad \theta = B\left(1 + D/\eta l_0\right)^{-1} \tag{3.48}$$

Here n_d, m, r, κ, η, A, B, and l_0 are constants to be obtained empirically from experiments, or by matching computational simulations. For small D, these simulations can be deterministic, based on the cohesive crack model or the crack band model. Because the finite nonlocal averaging zone emulates the finite-size FPZ, Eq. 3.48 can also be derived by asymptotic approximations of the nonlocal Weibull theory.

A difficult problem is the transition between type 1 and type 2 size effects. A more complicated explicit formula, called the universal size effect law, has been derived to describe this transition while preserving the type 1 and 2 size effect laws as the limiting cases. Chapters 12 and 13 of this book are devoted to this subject.

4 Failure Statistics of Nanoscale Structures

Damage and failure of a macroscale structure originate from the fracture of its nanoscale material elements, for example, a regular atomic lattice such as a single crystal grain of brittle ceramics, or a completely disordered structure such as a system of nanoparticles of the calcium silicate hydrate in concrete. Therefore, it is logical to begin with a study of the failure statistics of nanoscale structures to derive a probabilistic model for the failure of macroscale structures.

4.1 Background of Modeling of Nanoscale Fracture

With the advances in computational material sciences, multiscale analysis has been widely used to investigate the behavior of the macroscale structure from its meso- or nanoscale structures (Tadmor, Ortiz, & Phillips 1996; Ghoniem, Busso, Kioussis, & Huang 2003; Liu, Karpov, & Park 2005). Computational modeling of crack propagation in nanoscale structures, such as atomic lattices, largely relies on the use of empirical interatomic potentials (Omeltchenko, Yu, Kalia, & Vashishta 1997; Marder 2004; Khare et al. 2007, 2008; Xu et al. 2012). Recent research efforts have been directed to the coupling between the quantum mechanics (QM) and the molecular dynamics (MD) (Abraham, Broughton, Bernstein, & Kaxiras 1998; Broughton, Abraham, Bernstein, & Kaxiras 1999; Khare et al. 2007, 2008), which provides a more sound physical basis for the computational model.

However, these existing sophisticated numerical models are not suitable for studying the failure probability of nanoscale structures, especially when we are interested in the far left tail of the failure probability, which would require millions of numerical simulations. Meanwhile, the atomistic simulations are also not suitable for studying time-dependent fracture (e.g., static and cyclic fatigue), as the structural lifetime is many orders of magnitude longer than the vibration period of the atomic bond (Tadmor & Miller 2011). At the mesoscale level, there exists no physical law for the probability of microstructural failure, and only intuitive hypotheses and phenomenological models can be made. Therefore, the resulting failure statistics of macroscale structures strongly depends on these hypotheses and models about the mesoscale structural failure, which are often difficult to justify either physically or experimentally.

Compared to the proliferation of computational tools for simulating the failure behavior of nanoscale structures since the late 1990s, analytical modeling of breakage of interatomic bond has been pursued for a much longer time. A well established physical theory for the frequency of breakage of interatomic bond does exist. It is the rate process theory (Eyring 1936; Glasstone, Laidler, & Eyring 1941; Kramers 1941; Tobolsky & Erying 1943; Krausz & Krausz 1988; Kaxiras 2003), which theoretically justifies the Arrhenius thermal factor and has long been used to transit from the atomic scale to the material scale, providing the temperature and stress dependence of the rates of creep, diffusion, phase changes, adsorption, chemical reactions, and so forth. In this theory, the rates of breakage of interatomic bonds are characterized by the distribution of thermal energies among atoms and the frequency of passage over the activation energy barriers of the interatomic potential.

At the same time, it is also well known that the natural energy scale for chemical bonds, and thus also for activation barriers in molecular rearrangements between long-lived well-defined molecular states, is the electron-volt. This scale is at least an order of magnitude larger than the thermal energy scale ($kT = 0.25$ eV at room temperature, where $k =$ Boltzmann constant and $T =$ absolute temperature). In the case of a large free-energy barrier, the transition between two states is slow and so the process is quasi-stationary. The quasi-stationarity can further be justified by the observation that the interatomic bonds break at the rate of about 10^5/s in static fracture and about 10^{12}/s in fracture under missile impact. The frequency of thermal atomic vibrations is about 10^{14}/s. Therefore, one jump over the activation energy barrier, or one interatomic bond break, occurs only after every 10^9 or every 100 atomic vibrations, respectively.

Due to the quasi-stationarity of the breakage process of interatomic bonds, the failure probability of interatomic bonds must be proportional to their failure frequency. In this chapter, we derive the failure probability of the nanoscale structure by combining the classical transition rate theory and the atomistic fracture mechanics.

4.2 Stress-Driven Fracture of Nanoscale Structures

Consider crack propagation in a nanoscale structure, such as an atomic lattice block (Fig. 4.1a) or a disordered system of nanoparticles (Fig. 4.1b), under a remote stress τ. As a crack propagates through a nanoscale structure, the separation δ between the opposite atoms or nanoparticles across the nanocrack gradually increases by jumps. The force F_b transmitted between the opposite atoms or particles is governed by the interatomic or interparticle potential function $\Pi_b(\delta)$: $F_b = \partial\Pi_b(\delta)/\partial\delta$.

One may consider using various empirical potential functions, such as Morse potential (Morse 1929), Lennard-Jones potential (Lennard-Jones 1924), etc., to describe $\Pi_b(\delta)$. However, these pair-potentials cannot realistically represent the potential function for the atoms or nanoparticles across the nanocrack because they only include one-dimensional deformation and neglect the contribution of many surrounding atoms or nanoparticles. To capture the effect of the surrounding atoms or nanoparticles, one could use three-body potentials, such as Tersoff potential (Tersoff 1988) and Stillinger–Weber potential (Stillinger & Weber 1985), to model the entire nanoscale structure. A more physical approach is to use a coupled QM–MD method, which does not rely on any

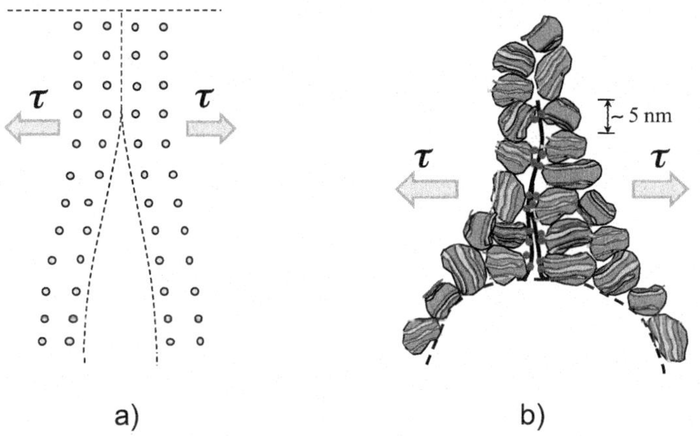

Figure 4.1. Fracture of a nanoscale structure. (a) Regular atomic lattice block and (b) disordered system of nanoparticles.

empirical potential functions. Nevertheless, for the purpose of the present formulation, it is not necessary to determine the interatomic potential functions because the aim here is to obtain only the *functional form* of the failure probability of a nanoscale structure.

Figure 4.2 shows the behavior of different pairs of atoms or nanoparticles along the crack. The pair with the peak value of the bond force, which corresponds to the point of maximum slope of the curve $\Pi_b(\delta)$ (state 3 in Fig. 4.2b and c), is normally defined as the front of a nanoscale cohesive crack. The real nanocrack ends at the atomic pair where the bond force F_b drops to 0 (state 5 in Fig. 4.2b and c) and the adjacent nanoscale elements may undergo a gross distortion. The atomic pairs between states 3 and 5 represent the cohesive zone [or the fracture process zone (FPZ)], in which the bond force decreases with increasing separation. This is the original idea due to Barenblatt, who proposed

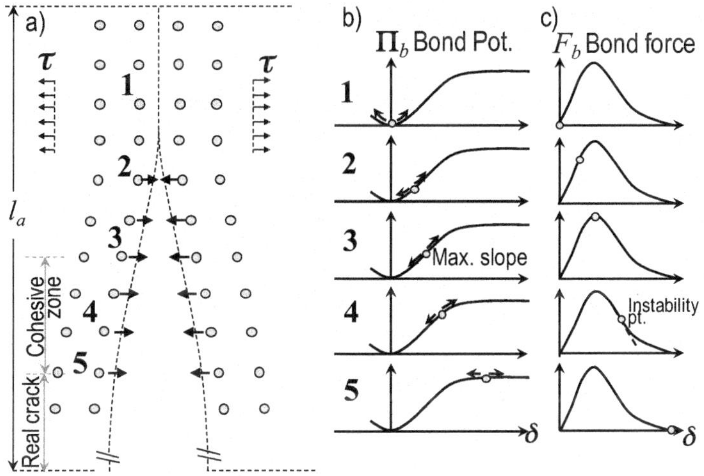

Figure 4.2. Mechanical behavior of interatomic bonds during nanocrack propagation.

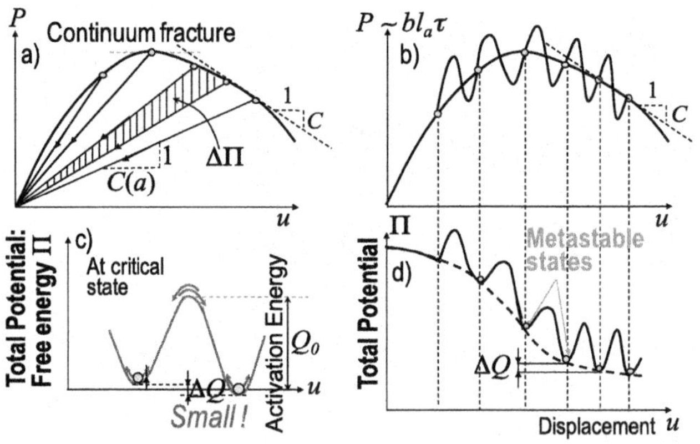

Figure 4.3. Continuum approximation of the fracture behavior of a nanoscale structure.

the concept of cohesive crack model (Barenblatt 1959). State 4 represents the limit of stability, at which the curve $\Pi_b(\delta)$ reaches a certain critical downward slope (equal in magnitude to the stiffness of the confinement within the surrounding solid). State 5, at which the separation of opposite atoms or nanoparticles greatly increases, must lie immediately next to state 4 because the transition from state 4 to state 5 is unstable, and thus dynamic and fast.

In a continuum approximation, the overall fracturing behavior of the nanoscale structure is characterized by a curve of equilibrium load P versus deflection u, in which hardening is followed by softening (Fig. 4.3a). The integral of this curve yields the curve of potential Π versus u. A crack in the nanoscale structure does not advance smoothly. Rather, it advances in numerous discrete jumps, which correspond to the jumps over the activation energy barriers of interatomic bonds (Fig. 4.1a) or nanoparticle connections (Fig. 4.1b). The length of these jumps is the spacing δ_a of the atomic bonds or the nanoparticle connections. The jumps make it necessary for an undulation to be superposed on the load-deflection curve, and a corresponding undulation on the potential curve; see Figure 4.3b and d in which $P = \tau b l_a =$ load, $\tau =$ applied remote stress, $u =$ displacement in the sense of P, $l_a =$ characteristic size of the nanoscale structure, and $b =$ width in the third dimension.

Consider, for the sake of simplicity, planar three-dimensional cracks that grow in a self-similar manner, expanding, e.g., in concentric circles or squares. According to the linear elastic fracture mechanics (LEFM), the stress intensity factor may generally be expressed as

$$K_a = \tau \sqrt{l_a} k_a(\alpha) \tag{4.1}$$

where $\alpha = a/l_a =$ relative crack length and $k_a(\alpha) =$ dimensionless stress intensity factor. In the context of linear elasticity, the remote stress τ applied on the nanoscale structure can be related to the macroscale stress σ by setting $\tau = c\sigma$, where

c = nano–macro stress concentration factor. Therefore, the energy release rate per unit crack front advance is

$$\mathcal{G}(\alpha) = \frac{K_a^2}{E_1} = \frac{k_a^2(\alpha) l_a c^2 \sigma^2}{E_1} \tag{4.2}$$

where E_1 = elastic modulus of the nanoscale structure. Let γ_1 = geometry constant such that $\gamma_1 a$ = perimeter of the radially growing crack front. The energy release along the entire perimeter, caused by crack advance δ_a, is

$$\Delta Q = \delta_a \left[\frac{\partial \Pi^*(P, a)}{\partial a} \right]_P = \delta_a(\gamma_1 \alpha l_a)\mathcal{G} = V_a(\alpha) \frac{c^2 \sigma^2}{E_1} \tag{4.3}$$

Here Π^* = complementary energy potential of the nanoscale structure, and $V_a(\alpha) = \delta_a(\gamma_1 \alpha l_a^2) k_a^2(\alpha)$ = activation volume [note that if the stress tensor is written as $\tau \boldsymbol{s}$ where τ = stress parameter, one could more generally write $V_a = \boldsymbol{s} : \boldsymbol{v}_a$ where $\boldsymbol{v}_a$ = activation volume tensor, as in the atomistic theories of phase transformations in crystals (Aziz, Sabin, & Lu 1991)].

A sharp LEFM crack is, of course, an idealization. In reality, there is always a finite FPZ, which exhibits a strain-softening behavior. However, for the global response, which is what matters here, a crack with a finite FPZ may be treated by an equivalent sharp LEFM crack giving the same energy release rate. Its tip is located roughly in the middle of the FPZ.

Now consider two states of a nanoscale structure: (1) before the nanocrack propagates, and (2) after the nanocrack propagates by one spacing δ_a. At the nanoscale, the transition between these two states is a thermally activated process. As a result of thermal activation, the energy states of the nanoscale structure fluctuate and lead to jumps over the activation energy barriers. The jumps occur both forward and backward, albeit with different frequencies (Fig. 4.3c). The energies required for the forward and backward jumps are $Q_0 - \Delta Q/2$ and $Q_0 + \Delta Q/2$, respectively, where Q_0 = activation energy at no stress. After overcoming many activation energy barriers (Fig. 4.3b and d), the nanocrack attains its critical length, at which the nanoscale structure loses its stability and the nanocrack begins to advance dynamically, with sound emission. So, the number of undulation waves on the $\Pi(u)$ curve and on the corresponding potential curve (Fig. 4.3b and d) must be large. Therefore, the barrier for the forward jump (crack propagation) must differ only little from the barrier for the backward jump (crack healing), i.e., ΔQ is small, and the forward and backward jumps must be happening with only slightly different frequencies.

The frequency of transition between two metastable states can be calculated based on the well-established transition rate theory (Phillips 2001; Kaxiras 2003). Kramers modeled the physical Brownian motion in the phase space and calculated the escape rate of a particle over an energy barrier (Kramers 1941). As mentioned before, the free-energy barrier must be at least one order of magnitude larger than the thermal energy ($Q_0 \gg kT$), and so the stress-induced energy bias between two states must be very small. For the limiting case of a large free-energy barrier ($Q_0 \gg kT \gg \Delta Q$), the transitions between local minima of free energy can typically be described by the asymptotic

Kramers formula of the first passage time, which predicts an exponential dependence on the barrier energy relative to kT (which is also an Arrhenius dependence on temperature). The net frequency of the forward crack front jumps is given by (Kramers 1941; Risken 1989)

$$f_1 = \nu_T \left(e^{(-Q_0+\Delta Q/2)/kT} - e^{(-Q_0-\Delta Q/2)/kT} \right) \tag{4.4}$$

$$= 2\nu_T e^{-Q_0/kT} \sinh[\Delta Q/2kT] = 2\nu_T e^{-Q_0/kT} \sinh[V_a(\alpha)/V_T] \tag{4.5}$$

where $V_T = 2E_1 kT/c^2\sigma^2$; ν_T is the characteristic attempt frequency for the reversible transition; $\nu_T = kT/h$, where $h = 6.626 \cdot 10^{-34}$ Js = Planck constant = (energy of a photon)/(frequency of its electromagnetic wave); T = absolute temperature; and $k = 1.381 \cdot 10^{-23}$ J/K = Boltzmann constant.

We have argued that $\Delta Q \ll kT$, based on the fact that there are numerous jumps between metastable states during the fracture process of the nanoscale structure. Based on Eq. 4.5, it is clear that $\Delta Q \ll kT$ implies $V_a/V_T \ll 1$. Therefore, we may also examine the V_a/V_T ratio to justify the assumption underpinning the Kramers formula. The atomic spacing or spacing of nanoparticle connection is typically on the order of 0.1 nm, and so $V_a \sim 10^{-26}\text{m}^3$. The volume V_T is a function of $\tau = c\sigma$. For example, for the nanoscale structure of hardened Portland cement gel, τ is on the order of MPa, and the elastic modulus E_1 of the nanoscale structure is doubtless larger, though not much larger, than the macroscopic elastic modulus E, which is on the order of GPa. Therefore, we estimate $V_T \sim 10^{-24} - 10^{-25}\text{m}^3$, and $V_a/V_T < 0.1$. Since $\sinh x \approx x$ for small x, Eq. 4.5 thus becomes (Bažant et al. 2009)

$$f_1 \approx e^{-Q_0/kT} [\nu_T V_a(\alpha)/kT] \, c^2\sigma^2/E_1 \tag{4.6}$$

One might object that, generally, there are multiple activation energy barriers: $Q_0 = Q_1, Q_2, \ldots$. However, the lowest one always dominates. The reason is that the factor $e^{-Q_1/kT}$ is very small, typically 10^{-12}. Thus, if, for example, $Q_2/Q_1 = 1.2$ or 2, then $e^{-Q_2/kT} = 0.004e^{-Q_1/kT}$ or $10^{-12}e^{-Q_1/kT}$, and so the higher barrier makes a negligible contribution. And if, for example, $Q_2/Q_1 = 1.02$, then Q_1 and Q_2 can be replaced by a single activation energy $Q_0 = 1.01Q_1$.

So far we have determined the rate of jumps over one activation energy barrier. For a nanoscale crack to propagate up to the critical crack length a_c at which stability of the nanocrack is lost, a certain number, n_f, of activation energy barriers must be overcome (up to point C in Fig. 4.3b). We do not know (and need not know) what number n_f is, but we know it must exist and be finite. Assuming that each jump is an independent process (Krausz & Krausz 1988) with the quasi-stationarity of the process, the failure probability of the nanoscale structure can be expressed as

$$P_f(\sigma) = \min\left[\nu_T^{-1} \int_{\alpha_0}^{\alpha_c} f_1(\alpha)\, \mathrm{d}\alpha, 1 \right] \tag{4.7}$$

$$= \min\left[C_T\sigma^2, 1 \right] \tag{4.8}$$

where $C_T = H_T c^2 \gamma_1 \int_{\alpha_0}^{\alpha_c} \alpha k_a^2(\alpha)\mathrm{d}\alpha$, $H_T = e^{-Q_0/kT}(\delta_a l_a^2/E_1 kT)$, $\alpha = a/l_a =$ relative crack length. The essential point here is that the strength distribution of a nanoscale structure follows a power-law function with a zero threshold. This zero threshold is the direct consequence of including both forward and backward jumps over the energy barrier.

The classical theory of structural lifetime (Zhurkov 1965; Zhurkov & Korsukov 1974; Kausch 1978) was derived by assuming that only forward jumps take place. In that case, as a generalization of the Arrhenius factor, the jump frequency is assumed to be an exponential function of the applied stress. In contrast, the present formulation shows that the jump frequency follows a power-law function of the applied stress because $\Delta Q \ll kT$, or equivalently $V_a/V_T \ll 1$. It has been consistently observed that the one-way jump model, which corresponds to the classical theory, underestimates the lifetime for small stresses by orders of magnitude. In Zhurkov (1965), it was documented that, for a low stress such as 20% of the short-time strength, the predicted lifetime was three orders of magnitude shorter than that observed experimentally on polymers, glass, and alumina. Another important point to be noted is that the classical theory of structural lifetime calculates the lifetime of a macroscale structure directly from the lifetime of the atomic bond, although the multiscale transition is lacking. This is discussed in detail in Chapter 6.

4.3 Probability Distribution of Fatigue Strength at Nanoscale

The foregoing analysis is now extended to time-varying stress histories. An important time-varying load case is cyclic loading. Consider a structure subjected to cyclic loading with a constant load amplitude. The loading history can be described by two quantities: the amplitude of nominal stress $\Delta\sigma = \sigma_{\max} - \sigma_{\min}$ and the stress ratio $R = \sigma_{\min}/\sigma_{\max}$, where $\sigma_i = c_n P_i/bD$ or $c_n P_i/D^2$ (i = max, min) for two- or three-dimensional scaling, $P_{\max}, P_{\min}$ = maximum and minimum applied loads of the structure or parameter of load system; c_n = a parameter chosen such that the nominal stress represents the maximum principal stress in the structure; b = structure thickness in the third dimension; and D = characteristic structure dimension or size. The fatigue strength of a structure is defined as the critical stress amplitude for a given number of cycles and stress ratio.

Given a stress history for the macroscale structure, the corresponding stress history for its nanoscale structures is hard to determine, especially for the beginning stage during which the residual stress builds up rapidly. However, for high cycle fatigue, the duration over which the residual stress is stabilized is short compared to the entire loading duration, and therefore the built-up of residual stress is not of particular interest here.

Figure 4.4b shows the stress profile of the nanoscale structure after the residual stress is stabilized. The stress amplitude at the nanoscale $\Delta\tau = \tau_{\max} - \tau_{\min}$ and the nanoscale stress ratio $R_\tau = \tau_{\min}/\tau_{\max}$ can thus be related to the stress amplitude $\Delta\sigma$ and the stress ratio R on the macroscale: $\Delta\tau = c_1\Delta\sigma$ and $R_\tau = c_2 R$. Parameters c_1 and c_2 are empirical but could conceivably be determined through a detailed micromechanical analysis of the build-up of residual stresses.

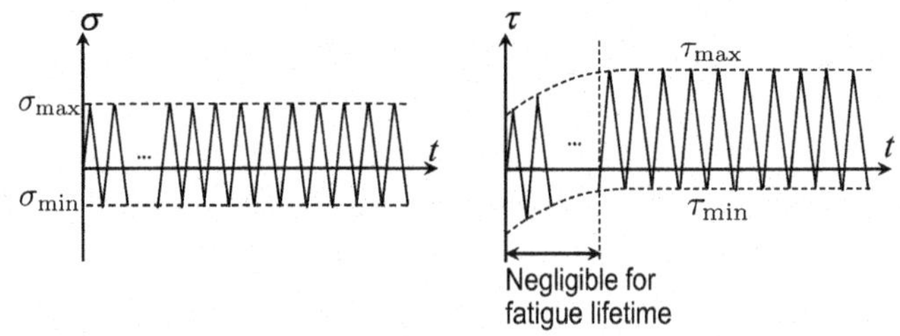

Figure 4.4. Fatigue loading histories at macro- and nanoscales.

Equation 4.7 gives the failure probability of a nanoscale structure under a constant stress. At the nanoscale, it may be assumed that the energy bias ΔQ depends only on the current stress, but not on the stress history. Therefore, for a given number of cycles N_0, the occurrence probability of a failure event can be obtained by simply integrating the entire stress history:

$$P_f = \min\left\{\frac{1}{\nu_T N_0 t_c}\int_{\alpha_0}^{\alpha_c}\int_0^{N_0 t_c} f_1[\alpha, \tau(t)]\,\mathrm{d}t\,\mathrm{d}\alpha,\ 1\right\} \tag{4.9}$$

$$= \min\left\{\frac{e^{-Q_0/kT}}{N_0 t_c}\left[\int_{\alpha_0}^{\alpha_c} V_a(\alpha)\,\mathrm{d}\alpha\right]\int_0^{N_0 t_c}\frac{\tau^2(t)}{E_1 kT}\,\mathrm{d}t,\ 1\right\} \tag{4.10}$$

where t_c = the duration of each cycle.

At the nanoscale, compressive stress cannot cause any breakage of atomic bonds or any rupture of nanoparticle connections. Therefore, the compressive stress does not contribute to the failure of the nanoscale structures. For the stress history depicted in Figure 4.4, Eq. 4.10 becomes

$$P_f = \min[C_T f(R_\tau)\Delta\tau^2, 1] = \min[C_T f(R_\tau)(c_1\Delta\sigma)^2, 1] \tag{4.11}$$

where $f(R_\tau) = (1 + \langle R_\tau\rangle + \langle R_\tau\rangle^2)/(1 - R_\tau)^2$, $C_T = H_T\left[\int_{\alpha_0}^{\alpha_c} V_a(\alpha)\mathrm{d}\alpha\right]$, $H_T = e^{-Q_0/kT}/3E_1 kT$, and $\langle x\rangle = \max(x, 0)$. Equation 4.11 shows that, similar to the statistics of static strength, the probability distribution of fatigue strength of a nanoscale structure follows a power law with a zero threshold.

4.4 Random Walk Aspect of Failure of Nanoscale Structures

The analysis presented so far focused purely on the stress-driven failure. When the applied stress is sufficiently small, the failure could be completely governed by the diffusive random walk of the tip of the nanocrack. The transition from the diffusion-governed failure and the stress-drift driven failure can be understood by the random walk analysis. The analytical solution for a complete 3D random walk analysis seems impossible. Here, a simplified 1D random walk model is presented as a qualitative analysis of drift-diffusion failure of a nanoscale structure.

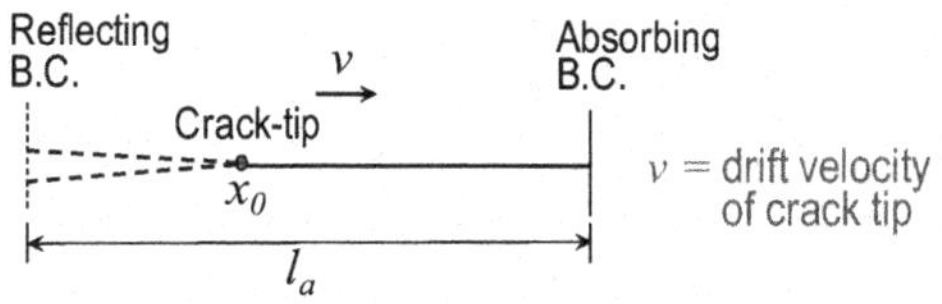

Figure 4.5. Simplified 1D analysis of random walk of the crack tip in a nanoscale structure.

The movement of the front of the cohesive crack (state 3 in Fig. 4.2a) represents a random walk biased by the energy release rate favoring the forward jumps and restricted by the tip (state 5 in Fig. 4.2a) of the real crack (this contrasts with a macroscale cohesive crack whose front cannot jump backward because of irreversible damage at the macro-level).

The boundary of the nanoscale structure at $x = l_a$, in front of the crack, is an absorbing boundary of the random walk of crack front. Before the real crack opens, the opposite boundary at $x = 0$ is a reflecting boundary of the random walk. However, the situation gets more complicated after the bonds of the atomic pair at crack mouth separate by several atomic spacings. This break, which produces the real crack (state 5), is irreversible. It is a sudden instability emitting a sound wave carrying energy, which cannot be recovered. The tip of the real crack (state 5) represents a reflecting boundary that moves forward as a function of the random walk. One thus faces a generalized random walk problem, which is beyond the scope of the present discussion.

The combination of random walk and stress driven propagation in a simplified 1D setting can be analyzed on the basis of the Fokker–Planck equation. To show the nature of the problem, let us consider here at least a simplified 1D random walk with fixed boundaries, in which the crack tip, initially located at x_0, moves at the drift velocity (Fig. 4.5):

$$v = \delta_a f_1 = 2\nu_T e^{-Q_0/kT} \delta_a V_a / V_T \tag{4.12}$$

For simplicity, the forcing (or bias) is assumed to be constant (uniform), even though in fracture mechanics it actually varies with the crack length.

When the front of a cohesive crack moves to the right, the FPZ grows; when to the left, it shrinks. At $x = 0$, there is a reflecting boundary, and at $x_1 = l_a$ an absorbing boundary. When the crack tip forms at $x = 0$, it will begin to move in the positive x direction. The random walk of the crack tip can at most reach $x = l_a$, which represents a complete fracture of the nanoscale structure (in reality, the nanoscale structure will fail earlier owing to instability, which is not captured by the random walk model). The probability $p(x, t)$ at time t of the crack tip being at position x, called the occupation probability, satisfies the Fokker–Planck equation (e.g., Redner 2001):

$$\frac{\partial p(x,t)}{\partial t} + v\frac{\partial p(x,t)}{\partial x} - \mathcal{D}\frac{\partial^2 p(x,t)}{\partial x^2} = 0 \tag{4.13}$$

where $\mathcal{D} = \frac{1}{2}\nu_T \delta_a^2 e^{-Q_0/kT}$ = diffusivity. The absorbing and reflecting boundary conditions are: $p(x = l_a, t) = 0$ and $vp(0, t) - \mathcal{D}[\partial p(x, t)/\partial x]|_{x=0} = 0$, respectively. The initial condition is: $p(x, 0) = \delta(x - x_0)$, where $\delta(x)$ = Dirac delta function.

To solve Eq. 4.13, it is convenient to apply the Laplace transform in the time domain (s = transform parameter):

$$sp(x,s) - p(x,0) + v\frac{\partial p(x,s)}{\partial x} = \mathcal{D}\frac{\partial^2 p(x,s)}{\partial x^2} \tag{4.14}$$

The general solution for Eq. 4.14 is simply given by

$$p(x,s) = A_1 \exp(\alpha_1 x) + A_2 \exp(\alpha_2 x) \quad (x \le x_0) \tag{4.15}$$

$$p(x,s) = A_3 \exp(\alpha_1 x) + A_4 \exp(\alpha_2 x) \quad (x > x_0) \tag{4.16}$$

where $\alpha_{1,2} = (v \pm \sqrt{v^2 + 4\mathcal{D}s})/2\mathcal{D} = l_a(\mathrm{Pe} + P_s)$, $\mathrm{Pe} = vl_a/2\mathcal{D} = 2(l_a/\delta_a)(V_a/V_T)$ and $P_s = l_a\sqrt{v^2 + 4\mathcal{D}s}/2\mathcal{D}$. The Péclet number, Pe, measures the relative dominance of stress-driven drift over stress-independent diffusion. The constants A_1, A_2, A_3 and A_4 are to be determined by the boundary and initial conditions: (1) $p(l_a, s) = 0$, (2)$-vp(0,t) + \mathcal{D}[\partial p(x,s)/\partial x]|_{x=0} = 0$, (3) continuity of $p(x,s)$ at $x = x_0$: $p(x = x_0^+, s) = p(x = x_0^-, s)$, and (4) discontinuity of spatial derivative of $p(x,s)$ at $x = x_0$: $[\partial p(x,s)/\partial x]|_{x=x_0^+} - [\partial p(x,s)/\partial x]|_{x=x_0^-} = -1/\mathcal{D}$. By solving these constants, we obtain the flux of occupation probability at the absorbing boundary as

$$j(l_a, s) = -\mathcal{D}\frac{\partial p(x,s)}{\partial x}\bigg|_{x=l_a} \tag{4.17}$$

$$= \frac{e^{\mathrm{Pe}(1-x_0/l_a)}[\mathrm{Pe}\sinh(x_0 P_s/l_a) + P_s\cosh(x_0 P_s/l_a)]}{\mathrm{Pe}\sinh P_s + P_s\cosh P_s} \tag{4.18}$$

The Laplace transform of the flux of occupation probability $j(l_a, s)$ can be written as

$$j(l_a, s) = \int_0^\infty j(l_a, t)e^{-st}\,\mathrm{d}t \tag{4.19}$$

$$= \int_0^\infty j(l_a, t)\left(1 - st + \frac{s^2t^2}{2!}\cdots\right)\mathrm{d}t \tag{4.20}$$

$$= \mathcal{E}\left[1 - s\langle t\rangle + \frac{s^2}{2!}\langle t^2\rangle\cdots\right] \tag{4.21}$$

where $\mathcal{E}$ is the exit probability at $x = l_a$, which equals 1 in this case, due to the absorbing boundary condition at $x = l_a$; $\langle t\rangle$ = the mean exit time, which is equivalent to the mean failure time of the nanoscale structure. The MacLaurin expansion of the Laplace transform of the occupation probability thus contains complete information about all the moments of the first passage time. In particular, we can easily compute the mean $\langle t\rangle$:

$$\langle t\rangle = -\frac{\partial j(l_a, s)}{\partial s}\bigg|_{s=0} \tag{4.22}$$

It should be noted that there are other analytical methods to arrive directly at the mean first passage time (Risken 1989; Redner 2001); but the present approach also provides the variance and all higher moments, if desired.

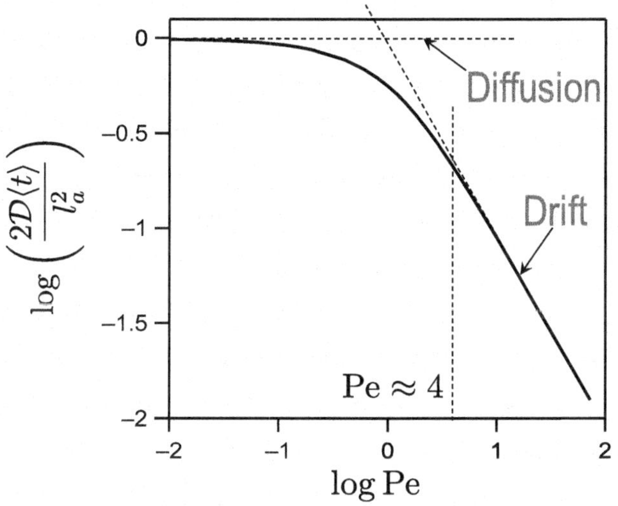

Figure 4.6. Relationship between the mean failure time and Péclet number Pe.

Substituting Eq. 4.17 into Eq. 4.22, we obtain the mean exit time, or the mean failure time, of the nanoscale structure as

$$\langle t\rangle = \frac{l_a^2}{2\mathcal{D}}\left[\frac{1-x_0/l_a}{\mathrm{Pe}} + \frac{\cosh \mathrm{Pe}}{\mathrm{Pe}^2 e^{\mathrm{Pe}}} - \frac{\cosh(\mathrm{Pe}x_0/l_a)}{\mathrm{Pe}^2 e^{x_0 \mathrm{Pe}/l_a}}\right] \tag{4.23}$$

Figure 4.6 shows the mean failure time $\langle t\rangle$ as a function of Péclet number Pe. For small Péclet numbers (Pe $\rightarrow 0$), the mean failure time is $l_a^2/2\mathcal{D}$. For large Pe, the mean failure time approaches $(l_a - x_0)/v$. As shown in Figure 4.6, the transition between the diffusion dominance and the drift dominance happens within a relatively narrow range of Pe values. The solution shows that, for Pe < 4, the fracture process is not governed purely by the stress-induced drift process. However, the condition Pe < 4 is not met in normal situations. Indeed, for $l_a \cong 100$ nm and the aforementioned values of V_a and V_T, one gets Pe ≈ 100.

Based on Figure 4.6, the diffusion completely dominates the failure at extremely low stresses, Pe ≈ 0.1. Then what would be the failure probability for which the applied stress corresponds to Pe ≈ 0.1? Consider that the strength of a material element follows a Gaussian distribution, and the mean strength μ_{G} corresponds to Pe $= 100$ and the coefficient of variation is 0.1. Based on the definition of the Péclet number, the stress that corresponds to Pe $= 0.1$ would be equal to $\mu_{\mathrm{G}}/\sqrt{1000}$, and the corresponding failure probability would be given by

$$P_f = \Phi\left(\frac{\sigma(\mathrm{Pe}=0.1) - \mu_G(\mathrm{Pe}=100)}{\delta_G}\right) \tag{4.24}$$

$$= \Phi\left(\frac{1}{\sqrt{1000}\times 0.1} - \frac{1}{0.1}\right) = 1.77\times 10^{-22} \tag{4.25}$$

where Φ = standard Gaussian (normal) distribution and δ_{G} = mean and standard deviation of the strength distribution of the material element.

It is clear that such a low failure probability is unlikely to be of interest in most practical engineering problems. So the random walk aspect can be safely be ignored for propagation driven purely by stress. Nevertheless, the diffusive fracture can play an important role for the corrosion-assisted fracture at very low applied stress.

5 Nano–Macroscale Bridging of Probability Distributions of Static and Fatigue Strengths

In Chapter 4 we showed that the probability distribution functions of static and fatigue strengths of a nanoscale structure, such as a regular atomic lattice or a random nanoparticle system, can be derived analytically based on the transition rate theory and atomistic fracture mechanics. As our main interest is the failure statistics of macroscale structures, it is now necessary to develop a multiscale framework relating the probability distributions of structural strength at macroscale and nanoscale.

The existing numerical stochastic multiscale approaches [e.g. (Xu 2007; Graham-Brady et al. 2006; Williams & Baxer 2006)] do not suffice for handling multiscale transition of probability distributions. In particular, they are incapable of dealing with the far-left tails, for two reasons:

1. The uncertainties in the physical laws for bridging various scales are difficult to quantify (Graham-Brady et al. 2006; Williams & Baxer 2006).
2. What is of interest is the far-out tail probability such as $<10^{-6}$, which is normally required as the maximum acceptable failure risk in engineering design of bridges, aircraft, and other structures. A numerical approach would require at least 10^8 numerical stochastic simulations with a numerical code based on nonlocal damage mechanics, which is currently beyond conventional computational means. Methods such as importance sampling would not help because they depend on a prior hypothesis about the tail (Bažant, Vořechovský, & Novák 2007).

So an analytical approach to the scale transitions is inevitable. Two fundamental models are available: (1) the chain model (or series coupling model) and (2) the fiber-bundle model (or parallel coupling model). These two models are mathematically tractable and, at the same time, they characterize, both mechanically and statistically, the basic failure mechanisms involved in scale bridging:

1. The chain model (Fig. 5.1a), in which the failure of one element triggers the failure of the entire chain, statistically represents, within one scale, the transverse *localization* of failure into a single fracture process zone (FPZ) among many potential FPZs (Fig. 5.1b).
2. The fiber-bundle model (Fig. 5.1c), which statistically models the condition of compatibility between one scale and its subscale. Specifically, it imposes the condition that the deformations of several cracked material subelements located side-by-side within the FPZ must be *compatible* with the overall deformation of this FPZ on the

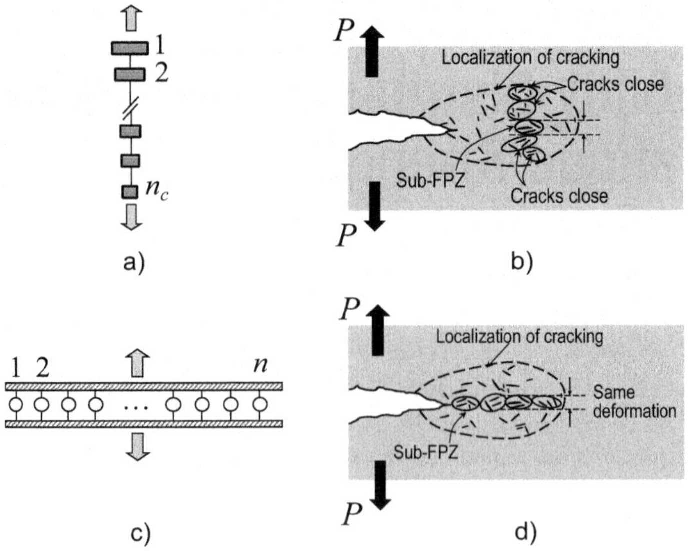

Figure 5.1. Two basic statistical models of structural failure. (a) Chain model, (b) localization in sub-FPZ, (c) bundle model, and (d) deformation compatibility of sub-FPZs.

next higher scale (Fig. 5.1d). At the same time, the fiber-bundle model also represents the load-redistribution mechanism in a partially damaged subelement at different scales.

This chapter presents the main important asymptotic properties of the chain and fiber-bundle models. By combining these two basic models, a model for the statistical multiscale transition is constructed and is then used to determine the analytical forms of the probability distribution functions of the static and fatigue strengths of a macroscale representative volume element, or RVE (see Section 1.4 for the definition of RVE for failure statistics, which differs from the definition in the theory of homogenization for non-softening materials).

5.1 Chain Model

The Weibull weakest-link model with its nonlocal generalization has been reviewed in detail in Chapters 2 and 3. Some of the key results are briefly summarized here as the starting point for developing the statistical multiscale model.

First consider a chain of material elements. If the strengths of these material elements are random, then the strength of the entire chain is determined by the strength of its weakest element. Since the cross-sectional areas of these elements might not be the same, the stress in each element can be different. Let $\sigma_N = P_{\max}/A_0$ denote the nominal strength of the chain, where $P_{\max}$ = maximum load that the chain can sustain and A_0 = reference cross-sectional area. The stress in the ith element is $\sigma_i = P_{\max}/A_i = \sigma_N A_0/A_i$, where A_i = cross-sectional area of the ith element. To determine the failure probability

P_f of a chain, we note that the whole chain survives if and only if all the elements survive. In other words, the survival probability, $1 - P_f$, of a chain is the joint probability of survival of all its elements. If the strengths of these elements are statistically independent, then the failure probability of a chain of n_c elements is given by the following well-known formula:

$$P_{f,\text{chain}}(\sigma_N) = 1 - \Pi_{i=1}^{n_c}[1 - \phi(\sigma_N A_0/A_i)] \tag{5.1}$$

where $\phi(x)$ = cumulative distribution function (cdf) of strength of one element.

The same concept can be directly applied to the statistics of fatigue strength of a chain. Consider a chain of elements (or links) subjected to cyclic loading with a prescribed number of cycles and stress ratio. The nominal fatigue strength $\Delta\sigma_c$ of the chain, i.e., the critical stress amplitude that leads to failure, is determined by the smallest fatigue strength of all the elements. Therefore, we may use the joint probability theorem to calculate the failure probability of the chain with n_c elements:

$$P_{f,\text{chain}}(\Delta\sigma_c) = 1 - \Pi_{i=1}^{n_c}[1 - \psi(\Delta\sigma_c A_0/A_i)] \tag{5.2}$$

where $\psi(x)$ = cdf of fatigue strength of one element.

Based on Eqs. 5.1 and 5.2, the strength distribution of the chain has the following simple but important asymptotic properties:

1. If the strength cdfs of all the elements have a power-law tail of exponent p, i.e., $\phi(x) = (x/s_\phi)^p$, $\psi(x) = (x/s_\psi)^p$, then the cdfs of static and fatigue strengths of the entire chain would also have a power-law tail and its exponent is also p.
2. For large enough n_c, we obtain the Weibull strength distribution, i.e., $P_f(\sigma) = 1 - e^{-n_{\text{eq},c}(\sigma/s_\phi)^p}$, or $P_f(\Delta\sigma) = 1 - e^{-n_{\text{eq},c}(\Delta\sigma_c/s_\psi)^p}$, where $n_{\text{eq},c}$ = equivalent number of elements that accounts for the nonuniform stress distribution. The detailed expression of $n_{\text{eq},c}$ can be found in Chapter 2.

5.2 Fiber-Bundle Model for Static Strength

In contrast to the chain model (or weakest-link model) which represents the failure localization mechanism, the fiber-bundle model (Fig. 5.1b) represents the distributed failure mechanism in which, after each element failure, the load gets redistributed among the surviving elements (usually called "fibers"). The load is reduced to zero when all fibers break, but the maximum load is reached when only a certain fraction of the fibers breaks. This fraction is a random number since the strength of each fiber is random.

The statistics of the static strength of the fiber-bundle has been studied for decades (Daniels 1945; Coleman 1958*a*, 1958*b*; Harlow & Phoenix 1978*a*, 1978*b*; Phoenix 1978*a*, 1978*b*; Harlow, Smith, & Taylor 1983). It has been shown that the strength distribution of the fiber-bundle depends on many factors, such as the strength cdf of fibers, the load-sharing rule, as well as the mechanical behavior of each fiber. Various intuitive hypotheses of load sharing after a fiber break have been proposed in the literature

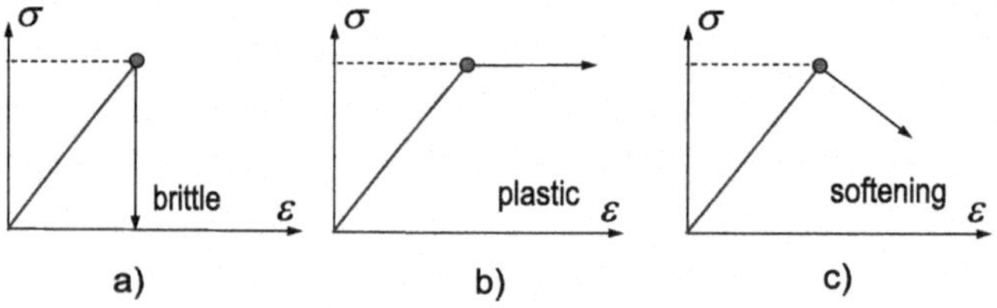

Figure 5.2. Different mechanical behaviors of fibers. (a) Brittle fibers, (b) plastic fibers, and (c) strain-softening fibers.

(Daniels 1945; Phoenix 1978*b*; Phoenix & Tierney 1983; Mahesh & Phoenix 2004). A more realistic approach, though, is to deduce the load-sharing rule from a mechanical model.

Since any fiber can be replaced by several fibers having different cross-sectional areas or different lengths but the same combined elastic stiffness, it is not unduly restrictive to assume that all the fibers have equal elastic stiffness E_f and equal cross-sectional area A_f, and are subjected to the same displacement. So, we consider initially elastic fibers spanning two parallel rigid plates. The load sharing rule is then fully determined by the mechanical behavior of the fibers. Three types of fiber behavior after reaching the strength limit are considered:

1. Brittle (Fig. 5.2a), in which case the stress drops suddenly to zero.
2. Plastic (Fig. 5.2b), in which case the fiber extends at constant stress.
3. Softening (Fig. 5.2c), in which case the fiber exhibits a gradual strain softening.

Clearly, the brittle and plastic bundles can be considered as the two limiting cases of the softening bundle. First we focus on the strength statistics for brittle and plastic bundles. Then we generalize it to softening bundles.

5.2.1 Brittle Bundle

Consider a bundle with n fibers. The fibers may be numbered as $k = 1, 2, \dots, n$ in the order of increasing random values of their strength σ_k. When the jth fiber is about to break, fibers $k = 1, 2, \dots, j-1$ have already broken and fibers $k = j, j+1, j+2, \dots, n$ are subjected to the same strain ϵ. The load applied on the entire bundle is shared equally among all the unbroken fibers and their stress is $\sigma_k = E_f\epsilon$. The average stress carried by the bundle at the moment when the jth fiber is about to break is $(n-j+1)\sigma_j/n$. The bundle strength is given by the maximization of this average stress among all the values of j:

$$\sigma_b = \frac{1}{n}\max_{\epsilon}\sum_{k=1}^{n} H(\sigma_k - E_f\epsilon)E_f\epsilon = \frac{1}{n}\max_{j}[(n-j+1)\sigma_j] \tag{5.3}$$

where σ_b = bundle strength = $P_{\max}/nA_f$, $P_{\max}$ = load capacity of the bundle, A_f = cross-sectional area of the fiber, and H = Heaviside step function. Alternatively, we may write Eq. 5.3 as

$$\sigma_b = n^{-1} \sum_{k=u}^{n} E_f \epsilon = [1 - (u-1)/n] E_f \epsilon \tag{5.4}$$

where $u = j$-value for which the last expression in Eq. 5.3 is maximized. It is clear that the randomness in fiber strength causes the randomness of the maximization ratio $\lambda = u/n$ and the corresponding strain.

Daniels (1945) derived the probability distribution of strength of a brittle bundle, which is summarized here. If the maximum load that the bundle can carry is less than a prescribed value S, then the following conditions must be satisfied (Let $A_f = 1$):

$$0 \le \sigma_1 \le S/n \tag{5.5a}$$

$$\sigma_1 \le \sigma_2 \le S/(n-1) \tag{5.5b}$$

$$\ldots\ldots$$

$$\sigma_{n-2} \le \sigma_{n-1} \le S/2 \tag{5.5c}$$

$$\sigma_{n-1} \le \sigma_n \le S \tag{5.5d}$$

where $\sigma_i (i = 1, \ldots, n)$ denote fiber strengths arranged in ascending order.

Assume that the fiber strengths are uncorrelated random variables. Then the strength cdf of the bundle, which is the probability that the foregoing conditions are met, can be written as

$$G_n(\sigma) == n! \int_0^{\sigma} f(\sigma_1)\,\mathrm{d}\sigma_1 \int_{\sigma_1}^{n\sigma/(n-1)} f(\sigma_2)\,\mathrm{d}\sigma_2 \ldots \int_{\sigma_{n-2}}^{n\sigma/2} f(\sigma_{n-1})\,\mathrm{d}\sigma_{n-1} \int_{\sigma_{n-1}}^{n\sigma} f(\sigma_n)\,\mathrm{d}\sigma_n \tag{5.6}$$

where $\sigma = S/n$, $f(\sigma_k) =$ probability density function (pdf) of fiber strength. Denoting $b(\sigma) = \int_0^{\sigma} f(u)\mathrm{d}u$, $b_r = b(n\sigma/r)$, and $x_r = b(\sigma_r)$, we can rewrite Eq. 5.6 in a simpler form:

$$G_n = n! \int_0^{b_n} \mathrm{d}x_1 \int_{x_1}^{b_{n-1}} \mathrm{d}x_2 \ldots \int_{x_{n-1}}^{b_1} \mathrm{d}x_n \tag{5.7}$$

Now consider a more general function:

$$\mathcal{G}_n(x) = n! \int_x^{b_n} \mathrm{d}x_1 \int_{x_1}^{b_{n-1}} \mathrm{d}x_2 \ldots \int_{x_{n-1}}^{b_1} \mathrm{d}x_n \tag{5.8}$$

It can easily be shown that the function $\mathcal{G}_n(x)$ has the following property:

$$\partial \mathcal{G}_n / \partial x = -n \mathcal{G}_{n-1} \tag{5.9}$$

with $\mathcal{G}_0(x) = 1$ and $\mathcal{G}_n(b_n) = 0$. Therefore, the Taylor series expansion of $\mathcal{G}_n(b_n)$ at $x = x_0$ leads to

$$\begin{aligned} 0 &= \mathcal{G}_n(x_0) - n(b_n - x_0)\mathcal{G}_{n-1}(x_0) \\ &+ \frac{n(n-1)}{2!}(b_n - x_0)^2 \mathcal{G}_{n-2}(x_0) - \cdots + (-1)^n (b_n - x_0)^n \end{aligned} \tag{5.10}$$

By setting $x_0 = 0$, we obtain Daniels' recursive equation for the strength cdf of brittle bundles:

$$G_n(\sigma) = \sum_{k=1}^{n} (-1)^{k+1} \frac{n!}{k!(n-k)!} F^k(\sigma_b) G_{n-k}\left(\frac{n\sigma}{n-k}\right) \tag{5.11}$$

where $F(\sigma_b)$ = strength cdf of fiber and $G_0(\sigma_b) = 1$. It has been shown that Eq. 5.11 is unusable in computations for $n > 75$, due to accumulation of round-off error.

For large bundles (n is large), the values of fiber strength are almost continuously distributed. Therefore, when $n \to \infty$, the area fraction occupied by the unbroken fibers when the fiber of strength σ is about to break is $1 - F(\sigma)$, and the stress carried by the bundle equals to $E_f\epsilon[1 - F(E_f\epsilon)]$. The strength of the bundle can be obtained by maximizing the bundle stress, which occurs at a certain critical strain ϵ^*. Daniels (1945) showed that the strength distribution of an infinitely large bundle is a Gaussian (or normal) distribution, which is symmetrical. Therefore, the maximum must be equal to the mean strength of the bundle, i.e.,

$$\mu_\sigma = E_f\epsilon^*[1 - F(E_f\epsilon^*)] \tag{5.12}$$

At the same time, Daniels also proved that, for large n, the variance of the total load converges to

$$\text{Var} = n(E_f\epsilon^*)^2 F(E_f\epsilon^*)[1 - F(E_f\epsilon^*)] \tag{5.13}$$

Therefore, the coefficient of variation (CoV) of strength of a large bundle has the following asymptotic expression:

$$\omega_{\sigma_b} \approx \rho_0 n^{-1/2} \tag{5.14}$$

where

$$\rho_0 \approx \sqrt{F(E_f\epsilon^*)/[1 - F(E_f\epsilon^*)]} = \text{constant} \tag{5.15}$$

Hence, ω_{σ_b} vanishes as $n \to \infty$. In other words, the strength of an infinite bundle is deterministic. Therefore, to model the scale transition of strength statistics, the number of fibers in the bundle must be finite.

A question crucial for the structural reliability is the shape of the far-left tail, which lies outside the Gaussian part of the strength cdf of a bundle with a finite number of fibers. It is evident that the far-left tail cannot be Gaussian because the Gaussian cdf has an infinite negative tail whereas the bundle strength cannot be negative. The distance from the mean μ_σ to the point $\sigma = 0$, which is a point sure to lie beyond the Gaussian core, may be written as

$$\Delta\sigma_{n\text{G}} = \mu_\sigma = (\delta_\sigma/\rho_0)\sqrt{n} \tag{5.16}$$

where δ_σ = standard deviation of bundle strength. The spread $\Delta\sigma_{n\text{G}}$ of the Gaussian core (i.e., the distance from the mean to the end of Gaussian core) must obviously be smaller than this; it is found to be also proportional to $\sqrt{n}$, i.e.,

$$\Delta\sigma_\text{G} = \gamma_G\delta_\sigma\sqrt{n} \tag{5.17}$$

where γ_G is some constant less than $1/\rho_0$. Smith (1982) showed that Daniels' Gaussian approximation to the cdf of bundle has a convergence rate of at least $O(n^{-1/6})$, which is a rather slow convergence, and proposed an improved Gaussian approximation with a mean depending on n, for which the convergence rate improves but still is not guaranteed to be better than $O(n^{-1/3}(\log n)^2)$.

Chapter 4 showed that the strength cdf of a nanoscale structure follows a power-law function. It would be interesting to explore the tail of the strength cdf of a bundle, in which all the fibers have strength distributions with power-law tails. By induction based on the set theory (Harlow et al. 1983; Phoenix, Ibnabdeljalil, & Hui 1997), it has been shown that, if the strength cdf of each fiber has a power-law tail of some exponent p, then the strength cdf of the bundle has also a power-law tail, and its exponent is equal to np. Here we present a simpler proof given in Bažant & Pang (2007), which is based on the asymptotic expansion of Eq. 5.11.

Let $F(\sigma) = \sigma^p$. For a one-fiber bundle, $n = 1$, this property obviously holds because $G_1(\sigma) = F(\sigma) = \sigma^p$. Now we assume this property to be true for the cdfs of all bundles with up to $n-1$ fibers, i.e., $G_k(\sigma) = (\sigma/s_k)^{kp}$ for $k = 1, 2, \ldots n-1$. Based on Eq. 5.11, we can obtain the strength cdf of the bundle with n fibers:

$$G_n(\sigma) = \sum_{k=1}^{n}(-1)^{k+1}\frac{n!}{k!(n-k)!}\sigma^{kp}G_{n-k}\left(\frac{n\sigma}{n-k}\right) \tag{5.18}$$

$$= (-1)^{n+1}\sigma^{np}G_0 + \sum_{k+1}^{n-1}\frac{(-1)^{k+1}n!}{k!(n-k)!}\sigma^{kp}\left(\frac{n\sigma}{(n-k)s_{n-k}}\right)^{(n-k)p} \tag{5.19}$$

$$= \left[(-1)^{n+1}G_0 + \sum_{k+1}^{n-1}\frac{(-1)^{k+1}n!}{k!(n-k)!}\left(\frac{n}{(n-k)s_{n-k}}\right)^{(n-k)p}\right]\sigma^{np} = (\sigma/s_n)^{np} \tag{5.20}$$

where s_n = scaling factor independent of σ. It is now clear that, with the parallel coupling, the power-law tail of strength cdf is preserved, and the exponent increases as the number of fibers in the bundle increases. But, what is the extent of the power-law tail of the bundle?

To check it, let us consider that the strength of each fiber follows the Weibull distribution (which has a power-law tail). By the Taylor series expansion, we have

$$F(\sigma) = (\sigma/s_1)^p(1 - \eta/2! + \eta^2/3! - \eta^4/4! + \cdots) \tag{5.21}$$

where $\eta = (\sigma/s_1)^p$. The dominant deviation from the power-law tail is represented by the first two terms of expansion, i.e.,

$$F(\sigma) = (\sigma/s_1)^p(1 - \sigma^p/t_1), \quad G_k(\sigma) = (\sigma/s_k)^{kp}(1 - \sigma^p/t_k) \tag{5.22}$$

where s_1, s_k, t_k ($k = 1, 2, \ldots n-1$) are constants and $t_1 = 2^{1/p}s_1$; $t_1^{1/p}$ and $t_k^{1/p}$ measure the vertical deviation from the power-law tail; $d_1 = t_1\varepsilon^{1/p}$ and $d_k = t_k\varepsilon^{1/p}$ describe the horizontal length of tail up to the (100ε)% from the power law. Now one may substitute

Table 5.1 Ratio of the horizontal length nd_n of power-law tail for total load of a brittle bundle to tail length d_1 of one fiber

	$n=2$	$n=3$	$n=4$	$n=6$	$n=12$	$n=24$
$p=1$	0.600	0.462	0.391	0.316	0.232	0.181
$p=2$	0.858	0.807	0.780	0.752	0.724	0.710
$p=4$	0.978	0.973	0.972	0.971	0.970	0.970
$p=6$	0.996	0.996	0.996	0.996	0.999	0.999
$p=8$	0.999	0.999	0.999	0.999	0.999	0.999
$p=12$	1.000	1.000	1.000	1.000	1.000	1.000

these expressions into Eq. 5.11 and, after binomial expansion, one gets

$$\begin{aligned}&\left(\frac{\sigma}{s_n}\right)^{np}\left[1-\left(\frac{\sigma}{t_n}\right)^{p}\right]\\&\approx\sum_{k=1}^{n-1}(-1)^{k+1}\frac{n!}{k!(n-k)!}\left(\frac{\sigma}{s_1}\right)^{kp}\left[1-k\left(\frac{\sigma}{t_1}\right)^{p}\right]\left(\frac{n\sigma}{js_j}\right)^{jp}\left[1-\left(\frac{n\sigma}{jt_j}\right)^{p}\right]\\&\quad+(-1)^{n+1}\left(\frac{\sigma}{s_1}\right)^{np}\left[1+n\left(\frac{\sigma}{t_1}\right)^{p}\right]\end{aligned}\tag{5.23}$$

where $j=n-k$. Since this equation must be valid for every small σ, the coefficient associated with the term σ^{np}, as well as σ^{np+1}, must vanish. The former yields Eq. 5.20, and the latter leads to a recursive equation for t_1/t_n:

$$\frac{t_1}{t_n}=\left\{\sum_{k=1}^{n-1}(-1)^{k+1}\frac{n!}{k!(n-k)!}\left(\frac{ns_1}{js_j}\right)^{jp}\left[k+\left(\frac{nt_1}{jt_j}\right)^{p}\right]+(-1)^{n+1}n\right\}^{1/p}\left(\frac{s_n}{s_1}\right)^{n}\tag{5.24}$$

Table 5.1 shows the numerical evaluation of nd_n/d_1 for bundles with different numbers of fibers while the exponent of the power-law tail of the strength cdf of each fiber is the same. Table 5.2 shows the extent of power-law tail of the failure probability for a brittle bundle with a tail exponent of 24, with different combinations of the number of fibers and the power-law exponent of each fiber, e.g., $((p=1, n=24); (p=2, n=12); \ldots)$.

Table 5.2 Extent of power-law tail probabilities for a brittle bundle with a tail exponent of 24

t_1/t_n	s_1/s_n	n	p	d_n/d_1	P_{tn}/P_{t1}	P_{tn}
1.00	1.00	1	24	1.000	1.00×10^{-0}	3.00×10^{-1}
2.00	1.46	2	12	0.500	1.00×10^{-4}	5.50×10^{-5}
3.00	1.78	3	8	0.333	1.00×10^{-7}	1.30×10^{-7}
4.02	2.05	4	6	0.249	1.00×10^{-9}	7.89×10^{-10}
6.18	2.53	6	4	0.162	1.00×10^{-12}	3.76×10^{-13}
8.74	2.98	8	3	0.114	1.34×10^{-15}	4.02×10^{-16}
16.57	3.83	12	2	0.060	9.43×10^{-22}	2.83×10^{-22}
132.79	6.23	24	1	0.008	1.21×10^{-44}	3.62×10^{-45}

Since d_n measures the strength corresponding to the deviation of $(100\varepsilon)\%$ from the power law, nd_n represents the total load of the bundle at such a deviation. As seen from Table 5.1, the load capacity of the bundle remains about constant for $p \geq 3$. This implies that d_n, the average strength of bundle at $(100\varepsilon)\%$ deviation, decreases roughly as $1/n$. From Table 5.2, we observe a drastic shortening of the power-law tail with an increasing number of fibers in terms of probability P_f:

$$P_{tn} \approx (d_n/s_n)^{np} = P_{t1}^n \left(\frac{t_n s_1}{t_1 s_n}\right)^{np} \tag{5.25}$$

where $P_{t1} = \varepsilon(t_1/s_1)^p =$ failure probability at the terminal point of power-law tail of one fiber. Based on Eq. 5.25, it has been shown that the power-law tail of a brittle bundle shortens as fast as $P_{t_n} \sim (P_{t_1}/n)^n - (P_{t_1}/3n)^n$ (Bažant & Pang 2006, 2007).

5.2.2 Plastic Bundle

For plastic bundles, in which each fiber sustains a constant peak load as it deforms, the maximum load of the bundle is the sum of n independent random variables $f_k(k = 1, 2, \ldots, n)$, which represent the fiber strengths. This classical problem can be dealt with by the Central Limit Theorem [e.g., Feller (1957); Bulmer (1979); Soong (2004)], which states that, if all f_k have a finite variance, the sum $Y = \sum_{k=1}^{n} f_k$ for $n \to \infty$ converges, except in the tails, to the Gaussian distribution with absolute error $O(n^{-1/2})$.

The distance from the mean of a sum to the tail of specified probability cutoff is known to be proportional to $\delta\sqrt{n}$, where δ is the standard deviation of the sum [e.g., Bouchaud & Potters (2000)]. To explain this, we note that the mean and variance of the maximum load of the bundle are $\mu_n = m\mu_\sigma$ and $s_n^2 = ns_\sigma^2$, where μ_σ and s_σ^2 are the mean and variance of f_k. If all f_k are nonnegative, μ must be nonnegative, even though the Gaussian pdf has an infinite negative tail. Of course, the Gaussian pdf of tensile strength cannot apply within the negative range; hence, the Gaussian core cannot reach farther from the mean μ_n than to the distance of rs_n, where

$$r = \frac{n\mu_\sigma}{\sqrt{ns_\sigma^2}} = \frac{\sqrt{n}}{\omega_0} \tag{5.26}$$

with ω_0 denoting the CoV of one fiber.

The tail outside the Gaussian core and the tails of f_k are known to be of the same type (Bouchaud & Potters 2000), i.e., if the tail of fibers is a power law, so is the tail of the mean. To show this, let us consider a bundle of two plastic fibers experiencing forces y and z. The tail cdf of strength can be written as

$$G(y) = \left(\frac{y}{y_0}\right)^{jp}\left[1 - \left(\frac{y}{t_j}\right)^{p}\right], \quad H(z) = \left(\frac{z}{z_0}\right)^{kp}\left[1 - \left(\frac{z}{t_k}\right)^{p}\right] \tag{5.27}$$

where j, k, p, y_0, z_0 are constants, and parameters t_k measure the power-law tail length corresponding to $(100\eta)\%$ deviations from each power-law tail. By differentiation, the pdf tails are

$$g(y) = \frac{jp}{y_0}\left(\frac{y}{y_0}\right)^{jp-1}\left[1-\left(\frac{y}{t'_j}\right)^p\right], \quad h(z) = \frac{kp}{z_0}\left(\frac{z}{z_0}\right)^{kp-1}\left[1-\left(\frac{z}{t'_k}\right)^p\right] \tag{5.28}$$

where

$$t'_j = t_j\left(\frac{jp}{jp+p}\right)^{1/p}, \quad t'_k = t_k\left(\frac{kp}{kp+p}\right)^{1/p} \tag{5.29}$$

The total load on the bundle is $x = y + z$. Load x can be obtained by all possible combinations of forces y and $z = x - y$ in the first and second fibers, both of which must be at their strength limits if the bundle load is maximum. According to the joint probability theorem, the pdf of the sum x is given by

$$f(x) = \int_0^x g(y)h(x-y)\,\mathrm{d}y \tag{5.30}$$

$$= jkp^2 y_0^{-jp} z_0^{-kp}\int_0^x y^{jp-1}(x-y)^{kp-1}\left[1-\left(\frac{y}{t'_j}\right)^p\right]\left[1-\left(\frac{x-y}{t'_k}\right)^p\right]\mathrm{d}y \tag{5.31}$$

Although the standard approach in the theory of probability is to take the Laplace transform of the convolution of integral and later invert it, for the present purpose, a simple power series approach will suffice. Here we perform the binomial expansion of $(x-y)^{kp-1}$ and $(x-y)^p$ and retain only two leading terms of the power series expansion of $f(x)$, which yields

$$f(x) \approx \frac{jkp^2 C'_0}{y_0^{jp} z_0^{kp}}\left[1 - C'_p\left(\frac{x}{t'_j}\right)^p - C'_q\left(\frac{x}{t'_k}\right)^p\right]x^{jp+kp-1} \tag{5.32}$$

where

$$C'_0 = \frac{\Gamma(jp)\Gamma(kp)}{\Gamma(jp+kp)}, \quad C'_p = \frac{\Gamma(jp+kp)\Gamma(p+jp)}{\Gamma(jp)\Gamma(jp+kp+p)}, \quad C'_q = \frac{\Gamma(jp+kp)\Gamma(p+kp)}{\Gamma(kp)\Gamma(jp+kp+p)} \tag{5.33}$$

Here $\Gamma(x)$ = Eulerian gamma function. The corresponding cdf of strength of a bundle consisting of two plastic fibers is

$$F(x) \approx \left(\frac{x}{x_0}\right)^{(j+k)p}\left[1-\left(\frac{x}{t^*}\right)^p\right] \tag{5.34}$$

where

$$x_0^{-(j+k)p} = \frac{C'_0}{y_0^{jp}z_0^{kp}}\frac{jkp}{(j+k)}, \quad (t^*)^{-p} = \frac{C'_p(j+1)(j+k)}{jt_j^p(j+k+1)} + \frac{C'_q(k+1)(j+k)}{kt_k^p(j+k+1)} \tag{5.35}$$

So we conclude that the exponents of fiber tails in a plastic bundle are additive, while the horizontal length of the power-law tail of the cdf of the total load of the bundle decreases from $d_1 = t_1\varepsilon^{1/p}$ to $d_{tn} = t^*\varepsilon^{1/p}$. In the case of fibers with $p = 1$, a bundle of three fibers is a coupling of one fiber with a bundle of two fibers, which gives $d_{tn}/d_1 = 0.667$; a

Table 5.3 Extent of power-law tail probabilities for a plastic bundle with a tail exponent of 24

t_1/t_n	s_1/s_n	n	p	d_n/d_1	P_{tn}/P_{t1}	P_{tn}
1.00	1.00	1	24	1.000	1.00×10^{-0}	3.00×10^{-1}
1.27	1.08	2	12	0.787	1.00×10^{-2}	3.00×10^{-3}
1.49	1.15	3	8	0.672	2.40×10^{-4}	7.20×10^{-5}
1.72	1.22	4	6	0.583	7.79×10^{-6}	2.34×10^{-6}
2.27	1.35	6	4	0.440	1.00×10^{-8}	3.01×10^{-9}
3.04	1.48	8	3	0.329	7.39×10^{-12}	2.22×10^{-12}
5.65	1.73	12	2	0.177	8.37×10^{-19}	2.51×10^{-19}
46.08	2.45	24	1	0.022	2.41×10^{-43}	7.23×10^{-44}

bundle of four fibers is a coupling of one fiber with a bundle of three fibers, which gives $d_{tn}/d_1 = 0.625$, etc., and for 24 fibers $d_{tn}/d_1 = 0.521$.

The cdf of the average strength of each fiber is simply a horizontal scaling of the cdf for the total load on the bundle, and so Eq. 5.34 can be written in terms of σ:

$$F(\sigma) \approx \left(\frac{n\sigma}{x_0}\right)^{(j+k)p}\left[1-\left(\frac{n\sigma}{t^*}\right)^{p}\right] = \left(\frac{\sigma}{s_n}\right)^{np}\left[1-\left(\frac{\sigma}{t_n}\right)^{p}\right] \tag{5.36}$$

where $s_n = x_0/n$, $t_n = t^*/n$ and $n = j + k$. So we see that the total load, as well as the average strength of the bundle, has a cdf tail with exponent np, which is the same as for a brittle bundle. The length of the power-law tail of the cdf of the strength of a bundle (i.e., the load per fiber), which is $d_n = d_{tn}/n$, gets changed, for $\varepsilon = 0.15$, by factors $0.667/3 = 0.222$ and $0.521/24 = 0.022$, respectively, with $p = 1$.

Table 5.3 shows the calculated extent of the power-law tail of strength cdf of a plastic bundle for different combinations of p and n. Compared to the brittle bundle (Table 5.2), the horizontal length of the power-law tail of the cdf for plastic bundle is about 2 to 3 times longer than the corresponding length for brittle bundle. In terms of the failure probability, the rate of shortening of the power-law tail for plastic bundle is slower than that for brittle bundle by one or two orders of magnitude. It has been shown that the power-law tail of a plastic bundle P_{t_n} shrinks as $(P_{t_1}/n)^n$ (Bažant & Pang 2007).

5.2.3 Softening Bundle with Linear Softening Behavior

In previous sections, we have shown two important properties of strength cdfs of both brittle and plastic bundles:

1. The power-law tails are always preserved and their exponents are additive. Specifically, if the strength cdf of each fiber in a bundle has a power-law tail of exponent $p_i (i = 1, \ldots, n)$, then the cdf of bundle strength has also a power-law tail and its exponent is $p = \sum_{i=1}^{n} p_i$.
2. As the number of fibers increases, the strength cdf of the bundle always approaches the Gaussian distribution, except for the far-left tail, where the power-law tail gets drastically shortened in terms of the failure probability.

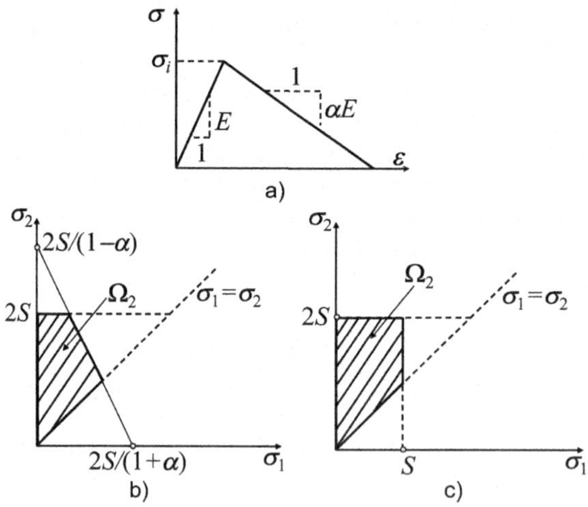

Figure 5.3. Analysis of softening fiber bundles. (a) Mechanical behavior of softening fibers, and (b, c) feasible domains of strength of individual fibers.

A more realistic model for quasibrittle materials is a softening bundle, in which the fiber experiences strain softening after reaching the peak load. Since the softening behavior represents an intermediate case between the brittle and plastic responses, we would expect that these properties should apply to softening bundle as well. Here we present a simple proof, first for a linearly softening bundle.

For brevity, let us limit attention to a bundle with two fibers having the same cross-sectional area, although a generalization to any number of fibers is straightforward. Assume that each fiber has a bilinear stress-strain curve (Fig. 5.3a), which has an elastic modulus E and softening modulus E_t ($E_t \leq 0$). Let the only random variable be the peak strength σ_i ($i = 1, 2$). Then the peak of average stress in the bundle can be written as

$$\sigma_b = \max_{\epsilon}[s_1(\epsilon) + s_2(\epsilon)] / 2 \tag{5.37}$$

where ϵ = strain in the fiber, and s_1, s_2 = stresses in fibers 1 and 2. We seek the critical strain ϵ^* at which the load on the bundle reaches its maximum. The critical strain depends on the ratio $\alpha = -E_t/E$ ($\alpha \geq 0$). Two cases must be distinguished, depending on whether the weaker fiber fails completely (1) before or (2) after the stronger one reaches its peak.

Let us number these two fibers such that $\sigma_1 \leq \sigma_2$. Then the peak stress of the bundle, σ_b, can be written as follows:

Case 1: $0 \leq \alpha \leq 1$

$$\text{if } (1+\alpha)\sigma_1/\alpha > \sigma_2: \quad \sigma_b = [(1+\alpha)\sigma_1 + (1-\alpha)\sigma_2] / 2 \tag{5.38}$$

$$\text{if } (1+\alpha)\sigma_1/\alpha \leq \sigma_2: \quad \sigma_b = \sigma_2/2 \tag{5.39}$$

Case 2: $\alpha > 1$

$$\text{if } (1+\alpha)\sigma_1/\alpha > \sigma_2: \quad \sigma_b = \sigma_1 \tag{5.40}$$

$$\text{if } (1+\alpha)\sigma_1/\alpha \leq \sigma_2: \quad \sigma_b = \max(\sigma_1, \sigma_2/2) \tag{5.41}$$

Note that the preceding expressions cover both cases of the plastic and brittle bundles. When $\alpha = 0$, the fiber deforms plastically and the peak average stress in the bundle is $(\sigma_1 + \sigma_2)/2$. When $\alpha \to \infty$, the fiber is brittle and the peak stress of the bundle is $\max(\sigma_1, \sigma_2/2)$.

If the average bundle strength is less than a prescribed value S, i.e., $\sigma_b \leq S$, then, based on Eqs. 5.38–5.41, the strength of each fiber must lie in the domain $\Omega_2(S)$, shown in Figures 5.3b and c. Since the strengths of these two fibers are considered to be independent random variables, we can use the joint probability theorem to calculate the cdf of the average bundle strength:

$$G_2(S) = 2\int_{\Omega_2(S)} f_1(\sigma_1)f_2(\sigma_2)\,\mathrm{d}\sigma_1\,\mathrm{d}\sigma_2 \tag{5.42}$$

where $f_i(x)$ = pdf of the strength of the ith fiber ($i = 1, 2$). For the limiting cases of brittle and plastic bundles, Eq. 5.42 becomes equivalent to Daniels' formulation for the brittle bundle (Daniels 1945) (Eq. 5.6), and to the convolution integral for the plastic bundle (Eq. 5.30).

Now we assume that the strength of each fiber has a cdf with a power-law tail, i.e., $F_i(\sigma) = (\sigma/s_0)^{p_i}$. Considering the transformation: $y_i = \sigma_i/S$, we can write the strength cdf of the bundle as

$$G_2(S) = 2S^{(p_1+p_2)}\int_{\Omega_2(1)} \frac{p_1p_2}{s_0^{p_1+p_2}}y_1^{p_1-1}y_2^{p_2-1}\,\mathrm{d}y_1\,\mathrm{d}y_2 \tag{5.43}$$

where $\Omega_2(1)$ denotes the feasible domain $\Omega_2(S)$ normalized by S. Since the integral in Eq. 5.43 results in a constant, the cdf of bundle strength has a power-law tail whose exponent is $p_1 + p_2$. Q.E.D.

It is evident that the foregoing analysis can directly be extended to a bundle with n fibers, for which the cdf of average bundle strength can be written as

$$G_n(S) = n!\int_{\Omega_n(S)} \prod_{i=1}^{n} f_i(\sigma_i)\,\mathrm{d}\sigma_1\,\mathrm{d}\sigma_2\ldots\mathrm{d}\sigma_n \tag{5.44}$$

$$= n!S^{p_1+p_2+\cdots+p_n}\int_{\Omega_n(1)}\left(\prod_{i=1}^{n}\frac{p_iy_i^{p_i-1}}{s_0^{p_i}}\right)\mathrm{d}y_1\,\mathrm{d}y_2\ldots\mathrm{d}y_n \tag{5.45}$$

where $\Omega_n(S)$ is the feasible domain of stresses in all the fibers, which defines an n-dimensional space, and $\Omega_n(1)$ is the corresponding feasible domain of normalized stresses, $y_i = \sigma_i/S$.

In principle, the reach of power-law tail of the strength cdf of softening bundle can be calculated from Eq. 5.45. However, for large bundles, it is difficult to handle the integral in Eq. 5.45 numerically. Previous sections showed that the reach of the power-law tail decreases with the number n of fibers rapidly, as $P_{tn} \sim (P_{t1}/n)^n - (P_{t1}/3n)^n$ for

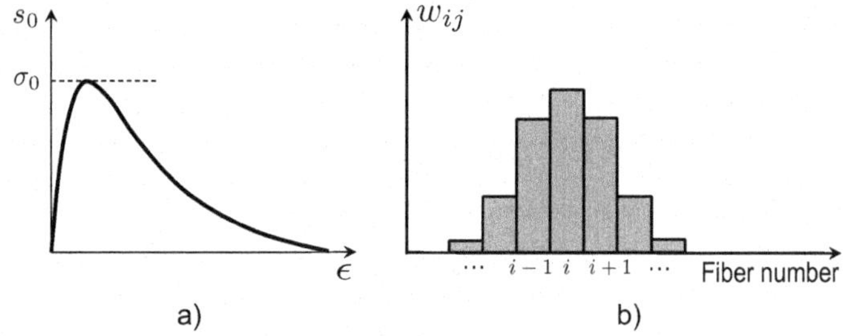

Figure 5.4. General stress-strain behavior of softening bundles. (a) Reference stress-strain relationship and (b) nonlocal weighting function.

brittle bundles, or $(P_{t1}/n)^n$ for plastic bundles. Since the behavior of softening bundles is bounded between these two extreme cases, the rate of shortening of the power-law tail of the softening bundles is expected to lie between them; i.e.,

$$P_{tn} \sim (P_{t1}/n)^n - (P_{t1}/3n)^n \tag{5.46}$$

5.2.4 Bundle with General Softening Behavior and Nonlocal Interaction

In the foregoing analysis of brittle, plastic, and softening bundles, we prescribed some particular stress-strain law for the individual fibers and the mechanical behavior of each fiber represents a local property. In this section, we extend the previous analysis to bundles with an arbitrary softening behavior. Furthermore, we also consider a nonlocal constitutive model for each fiber, to account for the nonlocal interactions between the fibers. In the context of failure of quasibrittle materials, the nonlocal interaction accounts for the microcrack interactions (Bažant & Ožbolt 1990; Bažant & Planas 1998; Bažant & Jirásek 2002). Furthermore, the nonlocal interaction also introduces, in a statistical sense, a spatial correlation of fiber strengths provided that the autocorrelation length is small compared to the system size.

Consider a reference stress-strain relationship shown as Fig. 5.4a, which is mathematically described as

$$s_0 = E_0\epsilon\,[1 - D(\epsilon)] \tag{5.47}$$

where E_0 = Young's modulus, $D(\epsilon)$ = damage function with a piecewise C^1-continuity, and the reference stress strain curve has a peak strength of σ_0. Now assume that the stress-strain curve of the ith fiber in a bundle exhibits a nonlocal behavior, and that it can be related to the reference stress-strain curve by the following affinity transformation:

$$s_i = \sigma_i\left(k\frac{\sigma_0\beta_i}{\sigma_i}\epsilon\right)\left[1 - \bar{D}_i\right] \quad (i = 1 \cdots n) \tag{5.48}$$

where $k = E_0/\sigma_0$, β_i = a positive constant, σ_i is the strength of the ith fiber, $\bar{D}_i$ = nonlocal damage parameter of ith fiber, and $\bar{D}_i$ is calculated from the weighted sum of the local damage parameters of all the fibers in the bundle, i.e.:

$$\bar{D}_i = \sum_{j=1}^{n} w_{ij} D\left(\beta_j \frac{\sigma_0}{\sigma_j}\epsilon\right) \tag{5.49}$$

Here w_{ij} = weights that describe the influence of the damage of the jth fiber on the nonlocal damage parameter of the ith fiber; the normalizing condition $\sum_j w_{ij} = 1$ must be met for every i. It is clear that by choosing an appropriate spatial weighting we can introduce a fixed nonlocal influence zone beyond which the damage of fibers would not affect the stress-strain curve of the ith fiber. Figure 5.4b shows a typical spatial weighting function w_{ij}.

Once the reference stress-strain relationship is defined, each fiber is, under the present assumptions, fully characterized by its strength, σ_i. Note that unlike the analysis of brittle, plastic, and linearly softening bundles, here no assumptions are made on the particular stress-strain behavior of each fiber.

Similar to the previous analysis of linearly softening bundles, let us first focus on a bundle of only two fibers, which are numbered by ascending strength, i.e., $\sigma_1 \leq \sigma_2$. Furthermore, we consider that the fiber strengths σ_i $(i = 1, 2)$ are the only random variables while the reference strength σ_0 and the constants β_1, β_2 are considered as deterministic quantities characterizing the transformation law.

According to Eqs. 5.47 to 5.49, the average stress in the bundle, σ_{avg}, can be written as

$$\sigma_{\text{ave}} = \frac{1}{2}\left[\sigma_1 f_1\left(\frac{\sigma_0\beta_1}{\sigma_1}\epsilon\right) + (\sigma_0\beta_1 k\epsilon)\, g_1\left(\frac{\sigma_0\beta_2}{\sigma_2}\epsilon\right) + \sigma_2 f_2\left(\frac{\sigma_0\beta_2}{\sigma_2}\epsilon\right) \right.$$
$$\left. + (\sigma_0\beta_2 k\epsilon)\, g_2\left(\frac{\sigma_0\beta_1}{\sigma_1}\epsilon\right)\right] \tag{5.50}$$

where:

$$f_1\left(\frac{\sigma_0\beta_1}{\sigma_1}\epsilon\right) = \left(\frac{\sigma_0\beta_1}{\sigma_1}k\epsilon\right)\left[1 - w_{11} D\left(\frac{\sigma_0\beta_1}{\sigma_1}\epsilon\right)\right] \tag{5.51}$$

$$f_2\left(\frac{\sigma_0\beta_2}{\sigma_2}\epsilon\right) = \left(\frac{\sigma_0\beta_2}{\sigma_2}k\epsilon\right)\left[1 - w_{22} D\left(\frac{\sigma_0\beta_2}{\sigma_2}\epsilon\right)\right] \tag{5.52}$$

$$g_1\left(\frac{\sigma_0\beta_2}{\sigma_2}\epsilon\right) = -w_{12} D\left(\frac{\sigma_0\beta_2}{\sigma_2}\epsilon\right) \tag{5.53}$$

$$g_2\left(\frac{\sigma_0\beta_1}{\sigma_1}\epsilon\right) = -w_{21} D\left(\frac{\sigma_0\beta_1}{\sigma_1}\epsilon\right) \tag{5.54}$$

Now, introducing for convenience the change of variable $\epsilon' = \epsilon\sigma_0/\sigma_1$, Eq. 5.50 can be written as

$$\sigma_{\text{avg}} = \frac{1}{2}\left[\sigma_1 f_1\left(\beta_1\epsilon'\right) + \sigma_1\left(\beta_1 k\epsilon'\right) g_1\left(\beta_2\frac{\sigma_1}{\sigma_2}\epsilon'\right) + \sigma_2 f_2\left(\beta_2\frac{\sigma_1}{\sigma_2}\epsilon'\right)\right.$$
$$\left. + \sigma_2\left(\beta_2\frac{\sigma_1}{\sigma_2}k\epsilon'\right) g_2\left(\beta_1\epsilon'\right)\right] \tag{5.55}$$

The critical value of ϵ', denoted by ϵ^*, which corresponds to the maximum value of σ_{avg}, is given by the condition $\partial\sigma_{avg}/\partial\epsilon' = 0$. Based on Eq. (5.55), it is easy to show that

$$\epsilon^* = f(\sigma_1/\sigma_2, \beta_1, \beta_2) \tag{5.56}$$

It is clear now that ϵ^* will be a function of the deterministic quantities β_1, β_2, and also of the ratio σ_1/σ_2 which, instead, is a random variable. Thus, the maximum average strength of the bundle, Eq. (5.55), can be rewritten in the following form:

$$\sigma_b = \frac{1}{2}\left[\sigma_1\Phi\left(\frac{\sigma_1}{\sigma_2}\right) + \sigma_2\Psi\left(\frac{\sigma_1}{\sigma_2}\right)\right] \tag{5.57}$$

where $\Phi(\sigma_1/\sigma_2)$ and $\Psi(\sigma_1/\sigma_2)$ are two generic functions of the random ratio σ_1/σ_2.

Now consider the case when the average bundle strength is less than some prescribed value S:

$$\frac{1}{2}\left[\sigma_1\Phi\left(\frac{\sigma_1}{\sigma_2}\right) + \sigma_2\Psi\left(\frac{\sigma_1}{\sigma_2}\right)\right] \le S \tag{5.58}$$

In this case, Eq. (5.58) describes a region $\Omega_2(S)$ in the plane $\sigma_2 - \sigma_1$ whose boundary is determined by the following curve:

$$\Gamma: \quad \sigma_1\Phi\left(\frac{\sigma_1}{\sigma_2}\right) + \sigma_2\Psi\left(\frac{\sigma_1}{\sigma_2}\right) = 2S \tag{5.59}$$

This can be written in parametric form as follows:

$$\Gamma: \quad \begin{cases} t = \dfrac{\sigma_1}{\sigma_2} \\ \sigma_2\left[t\Phi(t) + \Psi(t)\right] = 2S \end{cases} \qquad t \in [0, 1] \tag{5.60}$$

which can be rewritten in the following compact form:

$$\Gamma: \quad \begin{cases} \sigma_1 = StF(t) \\ \sigma_2 = SF(t) \end{cases} \qquad t \in [0, 1] \tag{5.61}$$

where

$$F(t) = \frac{2}{t\Phi(t) + \Psi(t)} \tag{5.62}$$

This system of equations describes the boundary of the region $\Omega_2(S)$, which is shown schematically in Figure 5.5a. The function $F(t)$ is not known a priori and depends on the particular family of the constitutive equations considered. However, since $F(t)$ need not be determined for the following analysis, it suffices to note that, after the parametrization, both σ_1 and σ_2 are expressed as the product of S with a function that does not depend on S.

According to Eq. 5.58, the strength of each fiber must lie in the domain $\Omega_2(S)$ whose boundary is now described by Eq. 5.61. Since the strengths of the individual fibers are considered to be statistically independent, we can use Eq. 5.42 to compute the strength distribution of the bundle. Furthermore, based on Eq. 5.61, it is clear that the failure

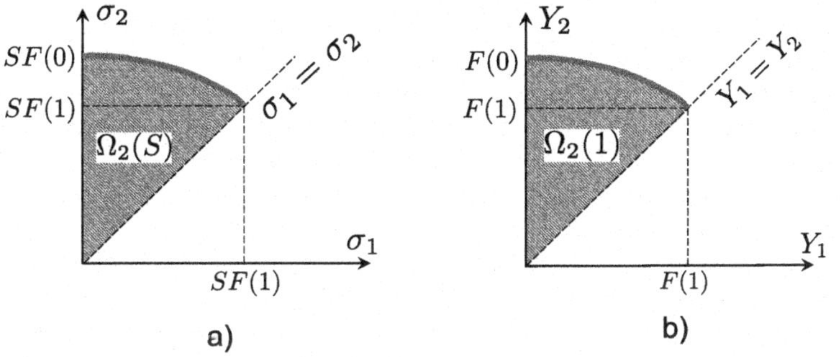

Figure 5.5. Schematic of feasible domain $\Omega_2(S)$ for general softening bundles.

domain $\Omega_2(S)$ can be normalized by S (Fig. 5.5b). This implies that, if each fiber has a strength cdf with a power-law tail, Eq. 5.43 can be used to calculate the strength cdf of the bundle. Clearly, this cdf has a power-law tail and the exponent is equal to the sum of the power-law exponents of the two fibers.

Similar to the case of linear softening bundles, the present analysis can be extended to a bundle with n fibers, and Eq. 5.44 is indeed valid. Thus it has been shown that the indestructibility of the power-law tail and the additivity of exponents are general properties of the fiber-bundle model with equal relative deformation, the affinity of the stress-strain laws of the individual fibers being the only restriction.

The other important property of strength cdf of the bundle is its asymptotic behavior as the number of fibers becomes very large. To determine this asymptotic strength cdf of softening bundles, let us consider a bundle of n fibers. The load carried by the bundle is given by $F(\epsilon) = \sum_{j=1}^{n} \sigma_j(\epsilon) A_f$, where σ_j = stress in jth fiber and ϵ = strain in each fiber. The mechanical behavior of each fiber can be random. This causes the randomness of the critical value ϵ^* of strain ϵ, at which F reaches its maximum value, $F_{\max}$.

Let us randomly separate these n fibers into two groups, A and B, and each group has the same number of fibers. The maximum load carried by the bundle is

$$F_{\max} = F_A(\epsilon^*) + F_B(\epsilon^*) \tag{5.63}$$

If n is large, the stress distribution over the fibers in these two groups will be similar to that in the bundle (Fig. 5.6). It follows that the cdf of $F_{\max}$ (i.e., the strength of bundle) and the cdfs of $F_A(\epsilon^*)$ and $F_B(\epsilon^*)$ must be of the same type. Then, to satisfy Eq. 5.63, the only possible distribution of $F_{\max}$ is the Gaussian distribution.

This asymptotic result for the softening bundle is expected because the strength cdf of large, both brittle and plastic, bundles asymptotically approaches the Gaussian distribution, and because the softening bundles should be bounded by these two limits. This observation indicates that the cdf of a softening bundle must also converge to the Gaussian cdf as the number of fibers increases. However, the rate of convergence to the Gaussian cdf is strongly dependent on the stress-strain behavior of the fibers. The slowest convergence, of the order of $O(n^{-1/3}(\log n)^2)$ (Smith 1982), occurs for the brittle bundles. The fastest convergence, of the order of $O(n^{-1/2})$, occurs for the plastic

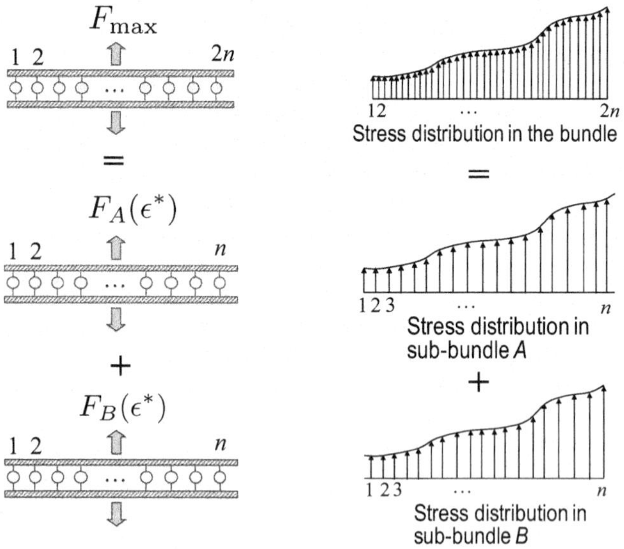

Figure 5.6. Stress distributions in a large bundle and its sub-bundles (fibers are arranged in the order of their breaking strengths).

bundles (Bažant & Pang 2007). The convergence rate of softening bundles is expected to be in between.

5.3 Fiber-Bundle Model for Fatigue Strength

So far we have analyzed the probability distribution of static strength of a bundle while disregarding the time scale. Following Chapter 4, let us now focus on the failure statistics of a bundle under fatigue loading, in particular the cdf of fatigue strength. Similar to the case of static strength, we are interested in the general formulation as well as the asymptotic properties of the cdf of fatigue strength of the bundle. Let us start with a bundle with two fibers having random fatigue strengths and the same cross-sectional areas, although present analysis can also be applied to bundles with any number of fibers.

Consider a bundle under cyclic loading with a prescribed stress ratio R. Here we define the fatigue strength of the fiber as the critical stress amplitude for a given stress ratio R and a specified number of cycles N_0. Denote $\Delta\sigma_i (i = 1, 2)$ = fatigue strengths of two fibers. The fibers are numbered so that $\Delta\sigma_1 < \Delta\sigma_2$. Figure 5.7 shows the loading histories of both the bundle and its two fibers. The bundle reaches its strength limit and fails at the N_0th cycle. After the first N_1 cycles, the first fiber fails and then the second fiber carries the entire load for the remaining $N_0 - N_1$ cycles.

The first fiber is subjected to a cyclic load with stress amplitude $\Delta\sigma_b$ and stress ratio R. Under this cyclic load, some subcritical crack inside the fiber grows from its original length a_0 to a critical length a_c at which the first fiber fails. To describe such a time-dependent failure process, a kinetics law is needed. A broadly applicable and commonly

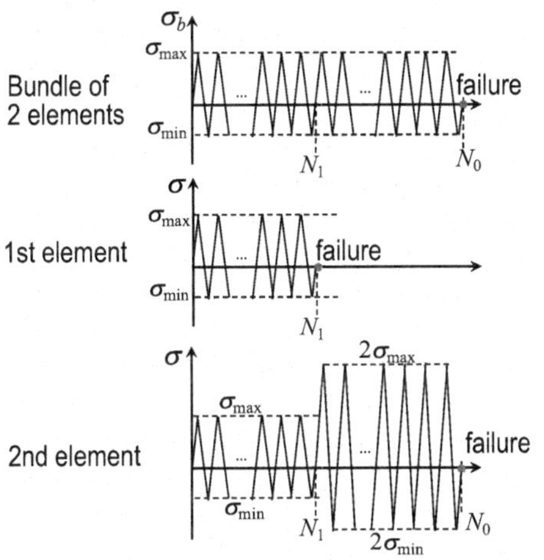

Figure 5.7. Loading histories of a two-fiber bundle and its fibers under cyclic loading.

used equation for fatigue kinetics is the Paris law (Paris & Erdogan 1963). It can be written as

$$\frac{\mathrm{d}a}{\mathrm{d}N} = Ae^{-Q_0/kT} \Delta K_e^{n_e} \tag{5.64}$$

where A and n_e are constants, $\Delta K_e = \Delta\sigma\sqrt{l}k_e(\alpha) =$ amplitude of the stress intensity factor of the fiber, $\Delta\sigma =$ stress amplitude, $l =$ characteristic size of the fiber, $k(\alpha) =$ dimensionless stress intensity factor, and $\alpha = a/l =$ relative crack length. The constant A is usually dependent on the stress ratio R. Though the Paris law has been widely applied to many quasibrittle materials, it is usually considered as an empirical law (Chapter 6, however, presents a physical justification of the Paris law based on a multiscale energetic analysis).

By separating the variables and integrating Eq. 5.64 from the original crack length a_0 to the length a_c, we obtain

$$\Delta\sigma_b^{n_e} N_1 = e^{Q_0/kT} I_1 \tag{5.65}$$

$$\text{where} \quad I_1 = A^{-1} l^{1-n_e/2} \int_{\alpha_0}^{a_c} k_{e1}^{-n_e}(\alpha)\, \mathrm{d}\alpha \tag{5.66}$$

where $k_{e1} =$ dimensionless stress intensity factor of the first fiber. It is clear that, for a particular fiber, I_1 must be a constant for different cyclic loads as long as the stress ratio is kept constant. Now, considering the fatigue strength tests for the first fiber and the bundle, we can write

$$\Delta\sigma_b^{n_e} N_1 = \Delta\sigma_1^{n_e} N_0 = e^{Q_0/kT} I_1 \tag{5.67}$$

$$\text{or:} \quad N_1 = N_0 \Delta\sigma_1^{n_e} / \Delta\sigma_b^{n_e} \tag{5.68}$$

The second fiber experiences the same load history as the first fiber does for the first N_1 cycles. After the first fiber fails, the stress in the second fiber doubles, i.e., the stress amplitude becomes $2\Delta\sigma_b$, because both fibers have the same elastic stiffness and the

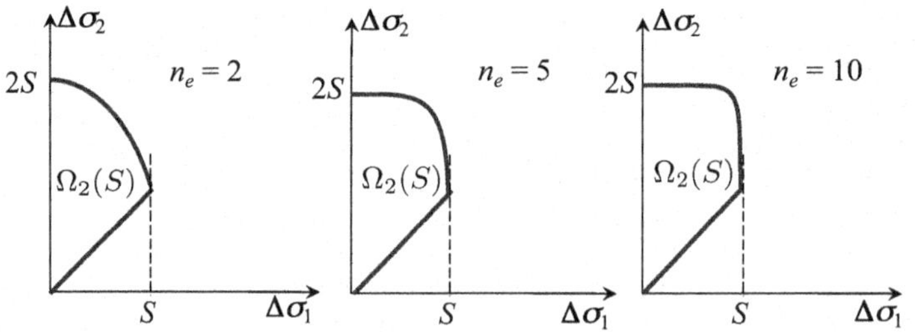

Figure 5.8. Feasible domains of fatigue strength of fibers for $n_e = 2,\ n_e = 5$, and $n_e = 10$.

same deformation. However, the stress ratio in the second fiber still remains R. The second fiber eventually fails at the N_0th cycle (Fig. 5.7). Therefore, the Paris law can be integrated for the second fiber taking into account its increased stress amplitude:

$$\Delta\sigma_b^{n_e} N_1 + (2\Delta\sigma_b)^{n_e}(N_0 - N_1) = e^{Q_0/kT} I_2 \tag{5.69}$$

$$\text{where} \qquad I_2 = A^{-1} l^{1-n_e/2} \int_{a_0}^{a_c} k_{e2}^{-n_e}(\alpha)\,\mathrm{d}\alpha \tag{5.70}$$

and k_{e2} = dimensionless stress intensity factor of the second fiber. Similar to the analysis for the first fiber, we have $e^{Q_0/kT} I_2 = \Delta\sigma_2^{n_e} N_0$, which yields

$$\Delta\sigma_b^{n_e} N_1 + (2\Delta\sigma_b)^{n_e}(N_0 - N_1) = \Delta\sigma_2^{n_e} N_0 \tag{5.71}$$

Substituting Eq. 5.68 into Eq. 5.71, we can then express the fatigue strength of the bundle as a function of the fatigue strengths of the individual fibers:

$$\Delta\sigma_b = \left[\Delta\sigma_1^{n_e}(1 - 1/2^{n_e}) + \Delta\sigma_2^{n_e}/2^{n_e}\right]^{1/n_e} \tag{5.72}$$

If the fatigue strength of the bundle does not exceed a certain value S, then the fatigue strengths of fibers are bounded by the region $\Omega_2(S)$ shown as Figure 5.8. Assuming that the fatigue strengths of the two fibers are independent random variables, then the cdf of fatigue strength of the bundle is given by

$$G_2(S) = 2\int_{\Omega_2(S)} f_1(\Delta\sigma_1) f_2(\Delta\sigma_2)\,\mathrm{d}\Delta\sigma_1\,\mathrm{d}\Delta\sigma_2 \tag{5.73}$$

where $f_i(x)$ = pdf of the fatigue strength of the ith fiber ($i = 1, 2$).

The foregoing analysis can readily be extended to a bundle with n fibers. Equation 5.72 can be generalized as

$$\Delta\sigma_b = \left\{\sum_{i=1}^{n} [\beta_i(n_e)\Delta\sigma_i]^{n_e}\right\}^{1/n_e} \tag{5.74}$$

$$\text{where} \quad \beta_i(n_e) = [(n - i + 1)^{n_e}/n^{n_e} - (n - i)^{n_e}/n^{n_e}]^{1/n_e} \tag{5.75}$$

It can easily be shown that

$$\beta_i = 1/n \qquad \text{for} \quad n_e = 1 \tag{5.76}$$

$$\beta_i = (n - i + 1)/n \qquad \text{for} \quad n_e \to \infty \tag{5.77}$$

So the cdf of fatigue strength of the bundle can be written as

$$G_n(S) = n! \int_{\Omega_n(S)} \prod_{i=1}^{n} f_i(\Delta\sigma_i)\, \mathrm{d}\Delta\sigma_1 \Delta\, \mathrm{d}\sigma_2 \ldots \mathrm{d}\Delta\sigma_n \tag{5.78}$$

Here $\Omega_n(S)$ is the feasible region of stresses in all the fibers defined by the following inequalities:

$$\left[\sum_{i=1}^{n} \beta_i(n_e)\Delta\sigma_i^{n_e}\right]^{1/n_e} \leq S \tag{5.79}$$

$$\Delta\sigma_1 \leq \Delta\sigma_1 \cdots \leq \Delta\sigma_{n-1} \leq \Delta\sigma_n \tag{5.80}$$

Similar to the statistics of static strength of the bundle, two important asymptotic properties of the cdf of fatigue strength of the bundle are of particular interest. The first property is the type of cdf of fatigue strength of large bundles. Consider the following two extreme values of n_e:

1. When $n_e = 1$, we have

$$\Delta\sigma_b = n^{-1} \sum \Delta\sigma_i \tag{5.81}$$

This is the same as the mathematical representation of the cdf of strength of a plastic bundle, in which each fiber deforms at constant stress after its strength limit is reached. By virtue of the Central Limit Theorem, the cdf of fatigue strength must follow the Gaussian distribution except for its far-left tail.

2. When $n_e \to \infty$, the fatigue strength of the bundle may be written as

$$\Delta\sigma_b = \max\left[\Delta\sigma_1, \frac{n-1}{n}\Delta\sigma_2, \ldots, \frac{1}{n}\Delta\sigma_n\right] \tag{5.82}$$

where $\Delta\sigma_1, \Delta\sigma_2, \ldots \Delta\sigma_n$ are the fatigue strengths of the fibers ordered by the sequence of their breaking, i.e., according to increasing strength. This is the same as the mathematical formulation of the cdf of strength of a brittle bundle, whose strength distribution approaches the Gaussian distribution as the number of fibers tends to infinity. In fact, Figure 5.8 shows that, when n_e is about 5, the feasible domain of fatigue strengths is very close to the feasible domain of monotonic strengths of a brittle bundle (Fig. 5.3c). Based on these two limiting cases, we expect that the cdf of fatigue strength of a large bundle should approach the Gaussian distribution for any value of $n_e \geq 1$.

The second important property is the tail of the cdf of fatigue strength of the bundle. Let us assume that the fatigue strength of each fiber has a cdf with a power-law tail, i.e., $F_i(\Delta\sigma) = (\Delta\sigma/s_0)^{p_i}$. Considering the transformation $y_i = \Delta\sigma_i/S$, we can rewrite Eq. 5.78 as

$$G_n(S) = n! S^{p_1+p_2+\cdots+p_n} \int_{\Omega_n(1)} \left(\prod_{i=1}^{n} \frac{p_i y_i^{p_i-1}}{s_0^{p_i}}\right) \mathrm{d}y_1\, \mathrm{d}y_2 \ldots \mathrm{d}y_n \tag{5.83}$$

where $\Omega_n(1)$ is the corresponding feasible region of the normalized fatigue strength.

Thus it is proven that, if the fatigue strength of each fiber has a cdf with a power-law tail, then the cdf of fatigue strength of the bundle will also have a power-law tail, and the power-law exponent will be the sum of the exponents of the power-law tails of the cdfs of fatigue strength of all the fibers in the bundle.

The reach of the power-law tail also needs to be identified (Bažant & Pang 2007). Based on the aforementioned two extreme cases ($n_e = 1$ and $n_e \to \infty$), the reach of the power-law tail of cdf of fatigue strength of the bundle can be estimated. The mathematical representation of the cdf of fatigue strength of bundle is the same as that of the cdf of strength of a plastic bundle for $n_e = 1$, or a brittle bundle for $n_e \to \infty$. Based on the previous section, we expect that, for $n_e \geq 1$, the reach of the power-law tail of fatigue strength would shrink with an increasing number of fibers at the rate between $(P_{t1}/n)^n$ and $(P_{t1}/3n)^n$.

5.4 Hierarchical Model for Static Strength

The essential question now is how to use the chain and bundle models to statistically represent one macroscale representative volume element (RVE) from the nanoscale structures, whose failure statistics has been derived in Chapter 4. One microcrack in an RVE, or too few of them, would not cause the RVE to fail. Rather, a certain number of microcracks must form and coalesce to cause the RVE to fail. This physically implies that one RVE cannot be represented by the chain model, and that the behavior of one RVE must be close to that of the fiber bundle. Further, it follows that the Weibull distribution cannot be used to represent the strength distribution of an RVE (Bažant & Pang 2007).

Another proof of this observation can be obtained by showing the impossibility of the opposite (Bažant & Pang 2007). Let a certain material element R_0 be considered as the RVE, and assume its strength distribution to be Weibullian. This would be possible only if element R_0 behaved as a chain. But then the failure would inevitably localize into just one subelement, R_1, of that chain. So, based on the definition of the RVE in Section 1.4, the true RVE would be subelement R_1 of the chain, rather than the originally considered element R_0. Q.E.D.

To show that the classical concept of an RVE is irrelevant for softening damage, note that in this concept the RVE is the smallest control volume for which the average elastic moduli (and possibly also the higher moments) remain constant when the control volume slides through the material. But the averaging consists of a summation and, according to the Central Limit Theorem (Section 2.7), it can never give information on the tails (Fig. 5.9). Yet it is solely the tail probability that governs the failure and size effect in large structures (of positive geometry) (Bažant & Pang 2007).

Does the impossibility of a Weibullian RVE mean that a pure fiber bundle would suffice to represent the behavior of one RVE? To answer this question, we first note that both the bundle and chain models will preserve the power-law tail of the cdf of their elements. In Chapter 4 we have shown that the strength cdf of the nanostructure follows a power-law function with an exponent equal to 2. Therefore, regardless of how we use the chains and bundles to represent one RVE, the strength cdf of one RVE must have a power-law tail.

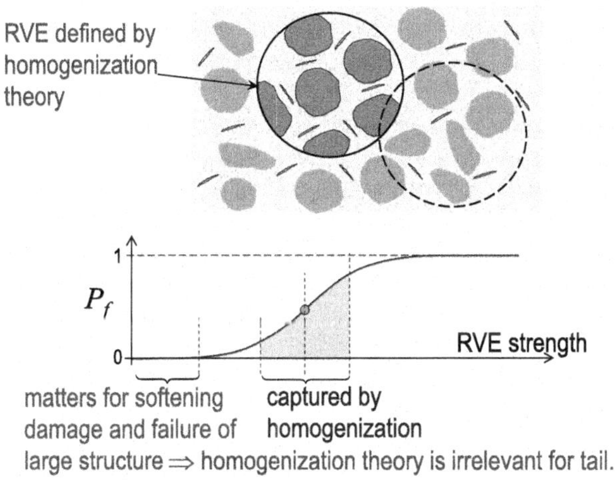

Figure 5.9. Application of the classical concept of RVE for reliability analysis of quasibrittle structures.

From Chapter 2 we may further recall that large quasibrittle structures with positive geometry [i.e., those that fail (under a controlled load) right at the initiation of a macrocrack from a damaged RVE] can be statistically modeled as a chain of RVEs. Therefore, it is clear that the Weibull modulus of the strength distribution of macroscale structures must be equal to the power-law tail exponent of the cdf of strength of one RVE.

Now, as an example, consider concrete. Its Weibull modulus is about 24 (Bažant & Novák 2000*a*; Bažant 2005), which means that the exponent of power-law tail of strength cdf of one RVE is equal to 24. If we statistically modeled the RVE as a bundle of brittle elements, each element representing one nanoscale structure, then the bundle would have 12 elements (because $12 \times 2 = 24$, where 2 is the tail exponent of a nanostructure). With a bundle of 12 elements, the power-law tail of the strength cdf of one RVE would reach, according to Table 5.2, only to the probability of about $P_f \approx 10^{-22}$. This implies that a chain that would statistically represent the entire macrostructure would have to consist of 10^{23} RVEs for the Weibull distribution to get manifested. But this is impossible.

Indeed, a chain with that many RVEs, with each RVE size being 0.1 m (as typical of concrete), would have a length of about 1 million light years, and thus the structure would have a galactic dimension (Bažant & Pang 2007). Therefore, if each RVE were statistically modeled by Daniels' bundle of 12 elements, then, on the scale of real structures, one would never observe Weibull distribution. Yet it is observed (Weibull 1939).

As a matter of fact, based on experimental observations, the cdf of strength of positive geometry structures is already much closer to the Weibull distribution than to the Gaussian distribution if the number of RVEs in the structure is about 10^3 or 10^4, but definitely not orders of magnitude more. Based on the weakest-link model (Chapter 2), this implies that the power-law tail of the strength cdf of one RVE must reach up to $P_f \approx 10^{-4}$ or 10^{-3}. Therefore, one RVE cannot be statistically represented by a bundle model, or else the power-law tail would be too short by orders of magnitude. To

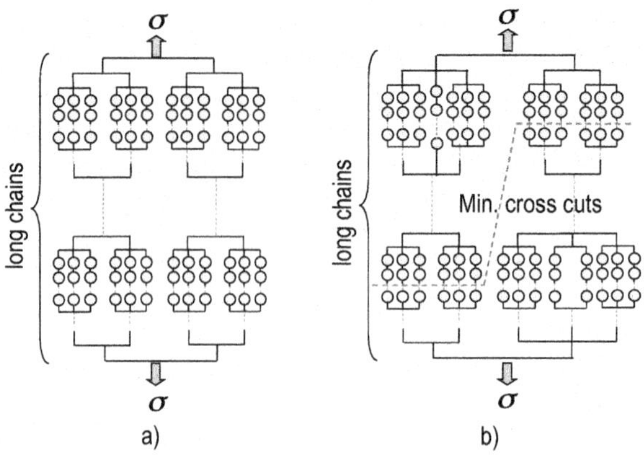

Figure 5.10. Hierarchical statistical model of failure of one RVE.

increase the reach of the power-law tail, we must rely on the chain model, since it has been shown that the chain model extends the power-law tail roughly by one magnitude for each 10-fold increase in the number of elements in the chain.

As a consequence of the foregoing analysis, we must introduce a statistical model consisting of both chains (i.e., series coupling) and bundles (i.e., parallel coupling). The model must satisfy two basic requirements (Bažant & Pang 2007):

1. The model must raise the tail exponent of the strength cdf from 2 at the nanoscale to about 10–50 at the macroscale, which is the range of Weibull modulus observed in real macroscale structures.
2. The model must yield a long enough power-law tail of correct length, reaching up to $P_f \approx 10^{-4}$ to 10^{-3}, so that a structure with more than about 10^4 RVEs would have a strength cdf close to the Weibull distribution, as revealed by analysis of experiments.

These requirements can be achieved only by a hierarchical statistical model involving both parallel and series couplings (Bažant & Pang 2007). A simple and adequate model of this kind is shown in Figure 5.10a. The first bundle (parallel coupling) must involve no more than two parallel elements, and each of them may then consist of a hierarchy of sub-chains of sub-bundles of sub-sub-chains of sub-sub-bundles, and so on. Each element in the hierarchical model represents a nanoscale structure.

As seen in Figure 5.10a, the elements, of identical power-law tails, are coupled in each sub-chain to extend the power-law tail to a realistic length (note that too long a sub-chain would eventually produce a Weibull cdf, but this would be unrealistic). On the next higher scale, the parallel coupling of two or three of these sub-chains in a sub-bundle will raise their tail exponent by summation but will shorten the power-law tail significantly. Then, on the next higher scale, a series coupling of many sub-bundles in a chain will again extend the power-law tail, and a parallel coupling of two of these chains will again raise the tail exponent and shorten the power-law tail significantly, until the macroscale RVE is reached. Thus, in each passage to a higher scale, the parallel coupling raises the tail exponent and the series coupling restores a realistic length of the tail.

To figure out the functional form of the cdf of RVE strength, we must specify the mechanical behavior of the bundles in the hierarchical model. Although different assumptions yield about the same results, here the following assumption is made: For the bundles at the lowest scale, three types of stress-strain behaviors, i.e., brittle, softening, and plastic, are considered for each element. Bundles at higher scales have brittle behavior, which would give the shortest power-law tail of the strength cdf of an RVE.

Note that the strength cdf needs to be calculated in a hierarchical manner. At the lowest scale, each element represents a nanoscale structure whose strength cdf has a power-law tail. We can then calculate the strength cdf of the sub-chain that connects these elements. At the next scale, the strength cdf of the sub-bundle, which consists of these sub-chains, can be calculated based on the strength cdf of the sub-chains. In this manner, we can move up through the scales, and finally obtain the strength cdf of one RVE.

As an example, we calculate the strength cdf of the hierarchical model shown in Figure 5.10a. Every element in the hierarchical model represents one nanoscale structure, whose strength cdf has a power-law form (Eq. 4.8 in Chapter 4). Three cases are considered:

1. Each element exhibits an elastic-brittle behavior.
2. Each element exhibits a local linear softening behavior, for which the softening modulus magnitude is 40% of the elastic modulus of the element.
3. Each element exhibits an elastic-plastic behavior.

Figure 5.11a shows the calculated strength cdfs of the hierarchical model for these three cases in the Weibull scale. For all these cases, the lower portion of the calculated strength cdf is a straight line on the Weibull plot, which indicates that it follows the Weibull distribution (which has a power-law tail). This property is, of course, expected since in the chain and bundle models the power-law tail of the strength cdf has been shown to be indestructible.

For the upper portion, the strength cdf deviates from the straight line in Figure 5.11a. Among the three cases considered, case 1 (i.e., elements with brittle behavior) gives the shortest Weibull tail, which terminates at the probability of about 0.7×10^{-4}, while case 3 (elements with plastic behavior) gives the longest Weibull tail, which terminates at the probability of about 0.7×10^{-3}. Although, in normal testing, the grafted Weibull tail for one RVE is too short to be observed, it is nevertheless needed to ensure that the strength cdf of large structure be predominantly Weibullian.

To identify the type of distribution for the upper portion of the cdf, the cdfs of RVE strength are plotted on the normal probability paper; see Figures 5.11b–d where the upper portion of the cdfs is seen to be fitted quite closely by a straight line. For case 1, the straight line is not too close if $P_f \geq 0.8$. For cases 2 and 3, the straight line fits quite closely, with a slight deviation occurring only for $P_f \geq 0.99$. This means that the upper portion of the strength cdf can be approximated as the Gaussian distribution. Since, for real quasibrittle structures, the sub-bundle is expected to have a softening behavior, the strength cdf of one RVE should be close to that of case 2.

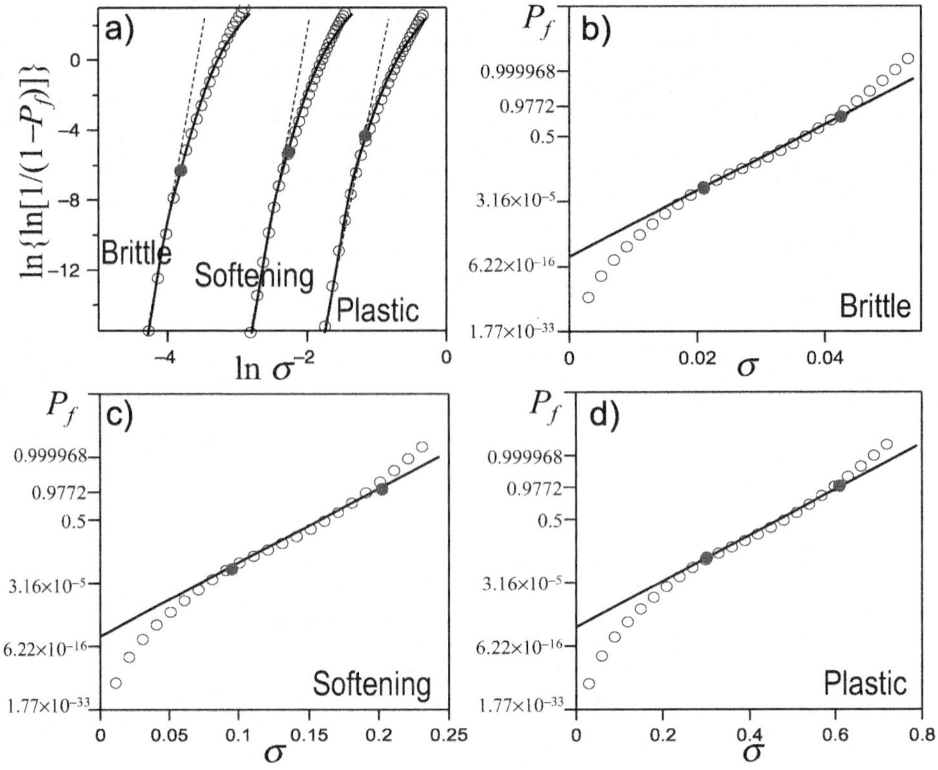

Figure 5.11. Calculated strength distributions of one RVE presented on (a) Weibull probability paper and (b–d) Normal probability paper.

Based on the aforementioned analysis, we may conclude that the strength distribution of one RVE can be approximately described by a Gaussian distribution with a Weibull tail grafted on the left at a probability of about 10^{-4} to 10^{-2}, with 10^{-3} as the mean estimate. The grafted cdf of strength of one RVE can be mathematically described as (Bažant & Pang 2006, 2007)

$$P_1(\sigma) = 1 - e^{-\langle\sigma/s_0\rangle^m} \qquad (\sigma \leq \sigma_{gr}) \tag{5.84a}$$

$$P_1(\sigma) = P_{\mathrm{gr}} + \frac{r_f}{\delta_G\sqrt{2\pi}} \int_{\sigma_{\mathrm{gr}}}^{\sigma} e^{-(\sigma'-\mu_G)^2/2\delta_G^2}\,\mathrm{d}\sigma' \qquad (\sigma > \sigma_{gr}) \tag{5.84b}$$

where m (Weibull modulus) and s_0 are the shape and scale parameters of the Weibull tail, and μ_G and δ_G are the mean and standard deviation of the Gaussian core if considered extended to $-\infty$; r_f is a scaling parameter required to normalize the grafted cdf such that $P_1(\infty) = 1$, P_{gr} = grafting probability, and $\langle x\rangle = \max(x, 0)$. Clearly, we have

$$P_{\mathrm{gr}} = 1 - e^{-(\sigma_{gr}/s_0)^m} \tag{5.85}$$

Finally, continuity of the pdf at the grafting point requires that

$$[\mathrm{d}P_1/\mathrm{d}\sigma]_{\sigma_{\mathrm{gr}}^+} = [\mathrm{d}P_1/\mathrm{d}\sigma]_{\sigma_{\mathrm{gr}}^-}. \tag{5.86}$$

The actual behavior of a RVE will, of course, correspond to some irregular hierarchical model, e.g., that in Figure 5.11b. In that case, according the aforementioned property of additivity of the power-law exponent for the bundle model, the exponent of the power-law tail for one RVE, and thus the Weibull modulus of a large structure, is determined by the minimum of the sum of strength powers among all possible cuts separating the model into two halves. Because random variations in couplings of the hierarchical model for extreme value statistics of the RVE must be expected, it would make hardly any sense to compute the structural failure probability directly from the failure probability of the nanoscale structure.

Nevertheless, the present statistical hierarchical model yields a physically based *functional form* of the strength distribution. The corresponding statistical parameters need to be calibrated from macroscale experiments, which are discussed in detail in Chapter 9.

Another noteworthy point is the statistical correlation of the element strength values within the hierarchical model. In the aforementioned calculations, the strength of each element is assumed to be statistically independent. Some statistical correlations, though, must exist. For the bundle model, they can be introduced indirectly by considering a nonlocal constitutive behavior for the fibers, provided that the autocorrelation length is small compared to the system size. As we have shown in the previous section, the nonlocal interaction between fibers does not change the qualitative behavior of the strength statistics of the bundle.

Similarly, the statistical correlation of element strengths will not alter the qualitative behavior of the chain. This is obvious from the fact that the stability postulate of extreme value statistics (Chapter 2) is also valid for correlated random systems [such as percolation models (Bazant 2000, 2002; van der Hofstad & Redig 2006)], as shown by renormalization group transformations, which homogenize the system recursively up to an element scale large enough for the correlations to become negligible. Therefore, it is expected that the strength correlation should have a negligible effect on the functional form of strength cdf of the RVE.

5.5 Hierarchical Model for Fatigue Strength

The cdf of fatigue strength of one RVE can also be calculated by using the same hierarchical model as shown in Figure 5.10a. As before, the cdf of fatigue strength of each sub-chain can be calculated from the joint probability theorem (Eq. 5.2). For the cdf of fatigue strength of sub-bundles (Eq. 5.78), we must specify the exponent of the Paris law. Based on the recent analysis of the multiscale transition of fracture kinetics, which will be described in Chapter 6, the Paris law exponent must be equal to 2 at the nanoscale and must then increase while passing to higher scales. Comparing the foregoing derivations of the cdfs of the fatigue and monotonic strengths of bundles, one would expect that the cdf of fatigue strength of a bundle at the nanoscale should be similar to the cdf of monotonic strength of a plastic bundle, while the cdf of fatigue strength of a bundle at an upper scale should be similar to the cdf of strength of a brittle bundle.

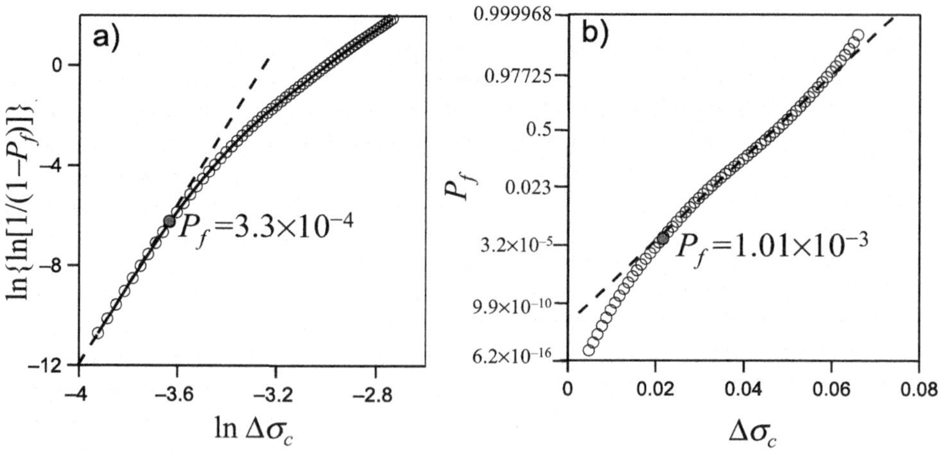

Figure 5.12. Calculated cdf of fatigue strength of one RVE presented on (a) Weibull probability paper and (b–d) Normal probability paper.

Figure. 5.12a and b displays the calculated cdf of fatigue strength of one RVE on the Weibullian and Gaussian probability papers, respectively. In the Weibull scale (Fig. 5.12a), the lower portion of the calculated cdf of fatigue strength is seen to be a straight line, which represents the Weibull distribution (whose tail is a power law). This result is consistent with the conclusion of the foregoing analysis: In the chain and bundle models, the power-law tail of cdf of fatigue strength is indestructible.

The upper part of the calculated cdf begins to diverge from the straight line at $P_f \approx 3 \times 10^{-4}$. Hence the core of the cdf does not follow the Weibull distribution. As shown in Figure 5.12b, the upper portions of the cdf can be fitted quite closely by a straight line on the Gaussian probability paper. Therefore, similar to the cdf of static strength of one RVE, the cdf of fatigue strength of one RVE is approximately described by the Gaussian distribution with a Weibull tail grafted on the left at a probability of about 10^{-4} to 10^{-2}. This can be mathematically described as follows:

$$P_1(\Delta\sigma_f) = 1 - e^{-(\Delta\sigma_f/s_{0f})^{m_f}} \qquad (\Delta\sigma_f \le \Delta\sigma_{gr}) \qquad (5.87a)$$

$$P_1(\Delta\sigma_f) = P_{\mathrm{gr},f} + \frac{r_{sf}}{\delta_{Gf}\sqrt{2\pi}}\int_{\Delta\sigma_{\mathrm{gr}}}^{\Delta\sigma_f} e^{-(\sigma'-\mu_{Gf})^2/2\delta_{Gf}^2}\,\mathrm{d}\sigma' \quad (\Delta\sigma_f > \Delta\sigma_{\mathrm{gr}}) \qquad (5.87b)$$

Here m_f is the Weibull modulus (equal to the tail exponent); s_{0f} = scale parameter of the Weibull tail; μ_{Gf} and δ_{Gf} are the mean and standard deviations of the Gaussian core if $\Delta\sigma_f$ is considered extended to $-\infty$; r_{sf} = a scaling parameter required to normalize the grafted cdf such that $P_1(\infty) = 1$; and $P_{\mathrm{gr},f}$ = grafting probability,

$$P_{\mathrm{gr},f} = 1 - e^{-(\Delta\sigma_{\mathrm{gr}}/s_{0f})^{m_f}} \qquad (5.88)$$

Finally, continuity of the pdf at the grafting point requires that

$$[(\mathrm{d}P_1/\mathrm{d}\Delta\sigma_f)]_{\Delta\sigma_{\mathrm{gr}}^+} = [(\mathrm{d}P_1/\mathrm{d}\Delta\sigma_f)]_{\Delta\sigma_{\mathrm{gr}}^-}. \qquad (5.89)$$

Note that, for one RVE, the cdf of fatigue strength has a form similar to that of the cdf of monotonic strength, as developed in the previous section. The locations of grafting points P_f of these two cdfs are similar, too. The reason is that the statistical chain models for fatigue strength and monotonic strength share the same equation, and that the mathematical formulations of the bundle models for the fatigue strength and for the monotonic strength are analogous.

6 Multiscale Modeling of Fracture Kinetics and Size Effect under Static and Cyclic Fatigue

In Chapter 5, we analyzed the nano–macro scale transition of probability distribution of static and fatigue strengths, and determined the strength distribution function of one macroscale representative volume element (RVE) of material. Aside from the strength requirement, another essential requirement is sufficient structural lifetime. To predict it, the macrocrack growth under constant or cyclic load must be taken into account. It turns out that the kinetics of macrocrack growth and the bridging of scales up to the RVE requires a different argument than the scale bridging of strength distribution. The argument may be based on energy dissipation rather than distribution tails.

6.1 Previous Studies of Fracture Kinetics

For many decades, extensive research has been devoted to fracture kinetics under both static and cyclic fatigue loading (e.g., Paris & Erdogan 1963; Wiederhorn 1967; Wiederhorn & Bolz 1970; Evans 1972; Thouless, Hsueh, & Evans 1983; Evans & Fu 1984; Krausz & Krausz 1988; Fett & Munz 1991; Fett 1991; Munz & Fett 1999). However, this research has been largely empirical in nature, and little physical justification for fracture kinetics of quasibrittle materials has been provided (until the recent findings to be described in this chapter).

For the case of static fatigue, i.e., structures subjected to constant loading (often combined with creep), the rate of crack growth was first experimentally studied in the context of a corrosive environment (Hillig & Charles 1964; Wiederhorn 1967; Wiederhorn & Bolz 1970), where the classical rate process theory (Glasstone et al. 1941) was adopted to explain the effect of chemical reactions on the crack propagation rate at low stresses. For the stress-driven crack growth in rock and ceramics, a simple power-law function, referred to as the Charles–Evans law, was proposed to describe the dependence of crack growth velocity on the applied stress intensity factor (Charles 1958*a*, 1958*b*; Evans 1972; Thouless et al. 1983; Atkinson 1984; Evans & Fu 1984):

$$\dot{a} = A_s e^{-Q_0/kT} K^{n_s} \tag{6.1}$$

where $\dot{a} = \mathrm{d}a/\mathrm{d}t$, A_s, n_s = constants, Q_0 = stress-free activation energy barrier, k = Boltzmann constant, T = absolute temperature, and K = stress intensity factor. Equation 6.1 is important for subcritical long-time crack growth in rock, especially on a geological time scale (Atkinson 1984), and has also been used as an empirical law for quasibrittle materials such as concrete and engineering ceramics (Bažant & Prat 1988; Fett & Munz 1991; Bažant & Planas 1998; Munz & Fett 1999; Lohbauer, Petchelt, & Greil 2002). A partial theoretical justification of the power-law form for creep crack growth rate for quasibrittle materials has also been suggested (Fett 1991; Munz & Fett 1999), based on the break frequency of a bond between a pair of two atoms. Such a justification, however, was rather limited, mainly because it did not consider any scale bridging between the atomic scale [and partly also because it did not consider nanoscale facture mechanics, and was limited to the simple Morse interatomic potential (Morse 1929), which neglects the contribution of surrounding atoms].

For the case of fatigue loading, Paris and Erdogan (1963) proposed expressing the fatigue crack growth rate as a power-law function of the amplitude of the stress intensity factor:

$$\frac{\mathrm{d}a}{\mathrm{d}N} = A_f e^{-Q_0/kT} \Delta K^{n_f} \tag{6.2}$$

where $\mathrm{d}a/\mathrm{d}N$ = crack growth rate per load cycle, A_f, n_f = constants, and ΔK = amplitude of the stress intensity factor. Various attempts have been made to justify Eq. 6.2, which is briefly called the Paris law (although the Paris–Erdogan law would be more appropriate). Weertman (1966) and Rice (1967) independently proposed damage accumulation models derived on the basis of tensile yielding or slip in the reversed plastic zone of metals ahead of the crack. By considering an elastic-perfectly plastic model with proportional plastic flow, Rice (1967) stated that the size of the reversed plastic zone, which determines the propagation of a fatigue crack, is proportional to $\Delta K^2/f_y^2$ (where f_y = yield strength). Another approach has been to model the cyclic plastic strain in the reverse yield zone at the crack tip (Coffin 1962).

The plasticity-based models for metals have predicted the exponent of Paris law to be 4, in agreement with the experimental observations for most metals. For quasibrittle materials, experiments show that the Paris law is also mostly applicable but the exponent n_f of the Paris law is significantly higher than it is for metals; specifically $n_f \approx 10$ for concrete (Bažant & Xu 1991), $n_f \approx 20$ for rock (Le, Manning, & Labuz 2014), and $n_f \approx 30$ for both alumina ceramics (Ogawa 1995) and zirconia dental ceramics (Studarta, Filser, Kochera, & Gauckler 2007).

An intriguing aspect of fatigue kinetics of quasibrittle materials has been its dependence on the specimen size. Understanding this size effect is essential because the actual engineering structure to be designed is often much larger than the specimens used in laboratory experiments. Although only limited attention has been paid to the scale effect on crack growth under static fatigue (Le & Bažant 2012), the scale effect under cyclic loading has been researched for decades (Barenblatt & Botvina 1981; Bažant & Schell 1993; Barenblatt 1996; Bažant 2005; Ritchie 2005; Ciavarella, Paggi, & Carpinteri 2008; Le, Manning, & Labuz 2014; Kirane & Bažant 2015*a*, 2015*b*).

The existing theoretical framework for the scale effect on the fatigue crack growth is based largely on the argument of incomplete similarity (Barenblatt & Botvina 1981; Barenblatt 1996, 2014; Ritchie 2005; Ciavarella et al. 2008), which does not directly provide the detailed form of the scaling function. Bažant and co-workers (Bažant & Xu 1991; Bažant & Schell 1993) proposed normalizing the Paris law by the apparent fracture toughness of the material, and deduced the size effect on the Paris law from the size dependence of the apparent fracture toughness. Nevertheless, it was found that the direct use of the size effect on the apparent fracture toughness cannot describe the observed size effect on the Paris law. Therefore, the rationale behind such normalization needs deeper examination.

This chapter, expanding the basic idea presented in Bažant et al. (2009), presents a justification of the large exponent of the power law for creep and fatigue crack growth rates. This justification is based on the fracture kinetics of the nanoscale structure and on the equality of energy dissipation at bridging the nanoscale and macroscale. This dissipation equality is also used to derive the size dependence of the fracture kinetics of quasibrittle structures under static and cyclic fatigue.

6.2 Fracture Kinetics at Nanoscale

It has been known that the kinetics of nanocrack propagation could be derived from the transition rate theory for the discrete thermal-activated crack jumps (Krausz & Krausz 1988). Consider a nanoscale structure depicted schematically in Figure 4.1. As discussed in Chapter 4, the frequency of each crack jump can be calculated by the transition rate theory (Eq. 4.5). If we multiply the frequency of crack jumps by the spacing of interatomic bonds or spacing of nanoparticle connections, δ_a, we obtain the velocity of nanocrack growth under a constant stress τ:

$$\dot{a}_{\text{nano}} = \delta_a f_1 = \nu_1 e^{-Q_0/kT} K_a^2 \tag{6.3}$$

$$\text{where} \quad \nu_1 = \delta_a^2 \frac{\gamma_1 \alpha l_a}{E_1 h} \tag{6.4}$$

$\dot{a}_{\text{nano}} = \mathrm{d}a_{\text{nano}}/\mathrm{d}t =$ nanocrack growth rate, and $K_a =$ stress intensity factor of the nanocrack. Therefore, we can conclude that the creep crack growth rate at the nanoscale is a power-law function of the stress intensity factor with exponent 2.

The same analysis can be applied to the case of cyclic loading. Consider a nanoscale structure subjected to cyclic loading, as depicted as Figure 4.4b. The distance by which a nanocrack would advance during each load cycle is given by $\Delta a_{\text{nano}} = \int_0^{t_c} \dot{a}(t)\,\mathrm{d}t$, where $\dot{a} = \mathrm{d}a/\mathrm{d}t$ and $t_c =$ period of one cycle. Since each crack jump can be considered as an independent process, the nanocrack growth rate per cycle can be written as

$$\frac{\mathrm{d}a}{\mathrm{d}N} = \nu e^{-Q_0/kT} \Delta K_a{}^2 \tag{6.5}$$

$$\text{where} \quad \nu = \frac{\nu_1}{3} t_c f(R_\tau), \quad f(R_\tau) = \frac{1 + \langle R_\tau \rangle + \langle R_\tau \rangle^2}{(1 - R_\tau)^2} \tag{6.6}$$

where $\langle x \rangle = \max(x, 0)$. Therefore, similar to the case of static fatigue, the nanocrack growth rate under cyclic loading also follows a power-law function of the amplitude of the stress intensity factor with the exponent equal to 2. Furthermore, Eq. 6.5 implies that the fatigue crack growth rate would depend on the nanoscale stress ratio R_τ.

Evidently, the crack growth rate at the nanoscale (defined by Eqs. 6.3 and 6.5) has the same form as the crack growth rate at the macroscale (defined by Eqs. 6.1 and 6.2), except for two differences:

1. Parameters ν_1 and ν of the fracture kinetics equations at the nanoscale depend on the crack length, but parameters A_s and A_f of the crack growth rate laws at the macroscale do not.
2. The exponent of the power law for both static and cyclic fracture kinetics at the nanoscale is 2, while the experiments show that, for brittle materials as well as quasibrittle materials such as concrete and ceramics, the exponent of the power law for static and cyclic fatigue crack growth at the macrocale is typically much higher; e.g., $n_f = 10$ and $n_s = 24$ for concrete (Bažant & Xu 1991; Bažant & Prat 1988), and $n_f, n_s = 10 - 30$ for engineering and dental ceramics (Ogawa 1995; Munz & Fett 1999; Studarta, Filser, Kochera, & Gauckler 2007).

The first difference can be explained by considering the averaging effect of nanocracks on the fracture process zone (FPZ) during the crack growth. As the macrocrack propagates under either the static or cyclic fatigue, there exists a FPZ at the crack tip (the size of which is different for creep loading and fatigue loading). In the FPZ, the external stress produces microcracks. The size of the FPZ does not change significantly as the crack propagates, this is the essential concept leading to the constancy of the fracture energy. This dictates the energy flux required to propagate the crack. Therefore, the constants A_s and A_f must characterize the average of the growth rates of all the nanocracks with different lengths inside the FPZ, and they should not depend on the size of the macrocrack.

The second difference, i.e., the difference between the exponents of the power law for crack growth rates at the macro- and nanoscales, requires analysis of the multiscale transition, which may be based on the idea of equality of nano- and macro-energy dissipations. This equality also leads to approximate equations for the size effect on the fracture kinetics under static and cyclic fatigue (Bažant et al. 2009; Le et al. 2014).

6.3 Multiscale Transition of Fracture Kinetics for Static Fatigue

Consider that the macrocrack propagates with a FPZ attached at the crack tip. This FPZ contains N_{a_s} active nanocracks. The nano–macro relation of fracture kinetics requires that the energy dissipation rate of the macrocrack must be equal to the sum of the energy dissipation rates of all the active nanocracks a_i ($i = 1, \ldots, N_{a_s}$) in the FPZ of the macrocrack (note that the dissipation includes not only the fracturing dissipation but also the dissipation by frictional slips, which is inevitably part of the fracture energy at each scale), i.e.,

$$U_s \dot{a} = \sum_{i=1}^{N_{as}} \dot{U}_{ai} \tag{6.7}$$

where U_s = energy required to advance the macrocrack by a unit length under static fatigue loading, and $\dot{U}_{ai}$ = net energy dissipation rate of the nanocrack propagation. According to Fig. 4.3c, it is clear that $\dot{U}_{ai}$ can be expressed as

$$\dot{U}_{ai} = (Q_0 - \Delta Q_i/2)\dot{a}_{fi} - (Q_0 + \Delta Q_i/2)\dot{a}_{bi} \tag{6.8}$$

where $\dot{a}_{fi}$ = rate of forward jumps and $\dot{a}_{bi}$ = rate of backward jumps. Based on the transition rate theory (Eq. 4.5), we have

$$\dot{a}_{fi} = \delta_a \nu_T e^{-(Q_0 - \Delta Q_i/2)/kT} \tag{6.9}$$

$$\dot{a}_{bi} = \delta_a \nu_T e^{-(Q_0 + \Delta Q_i/2)/kT} \tag{6.10}$$

Substitution of Eqs. 6.9 and 6.10 into Eq. 6.8 yields

$$\dot{U}_{ai} = 2Q_0\delta_a\nu_T e^{-Q_0/kT}\sinh(\Delta Q_i/2kT) - \Delta Q_i\delta_a\nu_T e^{-Q_0/kT}\cosh(\Delta Q_i/2kT) \tag{6.11}$$

Since $\Delta Q_i/kT \ll 1$, $\sinh(\Delta Q_i/2kT) \approx \Delta Q_i/2kT$ and $\cosh(\Delta Q_i/2kT) \approx 1$, Eq. 6.11 can be simplified as

$$\dot{U}_{ai} \approx Q_0\delta_a\nu_T e^{-Q_0/kT}\left(1 - \frac{kT}{Q_0}\right)\frac{\Delta Q_i}{kT} \tag{6.12}$$

$$\approx Q_0 \nu_i e^{-Q_0/kT} K_{ai}^2 \qquad (kT/Q_0 \ll 1) \tag{6.13}$$

where $\nu_i = \delta_a^2(\gamma_1\alpha_i l_i)/E_i h$.

It may be noted that a simpler (and simplified) derivation of Eq. 6.13 is to calculate $\dot{U}_{ai}$ by multiplying the nanocrack velocity $\dot{a}_i$ by the net energy dissipation rate U_a per forward jump, i.e.:

$$\dot{U}_{ai} = U_a \dot{a}_i \tag{6.14}$$

Since $\Delta Q_i \ll Q_0$, it is clear that $U_a \approx Q_0$. By substituting Eq. 6.3 into Eq. 6.14, we reach Eq. 6.13.

We can now substitute Eq. 6.13 into Eq. 6.7 to express the macrocrack growth rate as

$$\dot{a} = \frac{e^{-Q_0/kT}Q_0}{U_s}\phi(K) \quad \text{where} \quad \phi(K) = \sum_{i=1}^{N_{as}} \nu_i K_i^2 \tag{6.15}$$

where K_i is the stress intensity factor of nanocrack a_i of a nanoscale structure in the FPZ, and E_i = elastic modulus of each nanoscale structure. K_i depends on the nanoscale remote stress τ which, in the context of linear elasticity, may be assumed to be linearly proportional to the macroscale stress σ. Therefore, we can set $K_i = \omega_i K$, where ω_i = constants, and $\phi(K)$ can be rewritten as

$$\phi(K) = K^2 \sum_{i=1}^{N_{as}} \nu_i \omega_i^2 \tag{6.16}$$

We can calculate the number of active nanocracks in the FPZ of the macrocrack by adding them up through the hierarchy of FPZ scales (Fig. 6.1): The FPZ of the macrocrack contains q_1 mesocracks, each of which has a meso-FPZ at its tip. Each of the meso-FPZ contains q_2 microcracks, each of which contains a micro-FPZ with q_3

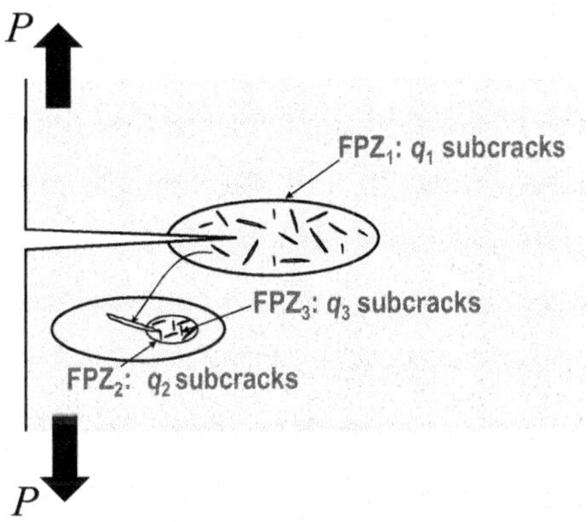

Figure 6.1. Hierarchy of fracture process zone scales.

submicrocracks,..., and so forth, all the way down to the nanoscale. If there are s different scales between the macroscale and nanoscale, then the total number of nanocracks in the macro-FPZ is simply given by

$$N_{a_s} = q_1 q_2 \cdots q_s \tag{6.17}$$

On scale μ, the number q_μ of activated cracks within the subscale FPZ must be a function of the stress intensity factor K_μ. Through dimensional analysis, we have

$$q_\mu = q_\mu(K_\mu^2/E_\mu U_\mu) \tag{6.18}$$

where E_μ = the elastic modulus of the subscale FPZ, and U_μ = critical energy dissipation for cracks on scale μ to propagate by a unit length.

It may be expected that there is no characteristic value of K_μ at which the behavior of function $q_\mu(K_\mu^2/E_\mu U_\mu)$ would qualitatively change. So, function $q_\mu(K_\mu^2/E_\mu U_\mu)$ should be self-similar. The only self-similar functions are power laws (Barenblatt 2003), and so we may write

$$q_\mu(K_\mu^2/E_\mu U_\mu) \propto (K_\mu^2/E_\mu U_\mu)^r \tag{6.19}$$

It appears plausible that function $q_\mu(K^2/E_\mu U_\mu)$ increases rapidly with increasing $K^2/E_\mu U_\mu$ while the ratios in $\phi(K)$ change far less. Therefore, we may replace ω_i and v_i by some effective mean values ω_a and v_a. It follows that function $\phi(K)$ can be rewritten as

$$\phi(K) = \frac{\eta v_a \omega_a^2}{\prod_{\mu=1}^{s} E_\mu^r U_\mu^r} \left(\prod_{\mu=1}^{s} K_\mu^{2r} \right) K^2 \tag{6.20}$$

where η = constant. We can further consider that $U_\mu = c_\mu U_s$ and $K_\mu = \kappa_\mu K$ where c_μ, κ_μ = constants. Substituting Eq. 6.20 into Eq. 6.15 and setting $2rs + 2 = n_s$, we

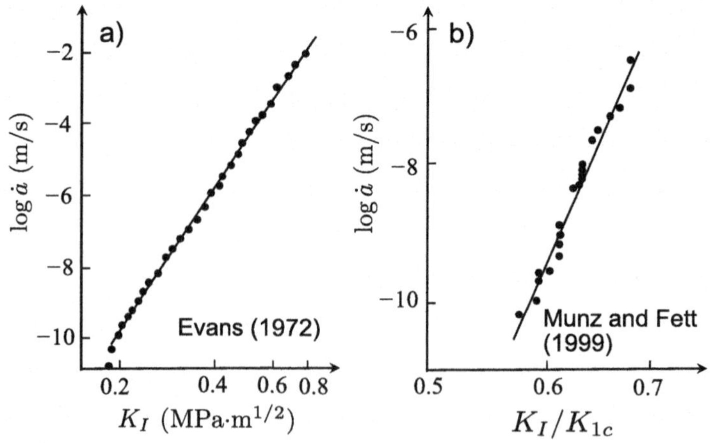

Figure 6.2. Measured fracture kinetics under static fatigue. (a) Glass (Evans 1972) and (b) 99.6% Al_2O_3 (Munz and Fett 1999).

obtain

$$\dot{a} = \frac{e^{-Q_0/kT} A'_s}{U_s^{n_s/2}} K^{n_s} \tag{6.21}$$

$$\text{where} \quad A'_s = \frac{\eta v_a \omega_a^2 Q_0 \prod_{\mu=1}^{s} \kappa_\mu^r}{\prod_{\mu=1}^{s} E_\mu^r c_\mu^r} \tag{6.22}$$

Clearly Eq. 6.21 has the same form as Eq. 6.1.

The present analysis proves that the growth rate of a nanocrack follows a power law with exponent 2, and further it shows that, under certain plausible simplifying assumptions, particularly the self-similarity of function $q_\mu(K_\mu^2/E_\mu U_\mu)$, the power-law form of crack growth rate at the macroscale can be physically justified. Nevertheless, the foregoing analysis does not present a rigorous mathematical proof of the power law for macrocrack growth rate. The experimental verification is still essential. Figure 6.2 shows the experimentally measured relationship between the crack growth rate and the stress intensity factor for different quasibrittle materials, such as glass (Evans 1972) and engineering ceramics (Munz & Fett 1999). It is clear that the power-law form of the fracture kinetics equation agrees well with these experimental observations.

6.4 Size Effect on Fracture Kinetics under Static Fatigue

In the foregoing analysis, we have demonstrated a plausible justification of the power-law form of the crack growth rate in quasibrittle materials, and of the fact that the power-law exponent increases while moving up from the nanoscale to the macroscale. But we have not yet addressed the size dependence of crack growth behavior in quasibrittle structures.

It is commonly accepted that, as the macrocrack propagates, the FPZ attached to the tip of the macrocrack has approximately a constant size l_s, which represents a characteristic length in the problem. Therefore, it is expected that the ratio of the structure size to the FPZ size, i.e., D/l_s, would influence the fracture kinetics. So far the size dependence of fracture kinetics under static fatigue has not been systematically studied. In this section, we derive this size effect based on a scaling model of the apparent fracture energy for static fatigue, which is analogous to the well-established size effect model on the apparent monotonic fracture energy.

One essential feature of the aforementioned analysis is that it uses the stress intensity factor to describe the loading on the macroscale FPZ, which is further related to the number of active nanocracks in the FPZ. This is clearly based on the concept of the linear elastic fracture mechanics (LEFM), which assumes that the FPZ is negligible compared to the structure size, i.e., $D/l_s \to \infty$. Now we extend the foregoing analysis to fracture kinetics of specimens of a finite size, i.e., D/l_s is finite. This may be done by introducing the size dependence of the critical energy dissipation for fatigue crack growth $U_s(D)$:

$$\dot{a} = \frac{e^{-Q_0/kT} A_s'}{U_s^{n_s/2}(D)} K^{n_s} \tag{6.23}$$

Note that a similar approach has been used successfully to derive an approximate size effect equation for the nominal strength of quasibrittle structures (Bažant 1984*b*, 2005) (see Eq. 3.37). Let us review it briefly. Based on the conventional LEFM, we can calculate the nominal strength of structure under monotonic loading as

$$G_{fa} = \frac{\sigma_N^2}{E} Dg(\alpha_0) \tag{6.24}$$

where σ_N = nominal structural strength as defined in Chapter 1, and G_{fa} = apparent fracture energy. To account for the finite FPZ size, we can introduce a scale effect on the apparent G_{fa} in the sense of LEFM (Bažant & Kazemi 1990*a*, 1990*b*) (which is equivalent to the size effect on the nominal strength):

$$G_{fa} = G_f \frac{D}{D + D_{0m}} \tag{6.25}$$

$$\text{where} \quad D_{0m} = c_f \frac{g'(\alpha_0)}{g(\alpha_0)} \tag{6.26}$$

Here G_f = fracture energy, which can be taken as the value of G_{fa} for an infinitely large specimen, $g(\alpha)$ = dimensionless energy release rate function such that $\sigma_N^2 Dg(\alpha)$ = energy release rate function, c_f = effective size of the FPZ under monotonic loading, $\alpha = a/D$ = relative crack length, a = actual crack length, and α_0 = original relative crack length. It has been shown that Eq. 6.25 can further be used to determine the crack resistance curve, aka the *R*-curve (Bažant & Kazemi 1991; Bažant & Planas 1998). By substituting Eq. 6.25 into Eq. 6.24, we reach the classical size effect on nominal structural strength (Eq. 3.37):

$$\sigma_N = \sigma_0 \left(1 + \frac{D}{D_{0m}}\right)^{-1/2} \tag{6.27}$$

$$\text{where} \quad \sigma_0 = \sqrt{\frac{EG_f}{g_0 D_{0m}}} \tag{6.28}$$

Now one may hypothesize that the size dependence of the critical dissipation rate for the macrocrack growth U_s under static fatigue has the same functional form as that of the apparent fracture energy for monotonic loading G_{fa}. Following Eqs. 6.25 and 6.26, we thus have

$$U_s = U_{s\infty} \left(\frac{D}{D + D_{0s}}\right) \tag{6.29}$$

$$\text{where} \quad D_{0s} = c_{fs} \frac{g'(\alpha_0)}{g(\alpha_0)} \tag{6.30}$$

where $U_{s\infty}$ = the value of U_s for an infinitely large specimen and c_{fs} = effective size of the static fatigue FPZ, which is proportional to the FPZ size l_s. Substitution of Eq. 6.29 into Eq. 6.23 yields

$$\dot{a} = C_s \left(1 + \frac{D_{0s}}{D}\right)^{n_s/2} K^{n_s} \tag{6.31}$$

$$\text{where} \quad C_s = e^{-Q_0/kT} A'_s U_{s\infty}^{-n_s/2} \tag{6.32}$$

As implied by Eq. 6.31, if the structure size is very large compared to the size of static fatigue FPZ, the crack growth rate would depend solely on the stress intensity factor. This must be expected because at the large-size limit LEFM must apply. On the other hand, if the structure size is comparable to the FPZ size, Eq. 6.31 indicates that the crack growth rate is determined by the applied nominal stress. This is expected as well, because at the small-size limit the cohesive stress in the FPZ is almost uniformly distributed and can be related directly to the applied nominal stress. Therefore, Eq. 6.31 agrees with the qualitative behavior of both small- and large-size asymptotes. However, no experiments have yet been performed to investigate the size dependence of static crack growth rate law. Future experimental verification of Eq. 6.31 is needed.

6.5 Multiscale Transition of Fracture Kinetics under Cyclic Fatigue

Let us now use the same mathematical framework to analyze the fracture kinetics of quasibrittle materials under fatigue loading. We limit our attention to fatigue loading with a fixed stress ratio R. Similar to the case of static fatigue, we expect that there exists a cyclic FPZ that is attached to the crack tip under cyclic fatigue and travels with the macrocrack (Fig. 6.3). Following the same energy equality condition as used for the static fatigue, we may assert that, during each load cycle, the energy dissipation due to the macrocrack growth must be equal to the sum of the energy dissipations due to the

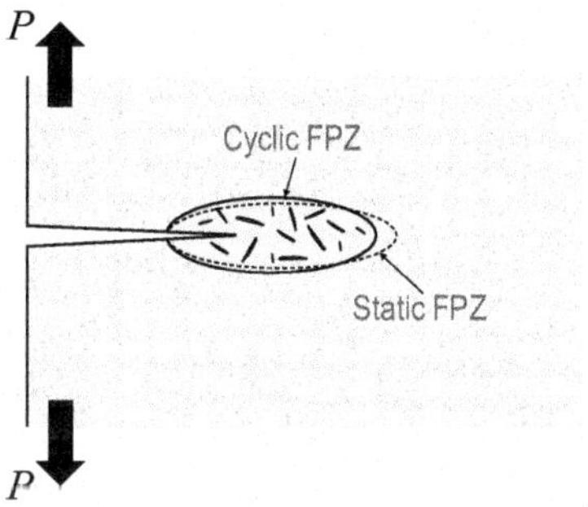

Figure 6.3. Cyclic fracture process zone in quasibrittle structures.

propagation of all the active nanoscale cracks inside the cyclic FPZ. Mathematically, this can be written as

$$U_c \frac{\mathrm{d}a}{\mathrm{d}N} = \sum_{i=1}^{N_{a_f}} U_a \frac{\mathrm{d}a_i}{\mathrm{d}N} \tag{6.33}$$

where U_c = critical energy dissipation per unit growth of the macrocrack, U_a = net energy dissipation associated with one forward jump of the nanocrack tip, a_i = length of the ith nanoscale crack, and N_a = number of active nanoscale cracks in the cyclic FPZ. As discussed in Section 6.2, $U_a \approx Q_0$ since $\Delta Q \ll Q_0$. With Eq. 6.5, we have

$$U_a \frac{\mathrm{d}a_i}{\mathrm{d}N} \approx Q_0 \nu_i e^{-Q_0/kT} \Delta K_{ai}{}^2 \tag{6.34}$$

$$\text{where:} \quad \nu_i = \frac{\delta_a^2 \gamma_1 a_i \left(1 + \langle R_{\tau i} \rangle + \langle R_{\tau i} \rangle^2\right)}{3E_i h (1 - R_{\tau i})^2} \tag{6.35}$$

It may be expected that, in the context of linear elasticity, the amplitude of the nanoscale stress intensity factors is proportional to the amplitude of the macroscale stress intensity factor, i.e., $\Delta K_{ai} = \omega_i \Delta K$, where ω_i = constants. Therefore, Eq. 6.33 can be rewritten as

$$\frac{\mathrm{d}a}{\mathrm{d}N} = \frac{Q_0 e^{-Q_0/kT}}{U_c} \left(\sum_{i=1}^{N_{af}} \nu_i \omega_i^2 \right) \Delta K^2 \tag{6.36}$$

Similar to the foregoing analysis of the fracture kinetics under static fatigue, we can calculate the number of active nanocracks N_{a_f} in the cyclic FPZ in a multiscale framework. In the cyclic FPZ at the tip of the macrocrack, there are p_1 mesocracks, which have their own cyclic meso-FPZs. In each of these meso-FPZs, there are p_2 microcracks, which contain their own cyclic micro-FPZs. In each of these micro-FPZs, there are p_3 sub-microcracks, and so on, all the way down to the nanoscale. If there are s scales between the macro- and nanoscales, then the total number of nanocracks in the macro-FPZ is

$$N_{a_f} = p_1 p_2 \ldots p_s \tag{6.37}$$

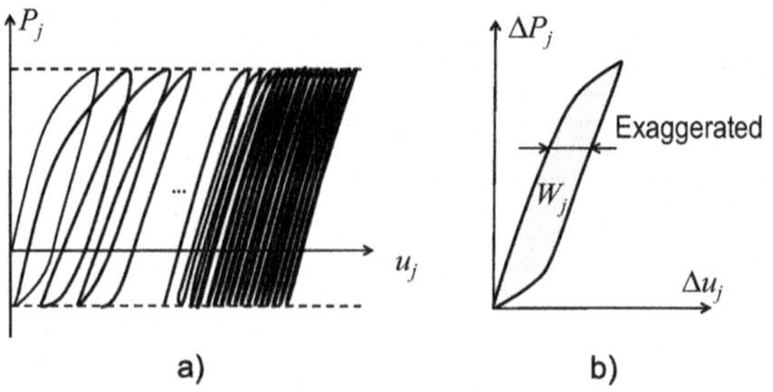

Figure 6.4. Load-displacement response of subscale FPZ.

Consider now the cyclic FPZ on the jth material scale. During one load cycle, the load-deformation curve of the subscale FPZ can be represented by Figure 6.4a. The energy dissipation during one load cycle is W_j, as shown by the area enclosed by the hysteresis loop in Figure 6.4a. It has been argued that the load-deformation hysteresis loop could be described by a constitutive law that does not depend on the maximum or the minimum loads (Fig. 6.4b) (Rice 1967). The area enclosed by the hysteresis loop governs the number of mesoscale cracks in the FPZ. Therefore, the damage in the cyclic FPZ, represented by the number of active mesoscale cracks inside the zone, must be a function of the amplitude of stress intensity factor ΔK_j. Using dimensional analysis, we infer that

$$q_j = \phi\left(\Delta K_j^2 / E_j U_{cj}, R_j\right) \tag{6.38}$$

where E_j = elastic modulus of the jth subscale FPZ, U_{cj} = critical value of energy dissipation for cracks on scale j to propagate by a unit length under fatigue loading, and R_j = stress ratio of the jth subscale FPZ. We may further write $R_j = \gamma_j R$, and $U_{cj} = \beta_j U_c$. Since we focus on the fatigue loading with a fixed stress ratio, for a fixed macroscale stress ratio R we can write rewrite Eq. 6.38 as

$$q_j = \phi_1\left(\Delta K_j^2 / E_j U_{cj}\right) \tag{6.39}$$

It may be expected that there is no characteristic value of ΔK_j at which the function ϕ changes its behavior. Therefore, the function ϕ should be self-similar, i.e., a power law (Barenblatt 2003). So,

$$q_j \propto \left(\Delta K_j^2 / E_j U_{cj}\right)^r \tag{6.40}$$

Noting that the amplitude of the stress intensity factor for the FPZs at all scales can be considered to be proportional to the macroscale stress intensity factor ΔK, we obtain the number of active nanocracks inside the macroscale cyclic FPZ as

$$N_{af} = \frac{C \Delta K^{2rs}}{\left(\prod_{j=1}^{s} E_j\right) U_c^{rs}} \tag{6.41}$$

where C is a constant. Then, upon replacing ν_i and ω_i by their effective mean values ν_a and ω_a, Eq. 6.36 becomes

$$\frac{\mathrm{d}a}{\mathrm{d}N} = \frac{A'_f e^{-Q_0/kT}}{U_c^{n_f/2}} \Delta K^{n_f} \tag{6.42}$$

$$\text{where} \quad A'_f = \frac{C \nu_a \omega_a^2 Q_0}{\prod_{j=1}^{s} E_j} \tag{6.43}$$

and $n_f = 2rs + 2$.

It is now clear that Eq. 6.42 matches the Paris law for fatigue crack growth, stated in Eq. 6.2. The present analysis shows that the exponent of Paris law on nanoscale is 2, and explains how the exponent can increase to $n_f \approx 10 - 30$ on macroscale. The reason simply is that the number of active nanocracks in the macro-cyclic FPZ rapidly increases with the applied stress amplitude. It was for this reason that, while using the statistical hierarchical model to formulate the probability distribution of fatigue strength of one macroscale representative volume element (Section 5.5 of Chapter 5), the Paris law exponent for each element of sub-bundle was considered to gradually increase as we moved from the nanoscale to the macroscale.

In a recent study of long-term cyclic creep behavior of concrete bridges (Bažant & Hubler 2014), it has been demonstrated that the macroscopic compressive cyclic creep behavior of concrete can be explained by applying the Paris law to the fatigue kinetics of microcracks. By matching experimental data, it was found that the exponent of this Paris law on the microscale of concrete is 4, which is much less than the Paris law exponent of concrete on the macroscale (where $n_f \approx 10$). This result supports the current conclusion about the increase of Paris law exponent while moving up through the material scales.

It has been documented in previous studies (Bažant & Planas 1998; Andersons, Hojo, & Ochiai 2004) that the growth rate of fatigue cracks depends also on the ratio of maximum to minimum stress (stress ratio), though only mildly, except for large amplitudes. The present analysis indicates such dependence, though not in an explicit manner. It must be emphasized that the present analysis is approximate. It is based on certain hypotheses about the self-similarity of the function ϕ_1, and does not represent a rigorous mathematical proof. For concrete, the scope of validity of Eq. 6.42 certainly needs deeper experimental verifications.

The characteristics of self-similarity are expected to hold for a wide range of medium-stress amplitudes but not for very small or very large stress amplitudes. Indeed, the growth rate of fatigue cracks is known to deviate from the power-law form for very small and very large stress amplitudes, as confirmed by many experiments (Ogawa 1995; Bažant & Planas 1998; Andersons et al. 2004; Studarta, Filser, Kochera, & Gauckler 2007). For small stress amplitudes, the reason may be that the distinction between the static and cyclic FPZs gets blurred. The static fatigue kinetics law (Eq. 6.1) is the limit of the Paris law for a vanishing stress amplitude, and the transitional region between these two laws is not well understood at the time of writing. For very large amplitudes, the reason for the deviations from Paris law may be that the maximum stress intensity factor during the load cycle is approaching the fracture toughness (Suresh 1998), at which the crack begins to propagate dynamically. However, practical interest usually does not include the extreme stress amplitudes.

6.6 Size Effect on Fatigue Crack Growth Rate and Experimental Evidence

Similar to the analysis of static fatigue kinetics in Section 6.3, the present formulation (Eq. 6.42) of fatigue crack growth uses the stress intensity factor to characterize the behavior of the cyclic FPZ. This is appropriate only if the structure size is much larger than the size of the cyclic FPZ.

Following the same approach as used for static fatigue kinetics (Section 6.4), we now extend Eq. 6.42 to the general case where the cyclic FPZ is not negligible compared to the structure size. To this end, we introduce a size effect on the critical energy dissipation rate U_c, and assume it to follow the same functional form as that of the apparent fracture energy for monotonic loading G_{fa} (Eq. 6.25), i.e.,

$$U_c = U_{c\infty} \frac{D}{D + D_{0c}} \tag{6.44}$$

$$\text{where} \quad D_{0c} = c_{fc} \frac{g'(\alpha_0)}{g(\alpha_0)} \tag{6.45}$$

Here c_{fc} = effective size of the cyclic FPZ, which is proportional to the actual size of the cyclic FPZ l_c. Substitution of Eq. 6.44 into Eq. 6.42 yields a size-dependent Paris law for cyclic loading with a constant stress ratio, i.e.:

$$\frac{da}{dN} = C_f e^{-Q_0/kT} \left(1 + \frac{D_{0c}}{D}\right)^{n_f/2} \Delta K^{n_f} \tag{6.46}$$

It may be noted that Bažant and co-workers (Bažant & Xu 1991; Bažant & Schell 1993) derived Eq. 6.46 in a different way, in which the Paris law was normalized with respect to a size-dependent apparent fracture toughness. However, it was found that the scaling of the apparent fracture toughness was insufficient to describe the observed size effect on the Paris law for concrete. To match the experimental data, it was found that the transitional size for scaling of apparent fracture toughness for cyclic loading must be different from the transitional size for monotonic loading (Bažant & Xu 1991).

It is also worth mentioning that recent reviews of the existing experimental observations on fatigue kinetics of some materials conclude that, on the macroscale, the power-law exponent n_f; too, can vary with the specimen size (Barenblatt & Botvina 1981; Ritchie 2005; Ciavarella et al. 2008). Unfortunately, the trend of such size dependence is found to be ambiguous. For some materials, n_f may increase, whereas for others it may decrease or remain constant with an increasing structure size. For concrete and rock, as will be shown later, experiments on fatigue kinetics of concrete and rock indicate that the Paris law exponent remains almost constant for different specimen sizes.

While the experimental data on the size effect on static fatigue kinetics of quasibrittle materials are scant, the test data on the size effect in cyclic fatigue kinetics are more abundant. Bažant and Schell (1993) performed a systematic set of size effect

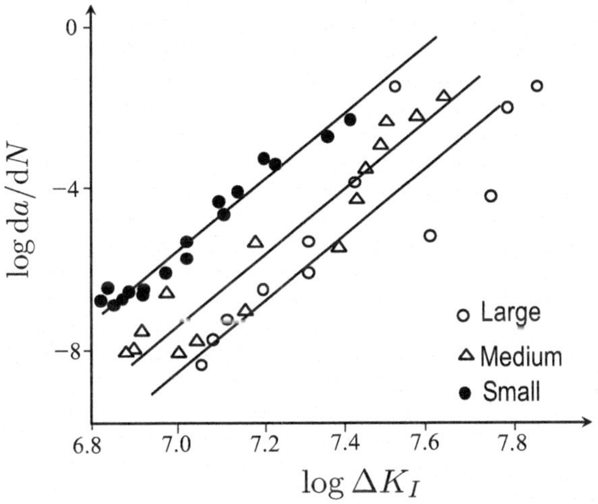

Figure 6.5. Measured size effect on fatigue kinetics of high strength concrete (Bažant & Schell 1993).

experiments on the monotonic strength and cyclic fatigue kinetics of high-strength concrete. In these experiments, a set of geometrically similar three point bend beams with a size ratio $1 : \sqrt{8} : 8$ was tested. The span-to-depth ratio was kept at 2.5, and the beam depths were $D = 38.1$, 107.8, and 304.8 mm. A notch of depth $a_0 = D/6$ was cut at the mid-span. The beams were tested under three point bending. In the fatigue tests, for each beam size, the maximum load was set to be about 77% to 84% of the load capacity measured in the strength tests, and the minimum load was set to be 7% of $P_{\max}$. The loading frequency was 10 Hz. Figure 6.5 shows the measured relationship between the fatigue crack growth rate and the amplitude of the stress intensity factor for these three beam sizes.

For each beam size, the plot of $\mathrm{d}a/\mathrm{d}N$ versus ΔK is seen to be a straight line. However, there is a clear horizontal shift among the three straight lines for the three beam sizes tested. This indicates that a size effect exists in the Paris law. On the other hand, the slopes of these three lines are approximately the same, which indicates that the size effect on the Paris law exponent is nonexistent or very weak. Figure 6.5 shows the optimum fits of these measured curves of $\mathrm{d}a/\mathrm{d}N$ versus ΔK using Eq. 6.46, from which we could determine the transitional size D_{0c} for cyclic loading. Based on Eqs. 6.45 and 6.26, it is clear that, for a fixed specimen geometry, the ratio between D_{0c} and D_{0m} is equal to the ratio between the monotonic and cyclic FPZ sizes, i.e.,

$$\frac{D_{0c}}{D_{0m}} = \frac{c_{f,c}}{c_f} \tag{6.47}$$

Based on the aforementioned optimum fitting of D_{0c} and D_{0m}, it is found that for HSC tested by Bažant and Schell the cyclic FPZ size is about 10% smaller than the monotonic FPZ size.

Recently, Le and co-workers investigated the size effects on monotonic strength and fatigue crack growth in Berea sandstone using also three point bend tests (Le, Manning,

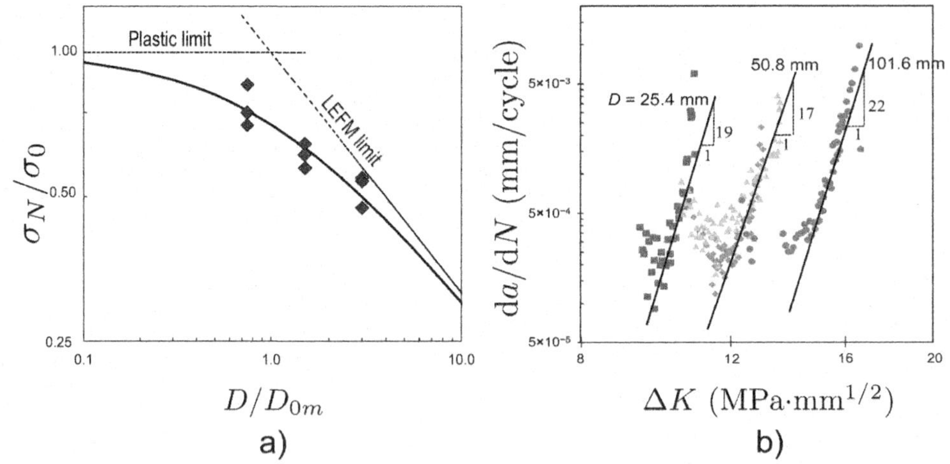

Figure 6.6. Measured size effects on monotonic strength and fatigue kinetics of Berea sandstone.

& Labuz 2014). All beams had a depth-to-span ratio of 1 : 2.5. A set of geometrically similar specimens of the size ratio 1 : 2 : 4 was prepared; the depths of the smallest and largest beams were $D = 25.4$ mm and 101.6 mm, respectively. The width of the beams was $b = 20$ mm, constant for all beam sizes. A notch of depth $0.2D$ was cut at the mid-span. For both monotonic and fatigue tests, the digital image correlation (DIC) technique was used to examine the fracture process at crack initiation and propagation, to allow experimental determination of the sizes of the monotonic and cyclic FPZs.

The monotonic short-time strength tests exhibited a marked size effect on the nominal strength $\sigma_N = P_{\max}/bD$, where $P_{\max}$ = maximum load capacity and b = beam width. Figure 6.6a shows the measured size effect on the nominal strength of the beam, with the optimum fit by the size effect law, Eq. 6.27. Based on this optimum fit, we obtain $D_{0m} = 34$ mm. In the fatigue tests, the maximum and minimum loads were kept as 75% and 3% of $P_{\max}$, respectively, and the loading frequency was 1 Hz. Figure 6.6b demonstrates, for the three beam sizes, a marked size effect on the measured plots of $\mathrm{d}a/\mathrm{d}N$ versus ΔK. The least-square optimum fitting with Eq. 6.46 gives $D_{0c} = 52$ mm. By comparing D_{0c} with D_{0m}, we can conclude that for this rock specimen the cyclic FPZ is about 59% longer than the static FPZ.

In the fatigue experiments, the DIC was used to track the movement of material points during the entire loading process. Figure 6.7a shows the contours of the incremental horizontal displacement for a medium size specimen in the strength test, in which the loading is increased from $0.9P_{\max}$ to $P_{\max}$. The tip of the preexisting notch is located at (0, 0). The contours clearly indicate that the material points on the two sides of the notch move in two opposite directions. Due to the symmetry of beam geometry, the horizontal displacement along the notch ligament should be zero provided that the material is not damaged. Therefore, the effective crack tip can be identified as the merging point of the displacement contours (Fig. 6.7a), which represents the point where the horizontal displacement jump becomes zero.

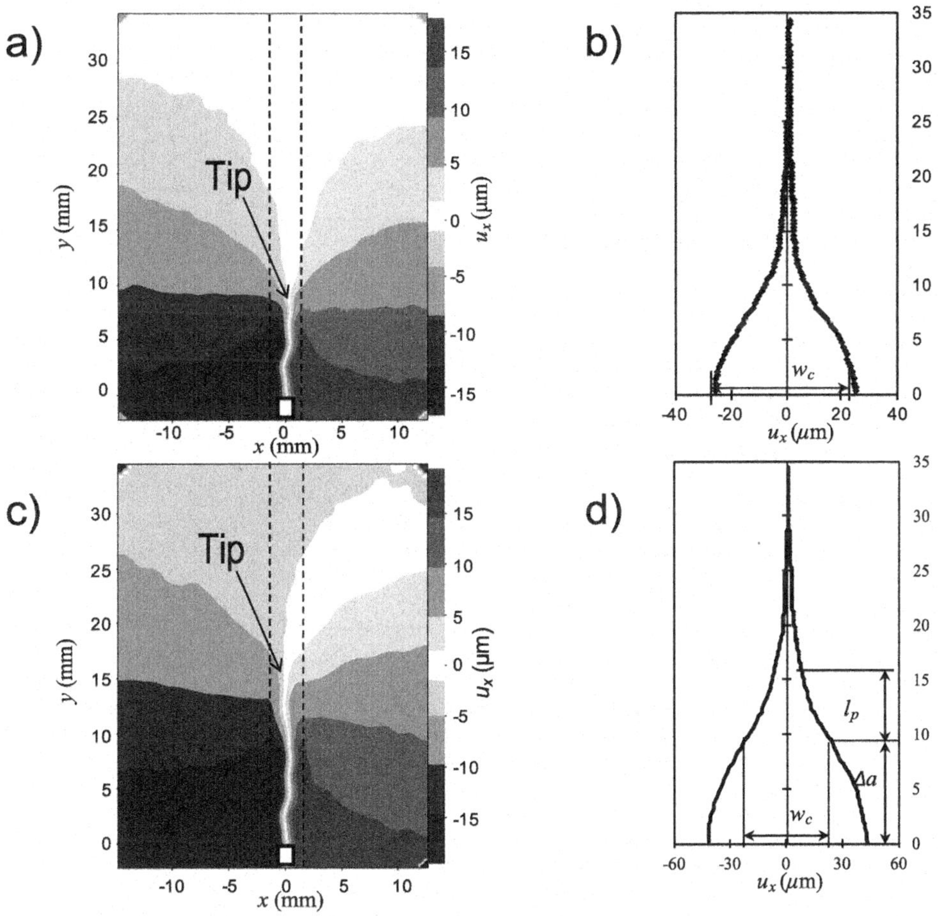

Figure 6.7. DIC measurements in strength test. (a) Incremental horizontal displacement contour for $P = 0.9P_{\max} - P_{\max}$; (b) total horizontal displacement jump at $P_{\max}$; (c) incremental horizontal displacement contour for $P = P_{\max} - 0.9P_{\max}$; and (d) total horizontal displacement jump at $0.9P_{\max}$ in the post-peak regime.

In the meantime, it is widely accepted that for, structures of positive geometry, the peak load is reached once the FPZ is fully developed (Bažant & Planas 1998; Bažant 2005). Therefore, it is reasonable to use the displacement jump at the notch tip at the peak load as the critical opening displacement w_c. Based on previous DIC application to Berea sandstones (Lin & Labuz 2013), the displacement jump was measured between two reference lines located at $x = \pm 1.6$ mm, which avoided the fluctuations due to the tortuosity of the crack propagation path caused by material inhomogeneities. Figure 6.7b presents the total horizontal displacement jump as the beam reaches its peak load, from which we can determine $w_c = 51$ μm for this particular specimen. By examining the displacement jump profile for all the medium and large specimens, it is found that the average value of w_c is about 40 μm. This value is then used to locate the end of the traction-free crack.

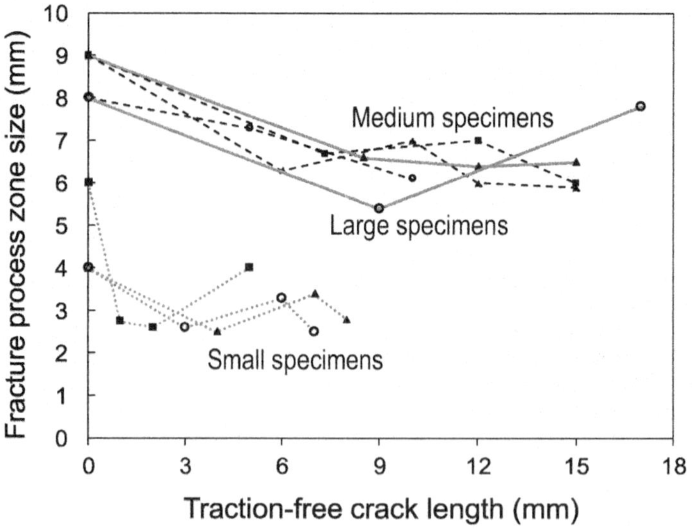

Figure 6.8. Evolution of fracture process zone size during crack propagation.

Figure 6.7c and d shows the contours of the horizontal displacement increments as the load is decreased from P_{max} to $0.9P_{max}$ in the post-peak region and of the total horizontal displacement jump between the reference lines at $0.9P_{max}$, from which the traction-free crack and the FPZ can be identified as mentioned previously. Figure 6.8 presents the evolution of the FPZ size as a function of the growth of the traction-free crack for all specimens based on the analysis of the displacement contours for the subsequent stages of the post-peak regime. It is clear that, as the macrocrack starts to propagate, the FPZ size remains approximately constant, which is a widely accepted feature of quasibrittle fracture. For small-size specimens, the average FPZ size is about 3.5 mm, whereas the average FPZ sizes for medium- and large-size specimens are both about 7.2 mm. Such a difference could be attributed to the fact that, for small-size specimens, the entire uncracked ligament length is only about 20 mm while the part of the ligament that is in tension is greatly restricted by flexural loading, the compressive part preventing the FPZ from developing fully. Since the measured FPZ sizes for medium- and large-size specimens are approximately the same, it can be reasonably expected that the FPZ is fully developed in those specimens and therefore we can conclude that the static FPZ size is about 7.2 mm.

In the fatigue tests, the DIC measurement was recorded at the maximum and minimum loads once every 20 cycles. Figure 6.9 shows the incremental horizontal displacement and the total horizontal displacement jumps between the two reference lines for the large-size specimen at various stages of the loading cycles (i.e., 12,000, 20,500, and 21,600 loading cycles). It is interesting to note that, after loading the specimen for 12,000 cycles, a significant cyclic FPZ is formed but the traction-free crack does not grow yet. At the 20,500th loading cycle, the traction-free crack starts to grow. By comparing the displacement jump for the 20,500 and 21,600 loading cycles, it is found that the cyclic FPZ size remains almost the same.

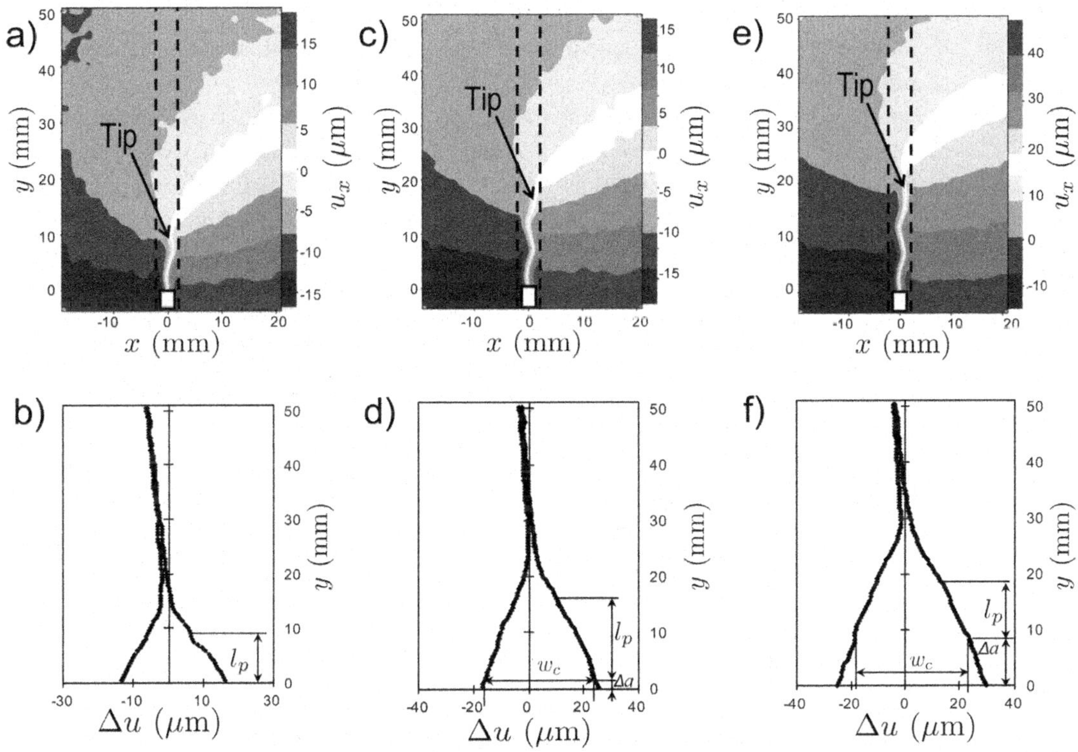

Figure 6.9. DIC measurements of displacement contours and jumps in the fatigue test.

This indicates that, similar to the static FPZ, a fully developed cyclic FPZ has a constant size, and it travels with the traction-free crack. From Figure 6.9, the cyclic FPZ size is estimated to be about 11.3 mm. The ratio between the cyclic FPZ size and the static FPZ size is thus about 1.57. This agrees well with the prediction of Eq. 6.47 and provides further support for the proposed size-dependent Paris law (Eq. 6.46).

6.7 Microplane Model for Size Effect on Fatigue Kinetics under General Loading

In parallel with the analytical and experimental investigations of scaling of fatigue kinetics, recent efforts have also been directed toward computational modeling of fatigue crack growth in quasibrittle structures that would be extendable to general triaxial loading of the material. In Kirane and Bažant (2015*a*, 2015*b*), a microplane model, initially developed to capture the nonlinear triaxial behavior of concrete under a great variety of loading paths, was extended by incorporating a new law for hysteresis and fatigue degradation.

In this model, the fatigue damage is related to the length of the path of the inelastic volumetric strain in the strain space. In the finite element implementation, the conventional crack band model is used to simulate propagation of the fatigue FPZ. Thus the fatigue crack with its wide and long process zone can be simulated as a damage band of a finite width.

For constant amplitude cycles, the model was shown to reproduce well, up to several thousands of cycles, the Paris law behavior with a high exponent that was previously identified for quasibrittle materials (Bažant & Schell 1993; Le, Manning, & Labuz 2014) with the size dependence of fatigue crack growth rate. In addition, the model could also be applied to load cycles of varying amplitudes, to residual strength under sudden overload, and to damage under nonproportional cyclic variation of the strain tensor.

This robust computational model has recently been used to investigate the monotonic and cyclic FPZs of high-strength concrete specimens tested in Bažant and Schell (1993). It was demonstrated that the model agrees well with the measured crack growth histories and lifetimes. In this study, the monotonic FPZ of a given specimen was defined to be the distance from the notch tip to the peak strength point along the ligament as the specimen reached the peak load. The cyclic FPZ was defined as the distance between the peak strength point and the point at which the stress is equal to the notch-tip stress at the monotonic peak load (Kirane & Bažant 2015*b*). Based on such a definition, the numerical simulations indicated that, for the aforementioned fatigue experiment on high-strength concrete, the cyclic FPZ is about 38% smaller than the monotonic FPZ. But this disagrees with the aforementioned analysis (Section 6.6).

Since the microplane model was shown capable of simulating the overall size effect on the fatigue kinetics measured in Bažant and Schell (1993), it is speculated that this discrepancy could be due to the difference in the FPZ definitions between the numerical simulation and the analytical model. However, it must also be pointed out that the experiments were limited to a rather small number of specimens, providing data of considerable random scatter. This could also affect the calibration of Eq. 6.46 and therefore the ratio between the monotonic and cyclic FPZs. Therefore, more comprehensive experimental studies of fracture processes under monotonic and fatigue loadings are needed.

Another way to clarify the behavior is to conduct simulations with the lattice discrete particle model (Cusatis et al. 2003*a*), which has recently achieved high fidelity in the works of Cusatis, Pelessone, and Mencarelli (2011). This approach should provide deeper insight into various damage mechanisms under fatigue and monotonic loading scenarios that are responsible for the difference between the FPZ sizes under cyclic and static loading.

7 Size Effect on Probability Distributions of Strength and Lifetime of Quasibrittle Structures

In Chapter 5, we derived the probability distribution of static and fatigue strengths of one material representative volume element (RVE), which can be approximated by a Gaussian distribution onto which a power-law tail (Weibull distribution) is grafted from its far left at a probability $P_f \approx 10^{-3}$ to 10^{-4}. This chapter focuses on the statistics of strength and lifetime of quasibrittle structures of positive geometry. First we derive the probability distribution of structural strength based on a finite weakest-link model. With the crack growth rate laws discussed in Chapter 6, we then study the lifetime distributions of structures subjected to both static creep and fatigue loading. The present probabilistic framework is further used to study the size effects on the probability distribution functions of structural strength and lifetime, on the mean structural strength and lifetime, and on the stress-life curve.

7.1 Probability Distribution of Structural Strength

For quasibrittle structures of positive geometry, a salient feature of the failure behavior is that the peak load is attained as soon as a macrocrack initiates from one RVE. Therefore, the failure statistics of the structure can be formulated by the weakest-link model, where each link corresponds to one RVE. As mentioned in Chapter 1 (Section 1.4), here the RVE is defined as the smallest material volume whose failure causes the whole structure to fail, which is necessarily different from the conventional definition of RVE in the homogenization theory [e.g., Hill (1963)]. The survival probability of the structure is the joint probability of survival of all the RVEs, numbered $i = 1, \dots n_s$. Therefore, under the assumption of statistical independence of the random strengths of the RVEs (which requires the autocorrelation length of material strength not to exceed the FPZ size), we have

$$1 - P_f = \prod_{i=1}^{n_s} P_{si} \tag{7.1}$$

$$\text{or} \quad P_f(\sigma_N) = 1 - \prod_{i=1}^{n_s} \{1 - P_{1\sigma}[\langle s(\boldsymbol{x}_i)\rangle \sigma_N]\} \tag{7.2}$$

where P_{si} = survival probability of the ith RVE, $P_{1\sigma}(x)$ = strength cdf of one RVE given by Eqs. 5.84a and 5.84b, P_f = failure probability of the entire structure, σ_N = nominal

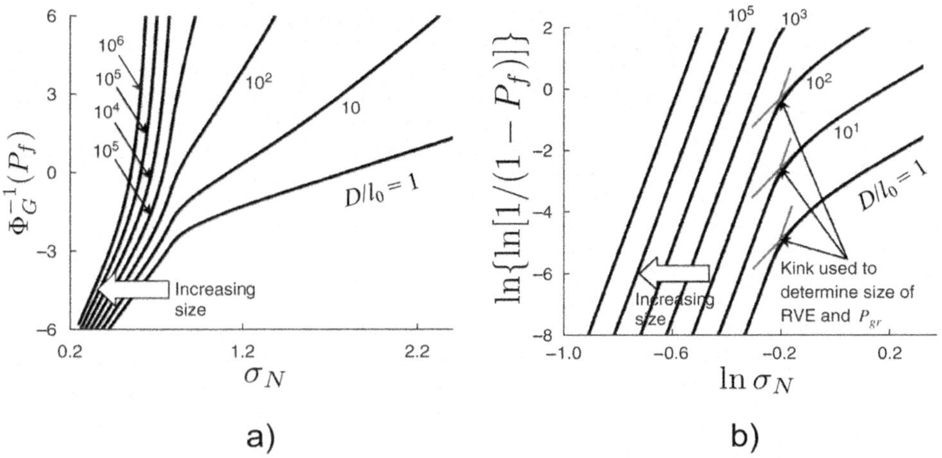

Figure 7.1. Strength distributions of geometrically similar structures of different sizes presented in (a) Normal probability paper and (b) Weibull scale. Weibull probability paper.

strength of the structure as defined in Chapter 1, and $s(\boldsymbol{x}_i)$ = dimensionless stress field such that $\sigma(\boldsymbol{x}_i) = \sigma_N s(\boldsymbol{x}_i)$ is equal to the maximum principal stress at the center of the ith RVE with the coordinate $\boldsymbol{x}_i$. Here we assume that the principal stresses in each RVE are fully statistically correlated to the maximum one, which seems realistic. If they were uncorrelated, each principal stress would require one element in the chain.

Equation 7.1 is further contingent upon the hypothesis that the strengths of different RVEs are statistically uncorrelated. This is certainly a simplification, though probably quite realistic. Strictly speaking, the strength field within the structure should be represented by an autocorrelated random process. But previous studies of random lattice-particle models (Grassl & Bažant 2009) showed that the autocorrelation length is approximately equal to the size of the RVE, l_0, which is about two to three times the maximum size of the material inhomogeneity (Bažant & Pang 2007). Therefore, though strength correlations exist, they can be neglected for distances larger than the RVE size.

Equation 7.2 directly implies that the probability distribution of structural strength would depend on the structure size, or equivalently, on the number of RVEs n_s. Figure 7.1 presents the calculated cumulative distribution function (cdf) of structural strength for different structure sizes on both Weibull and Gaussian probability distribution papers. For small-size structures (small n_s), the strength cdf is predominantly Gaussian, which corresponds to the case of quasi-plastic behavior. When the structure size increases but is not too large, the core of the cdf of structural strength is still predominantly governed by the lower part of the Gaussian core of the strength cdf of one RVE lying close to the grafting point. According to the stability postulate (Fisher & Tippett 1928) or a renormalization group analysis (Bazant 2000; van der Hofstad & Redig 2006), the core of cdf of structural strength should thus approach the Gumbel (or Fisher–Tippett) distribution, in the sense of intermediate asymptotics (Barenblatt 1979). For sufficiently large structures (large n_s), what matters for P_f is the tail of the strength cdf of one generic RVE, i.e., $P_{1\sigma}(\sigma) = (\sigma/s_0)^m$. Therefore, Eq. 7.2 can be rewritten as

$$P_f(\sigma_N) = 1 - \exp\left[-\left(\int_V \langle s(\boldsymbol{x}_i)\rangle^m \frac{\mathrm{d}V(\boldsymbol{x})}{V_0}\right)\left(\frac{\sigma_N}{s_0}\right)^m\right] \tag{7.3}$$

where $V_0 = l_0^{n_d}$, n_d = dimension of the problem, and l_0 = size of one RVE, which is a material property (material length). As shown in Eq. 7.3, the strength cdf of large-size structures approaches the Weibull distribution, which corresponds to the case of perfectly brittle behavior (Weibull 1939).

As mentioned in Chapter 2 (Section 2.4), it is convenient to define the equivalent number $N_{\mathrm{eq},\sigma}$ of RVEs, i.e., $N_{\mathrm{eq},\sigma} = \int_V \langle s(\boldsymbol{x}_i)\rangle^m \,\mathrm{d}V(\boldsymbol{x})/V_0$ (Bažant & Pang 2007). $N_{\mathrm{eq},\sigma}$ represents the number of RVEs under uniform stress for which σ_N gives the same cdf of structure strength, as does Eq. 7.2 under the assumption that the strength cdf follows the Weibull distribution. The concept of $N_{\mathrm{eq},\sigma}$ essentially converts the geometry effect (stress distribution) on the strength cdf to the size effect on the strength cdf. For the grafted distribution, the calculation of $N_{\mathrm{eq},\sigma}$ becomes cumbersome, because $N_{\mathrm{eq},\sigma}$ is a function of both σ_N and stress field $s(\boldsymbol{x}_i)$.

Here it should be pointed out that, in the literature, the strength distribution of one RVE was sometimes erroneously assumed as Weibullian. However, it is simple to prove that this is impossible. Consider that the strength of a presumed "RVE" has the Weibull distribution. But this distribution can arise only from the weakest-link model for a chain, described by Eq. 7.1. In a chain, the fracture must always localize into one link. So the presumed "RVE" cannot be the true RVE. Rather, the failing link must represent the true RVE, which is the smallest material volume whose failure triggers the failure of the entire structure.

To summarize, based on the present theory, the statistics of structural strength must evolve with increasing size D through three asymptotic regimes:

1. For small sizes [roughly $N_{\mathrm{eq},\sigma} \in (1, 20)$], the cdf of strength is virtually Gaussian.
2. In the intermediate range of not too large sizes [roughly $N_{\mathrm{eq},\sigma} \in (50, 500)$], the chance that the strength of the weakest RVE would fall into the power-law tail $P_f < 10^{-3}$ is still very small, and so the weakest RVE is likely to be in the Gaussian core. This means, in theory, that the strength should approach the extreme value distribution of Gaussian variables, which is the Gumbel distribution (Fisher & Tippett 1928; Gumbel 1958; Ang & Tang 1984; Haldar & Mahadevan 2000*a*; Bazant 2000), representing the so-called intermediate asymptotic regime in the sense of Barenblatt (Barenblatt 1979, 2003).
3. For large sizes (roughly $N_{\mathrm{eq},\sigma} > 5000$), there is a very high chance that the strength of the weakest element would be in the power-law tail $P_f < 10^{-3}$. This means that the strength distribution must converge to the extreme value distribution for power-law tails, which is the Weibull distribution (Fisher & Tippett 1928; Ang & Tang 1984; Haldar & Mahadevan 2000*a*).

Computations show, however, that the Gumbel statistics is not approached closely, and that the cdf for structures of any size may be well approximated as a Gauss–Weibull

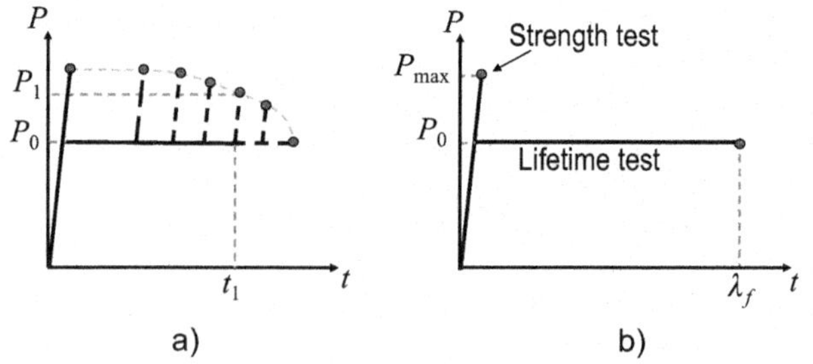

Figure 7.2. Loading histories of strength and creep-rupture (lifetime) tests.

grafted distribution, with the grafting point moving to from left to right as the structure size increases.

It is clear that we can also use the finite weakest-link model to derive the probability distribution of fatigue strength, which measures the critical stress amplitude for a prescribed number of loading cycles and a fixed stress ratio. Since the strength cdf of one RVE has the same form as the cdf of fatigue strength of one RVE (Eqs. 5.84a and 5.84b and Eqs. 5.87a and 5.87b), it is expected that the cdf of fatigue strength of the structure would also vary with the structure size, in the same way as the cdf of the monotonic strength does. However, it is not of practical interest to study the cdf of fatigue strength because it is not feasible to obtain the histogram of fatigue strength. (Before testing a structure, we would not know the critical stress amplitude for a prescribed number of cycles.) Instead, for fatigue loading the main design concern is the structural lifetime.

7.2 Probability Distribution of Structural Lifetime

7.2.1 Creep Lifetime

Many engineering structures, such as bridges and dams, are subjected to a large dead load throughout their service life, and so the structural lifetime under constant loads (i.e., creep lifetime) becomes an important design aspect. It has long been recognized that an easy way to study the structural lifetime is through the fracture kinetics [e.g., Fett (1991); Bažant & Xu (1991); Fett Muuz (1991); Munz & Fett (1999)].

Here we use the crack growth equation derived in Chapter 6 to relate the strength and creep lifetime of one RVE. Consider the load history in the creep-rupture test (or lifetime test), in which the load, P, is rapidly raised to some value P_0; then is held constant for various lengths of time, t_1; and finally is rapidly increased to some random value P_1 at which the failure occurs (Fig. 7.2a). When $t_1 \to \lambda_f =$ lifetime, we have $P_1 \to P_0$. For $t_1 = 0$, $P_1 = P_{\max}$, which is the strength test. In between there must be a continuous transition, and so the statistics of failure load $P_{\max}$ must be related to the statistics of lifetime λ_f.

The focus of this chapter is on monotonic strength and structural lifetime. The general transition between the strength and lifetime tests will be dealt with in Chapter 10.

As the RVE is subjected to a sustained load, damage develops and eventually it turns into a macrocrack. Within the framework of equivalent linear elastic fracture mechanics (LEFM), we can replace the distributed damage zone by an equivalent subcritical LEFM crack. Therefore, the failure kinetics of the RVE can be described by the crack growth law. Consider the growth of a subcritical crack with initial length a_0, which represents the initial flaws in the RVE. Under a certain loading history, this crack grows to its critical value a_c, at which the RVE fails. The growth rate of the subcritical crack can be described by the Charles–Evans law (Eq. 6.1), in which we may further write the stress intensity factor of one RVE as

$$K_R = \sigma \sqrt{l_0} k_R(\alpha) \tag{7.4}$$

where σ is a load parameter of the dimension of stress, called the nominal stress and defined as $\sigma = P/bl_0$; l_0 = RVE size; b = thickness of the RVE in the transverse direction; and $\alpha = a_R/l_0$ = relative crack length.

For the case of strength test, the load is linearly increased until the failure of RVE with loading rate r (i.e., $\sigma = rt$); Figure 7.2b. Denoting $\sigma_N = P_{\max}/bl_0$ and integrating Eq. 6.1, we obtain the nominal strength σ_N:

$$\sigma_N^{n_c+1} = r(n_c+1)e^{Q_0/kT} \int_{\alpha_0}^{\alpha_c} \frac{\mathrm{d}\alpha}{A_s l_0^{(n_c-2)/2} k_R^{n_c}(\alpha)} \tag{7.5}$$

For the case of lifetime test (Fig. 7.2b), the load is rapidly increased to P_0, which is smaller than the load capacity of the RVE. By letting the applied nominal stress be $\sigma_0 = P_0/bl_0$ and integrating Eq. 6.1 for the entire loading history of lifetime, we get

$$\frac{1}{r}\int_0^{\sigma_0} \sigma^{n_c}\,\mathrm{d}\sigma + \int_{t_0}^{\lambda_f} {\sigma_0}^{n_c}\,\mathrm{d}t = e^{Q_0/kT} \int_{\alpha_0}^{\alpha_c} \frac{\mathrm{d}\alpha}{A_s l_0^{(n_c-2)/2} {k_1}^{n_c}(\alpha)} \tag{7.6}$$

where $t_0 = \sigma_0/r$. This yields

$${\sigma_0}^{n_c+1} + r(n_c+1){\sigma_0}^{n_c}(\lambda_f - t_0) = r(n_c+1)e^{Q_0/kT} \int_{\alpha_0}^{\alpha_c} \frac{\mathrm{d}\alpha}{A_s l_0^{(n_c-2)/2} {k_1}^{n_c}(\alpha)} \tag{7.7}$$

Since the lifetime of interest is typically far longer than the duration of laboratory strength tests, the initial rapidly increasing portion of the load history makes a negligible contribution to the overall structural lifetime (i.e., $\lambda_f \gg t_0$). Since the mechanisms of crack growth can be considered to be the same for both load histories, one may eliminate the integrals from Eqs. 7.5 and 7.7 to get a very simple relation between σ_N, λ_f, and σ_0.

$${\sigma_0}^{n_c+1} + r(n_c+1){\sigma_0}^{n_c}(\lambda_f - t_0) = {\sigma_N}^{n_c+1} \tag{7.8}$$

Equation 7.8 can equivalently be written as follows:

$$\sigma_N = \sigma_0^{n_c/(n_c+1)}[r(n_c+1)\lambda_f - n_c\sigma_0]^{1/(n_c+1)} \tag{7.9}$$

where it should be noted that for $\sigma_0 \to \sigma_N$, $\lambda_f \to \sigma_N/r$, while for $\sigma_N \gg \sigma_0$ (i.e., $\lambda_f \gg t_0$):

$$\sigma_N \approx \sigma_0^{n_c/(n_c+1)}[r(n_c+1)]^{1/(n_c+1)}\lambda_f^{1/(n_c+1)} \tag{7.10}$$

$$\text{or } \sigma_N = \beta\sigma_0^{n_c/(n_c+1)}\lambda_f^{1/(n_c+1)} \tag{7.11}$$

where $\beta = [r(n_c+1)]^{1/(n_c+1)}$ = a constant.

Note that although, in Chapter 6, the power law for creep crack growth rate was physically justified as the mean structural behavior, it is now used to relate the randomness of strength and of lifetime in one RVE. This is a simplification. Equations 7.5 and 7.7 clearly indicate that the randomness of strength and lifetime of RVE is caused by the geometrical randomness of the dominant subcritical crack, which is determined by both the crack size and the local fracture energy. This agrees well with the stochastic lattice-particle model (Grassl & Bažant 2009). Furthermore, it should be emphasized that we do not need to consider the scale effect on the static fatigue kinetics (i.e., Eq. 6.21) because the focus here is on the static fatigue crack growth in one RVE, whose size is independent of the structure size.

Furthermore, it should be noted that Eqs. 7.10 and 7.11 neglect the initial loading portion of the loading history of the lifetime test. This simplification is valid only if the sustained stress σ_0 is so low compared to the strength that the initial short-time rising segment of load history has a negligible contribution to the failure probability. This is generally true for sustained loads in the service stress range.

If, however, σ_0 is not low enough, then Eq. 7.7 needs to be used, which will introduce a certain threshold into the statistics of lifetime. The physical justification of the threshold consists in the fact that there is a nonzero probability of failure during the initial short-time increase of the load, as some specimens may have a strength even lower than σ_0. This general case will be discussed in Chapter 10. For common n_c values [$n_c = 20$–30, as indicated in Kawakubo (1995), for example], Eq. 7.11 gives a very good approximation when $\sigma_0 \leq 0.85\sigma_N$, and then the threshold is important only above $0.85\sigma_N$.

Because of the relationship between the monotonic strength and the creep lifetime (Eq. 7.11), the cdf of creep lifetime can be written as

$$\Pr(\lambda' \leq \lambda) = \Pr\left(\sigma' \leq \lambda^{1/(n_c+1)}\beta\sigma_0^{n_c/(n_c+1)}\right) \tag{7.12}$$

where λ', σ' = random realizations of creep lifetime and strength of one RVE, respectively. Based on the probability distribution of RVE strength given by Eqs. 5.84a and 5.84b, the cdf of creep lifetime of one RVE can be obtained as follows:

$$\text{for } \lambda < \lambda_{\text{gr}}: \quad P_{1\lambda}(\lambda) = 1 - \exp[-(\lambda/s_\lambda)^{\bar{m}}]; \tag{7.13}$$

$$\text{for } \lambda \geq \lambda_{\text{gr}}: \quad P_{1\lambda}(\lambda) = P_{\text{gr}} + \frac{r_f}{\delta_G\sqrt{2\pi}}\int_{\gamma_c\lambda_{\text{gr}}^{1/(n_c+1)}}^{\gamma_c\lambda^{1/(n_c+1)}} e^{-(\lambda'-\mu_G)^2/2\delta_G^2}\,d\lambda' \tag{7.14}$$

where $\gamma_c = \beta\sigma_0^{n_c/(n_c+1)}$ $\lambda_{\text{gr}} = \beta^{-(n_c+1)}\sigma_0^{-n_c}\sigma_{N,\text{gr}}^{n_c+1}$, $s_\lambda = s_0^{n_c+1}\beta^{-(n_c+1)}\sigma_0^{-n_c}$, and $\bar{m} = m/(n_c+1)$.

Similar to the strength distribution of one RVE, the lifetime cdf of one RVE, too, has a Weibull tail (power-law tail). The grafting probability P_{gr} for the lifetime distribution of one RVE is the same as that for the strength cdf of one RVE. However, as Eq. 7.14 suggests, the rest of the lifetime cdf of one RVE does not follow the Gaussian distribution exactly.

According to Eqs. 7.13 and 7.14, it is clear that the statistical parameters of lifetime cdf of one RVE depend on the stress, σ_0, applied to the RVE. To facilitate the later calculation of lifetime cdf of the entire structure, we rewrite Eqs. 7.13 and 7.14 by introducing a new random variable $\Lambda' = \sigma_0^{n_c}\lambda'$:

$$\text{for } \Lambda < \Lambda_{\text{gr}}: \quad P_{1\Lambda}(\Lambda) = 1 - \exp[-(\Lambda/s_\Lambda)^{\bar{m}}]; \tag{7.15}$$

$$\text{for } \Lambda \geq \Lambda_{\text{gr}}: \quad P_{1\Lambda}(\Lambda) = P_{\text{gr}} + \frac{r_f}{\delta_{\text{G}}\sqrt{2\pi}} \int_{\beta\Lambda_{\text{gr}}^{1/(n_c+1)}}^{\beta\Lambda^{1/(n_c+1)}} e^{-(\Lambda'-\mu_{\text{G}})^2/2\delta_{\text{G}}^2}\, \mathrm{d}\Lambda' \tag{7.16}$$

where $\Lambda_{\text{gr}} = \beta^{-(n_c+1)}\sigma_{N,\text{gr}}^{n_c+1}$, $s_\Lambda = s_0^{n_c+1}\beta^{-(n_c+1)}$.

Since the lifetime of a chain is the shortest individual lifetime of its links, the weakest-link model may still be used to compute the lifetime cdf of a structure. Similar to the definition of nominal strength, here we can define the nominal stress, $\sigma_{n0} = c_n P/bD$ or $= c_n P/D^2$, for two- or three-dimensional scaling (P = applied load) and further denote $\Lambda_f = \sigma_{n0}^{n_c}\lambda_f$ (λ_f = creep lifetime of the structure). According to the joint probability theorem, the lifetime distribution of structure can be expressed as

$$P_f(\Lambda_f) = 1 - \prod_{i=1}^{n_s}\{1 - P_{1\Lambda}[\langle s(\boldsymbol{x}_i)\rangle^{n_c}\Lambda_f]\} \tag{7.17}$$

Similar to the chain model for the cdf of the structural strength, the lifetime of the ith RVE is here assumed to be governed by the maximum average principal stress $\sigma_{n0}s(\boldsymbol{x}_i)$ within the RVE, provided that the other principal stresses are fully statistically correlated to the maximum one.

As shown in Eq. 7.16, the core of lifetime cdf of one RVE, denoted as $\Phi_u(\Lambda)$, is a Gaussian cdf transformed by the power law for crack growth such that the lifetime power $\xi = (r/\kappa)^\kappa\Lambda^\kappa$ ($\kappa = 1/(n_c+1)$) would have the Gaussian distribution:

$$\Phi_u(\xi) = \int_{-\infty}^{\xi} \left(2\pi s_u^2\right)^{-1/2} e^{-(t'-\mu_u)^2/2s_u^2}\, \mathrm{d}t' \tag{7.18}$$

When κ is very small, the foregoing transformation can be approximated by the logarithmic transformation for the region of Λ that is near some prescribed lifetime Λ_0 (Ibnabdeljalil & Phoenix 1995):

$$\xi = (r/\kappa)^\kappa\Lambda_0^\kappa\,[\kappa\ln(\Lambda/\Lambda_0) + 1] \tag{7.19}$$

This approximation rests on the fact that, when κ is small enough, $(\Lambda/\Lambda_0)^\kappa$ is close to 1. Therefore, $\ln[(\Lambda/\Lambda_0)^\kappa] \approx (\Lambda/\Lambda_0)^\kappa - 1$ and $\Lambda_0^\kappa[\kappa\ln(\Lambda/\Lambda_0) + 1] \approx \Lambda^\kappa$.

To ensure the continuity of the lifetime cdf at the grafting point, we choose $\Lambda_0 = \Lambda_{\text{gr}}$. From Eqs. 7.18 and 7.19, we find that the core of lifetime distribution can be approximated as the lognormal distribution for the range of Λ that is near Λ_{gr}. As κ decreases

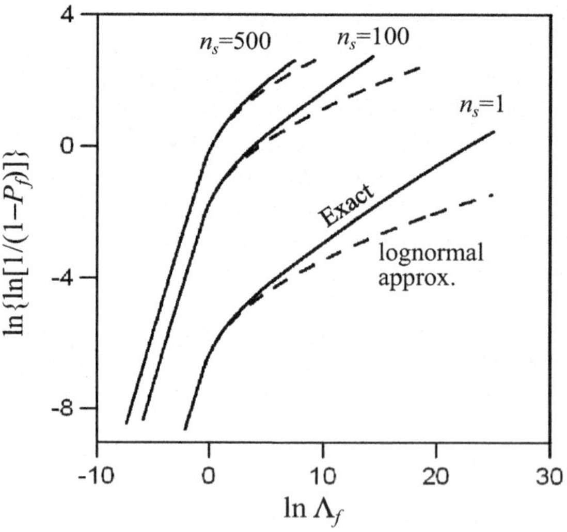

Figure 7.3. Probability distribution of structural lifetime and its logarithmic approximation.

(n_c increases), we would expect a wider range of applicability of the logarithmic transformation. But, interestingly, this is not really the case. The reason is as follows: A larger value of n_c would doubtless make $(\Lambda/\Lambda_{gr})^\kappa$ converge to 1. At the same time, however, a larger value of n_c, which implies a lower Weibull modulus, means a much broader scatter of the lifetime. This would greatly increase the ratio Λ/Λ_{gr} for the main part of the core; thus it would make $(\Lambda/\Lambda_{gr})^\kappa$ deviate further from 1. Calculations show that these two effects cancel each other. As a result, the logarithmic transformation gives only an approximation of the core near Λ_{gr}, albeit a good one, and does so even for a large value of n_c.

Figure 7.3 shows the lifetime distributions of structures consisting of different numbers of RVEs for $n_c = 20$ (or $\kappa = 1/21$), calculated by both the exact transformation from the power law of crack growth and its logarithmic approximation. For the lifetime distribution of one RVE, the logarithmic approximation works well only for a very narrow range of Λ_f. When the number of RVEs increases, the logarithmic approximation gets closer to the exact solution. This is because, if the structure is sufficiently large, the portion of the core of lifetime cdf of one RVE that matters for the lifetime cdf of the entire structure is small and is very close to the grafting lifetime. Since that part of the core can be modeled by the lognormal distribution, the deviation of lifetime cdf from the Weibull distribution can be approximated as a chain of elements with lognormal distribution provided that the structure is large enough.

For very large structures, the lifetime cdf is determined by the far-left tail of the lifetime cdf of one RVE: $P_{1\Lambda}(\Lambda_f) = (\Lambda_f/s_\Lambda)^{\bar{m}}$. By using an analysis similar to that for the cdf of strength of large-size structure, we obtain the lifetime cdf as

$$P_f(\Lambda_f) = 1 - \exp\left[-\left(\int_V \langle s(\boldsymbol{x}_i)\rangle^{n_c\bar{m}} \frac{\mathrm{d}V(\boldsymbol{x})}{V_0}\right)\left(\frac{\Lambda_f}{s_\Lambda}\right)^{\bar{m}}\right] \tag{7.20}$$

Clearly, the lifetime cdf of large-size structures asymptotically approaches the Weibull distribution. Here we may also define the equivalent number of RVEs for the lifetime distribution of large-size structure: $N_{\mathrm{eq},\Lambda_f} = \int_V \langle s(\boldsymbol{x}_i)\rangle^{n_c\bar{m}}\,\mathrm{d}V(\boldsymbol{x})/V_0$, which physically means the number of RVEs for which a chain of $N_{\mathrm{eq},\Lambda_f}$ elements subjected to a uniform stress σ_{n0} gives the same cdf of lifetime as Eq. 7.17 does. Note that the equivalent numbers of RVEs for the strength and lifetime distributions are different. The reason is that the corresponding loading histories for strength and lifetime tests are different. However, the difference is very small since, due to the large value of n_c, $mn_c/(n_c + 1) \approx m$.

In the present theory, Eqs. 5.84a and 5.84b and Eq. 7.11 yield, for the Weibull moduli of strength and lifetime distributions, a strikingly simple relation involving the exponent of the power law for crack growth:

$$\bar{m} = \frac{m}{n_c + 1} \tag{7.21}$$

where $\bar{m}$ = the Weibull modulus of lifetime distribution. This relation indicates an efficient way to obtain the Weibull modulus of lifetime distribution without any testing of lifetime histograms, which is time consuming and costly, and for long lifetimes virtually impossible. Aside from the exponent n_c, we merely need the Weibull modulus m of strength distribution, which can be most easily determined by size effect tests (Pang, Bažant, & Le 2008). Exponent n_c of power law for crack growth can be obtained by the standard test, of not too long duration, that measures the mean crack growth velocity.

It is noted from Eq. 7.21 that the Weibull modulus for lifetime would typically be 10 to 30 times smaller than that for strength. This means that, for lifetime, the size effect is far stronger than it is for structural strength. Regrettably, there seem to be no experiments in the literature to document it.

7.2.2 Fatigue Lifetime

Besides creep lifetime, structural lifetime under fatigue loading is also of great interest to the design of many engineering structures such as aircraft, dental implants, prosthetic joints, and so on. Evidently, we can use the present framework to derive the probability distribution of fatigue lifetime from the cdf of fatigue strength (Eqs. 5.87a and 5.87b). Similar to the analysis of creep lifetime, we may consider the tests of fatigue strength and fatigue lifetime made on the same RVE. In the fatigue strength test, the RVE is subjected to a cyclic load with a prescribed number of cycles, N_0, and a given stress ratio, R, and the critical load amplitude (i.e., the fatigue strength ΔP_m), at which the RVE fails, is recorded. In the fatigue lifetime test, the load amplitude ΔP_0 and the stress ratio R are prescribed, and the critical number of cycles N at which the RVE fails is recorded.

The kinetics of cyclic failure of one RVE can also be described by the growth of a subcritical crack, which represents the distributed damage in the RVE. The RVE fails when the dominant subcritical crack grows from its original length a_0 to a critical length a_c.

The growth rate of this subcritical crack follows the Paris law (Eq. 6.2). By separation of variables, we have

$$\Delta\sigma^{n_f} N_c = \int_{\alpha_0}^{\alpha_c} \frac{d\alpha}{A_c k^{n_f}(\alpha) l_0^{n_f - 1}} \tag{7.22}$$

where $\Delta\sigma = (P_{\max} - P_{\min})/bl_0$ = nominal stress amplitude, $\alpha = a/l_0$ = dimensionless crack size, $k(\alpha)$ = dimensionless stress intensity factor of the RVE, l_0 = RVE size, and N_c = number of loading cycles to failure. By applying Eq. 7.22 to the tests of both fatigue strength and fatigue lifetime, we can relate the fatigue strength for the given number of cycles N_0 to the fatigue lifetime N_c for the given load amplitude:

$$\Delta\sigma_f = \Delta\sigma_0 (N/N_0)^{1/n_f} \tag{7.23}$$

where $\Delta\sigma_f = \Delta P_m/bl_0$ and $\Delta\sigma_0 = \Delta P_0/bl_0$. Similar to the derivation of the cdf of creep lifetime, we assume that both the fatigue strength and the fatigue lifetime are random variables, but their relationship, which follows Eq. 7.23, is not random. Furthermore, the size effect on the Paris law is not needed here because we are interested in the fatigue crack growth in one RVE.

With Eq. 7.23, we can relate the probability distributions of fatigue lifetime and fatigue strength as

$$\Pr(N' \le N) = \Pr\left(\Delta\sigma_f' \le \Delta\sigma_0 N_0^{-1/n_f} N^{1/n_f}\right) \tag{7.24}$$

and further, with Eqs. 5.87a and 5.87b, we obtain the probability distribution of fatigue lifetime of one RVE:

$$\text{for } N < N_{\text{gr}}:\ \ P_{1N}(N) = 1 - \exp[-(N/s_N)^{\bar{m}_f}]; \tag{7.25}$$

$$\text{for } N \ge N_{\text{gr}}:\ \ P_{1N}(N) = P_{\text{gr},f} + \frac{r_{ff}}{\delta_{Gf}\sqrt{2\pi}} \int_{\gamma_f N_{\text{gr}}^{1/n_f}}^{\gamma_f N^{1/n_f}} e^{-(N' - \mu_{Gf})^2/2\delta_{Gf}^2} dN' \tag{7.26}$$

where $\gamma_f = \Delta\sigma_0 N_0^{-1/n_f}$, $N_{\text{gr}} = (\Delta\sigma_{\text{gr}}/\Delta\sigma_0)^{n_f} N_0$, $s_N = N_0 \Delta\sigma_0^{-n_f} s_{0f}^{n_f}$, and $\bar{m}_f = m_f/n_f$. Similar to the creep lifetime, it is convenient to define a new random variable $\mathcal{N}' = \Delta\sigma_0^{n_f} N'$, and Eqs. 7.25 and 7.26 can be rewritten as:

$$\text{for } \mathcal{N} < \mathcal{N}_{\text{gr}}:\ \ P_{1\mathcal{N}}(\mathcal{N}) = 1 - \exp[-(\mathcal{N}/s_{\mathcal{N}})^{\bar{m}_f}]; \tag{7.27}$$

$$\text{for } \mathcal{N} \ge \mathcal{N}_{\text{gr}}:\ \ P_{1\mathcal{N}}(\mathcal{N}) = P_{\text{gr},f} + \frac{r_{ff}}{\delta_{Gf}\sqrt{2\pi}} \int_{\gamma_{\mathcal{N}} \mathcal{N}_{\text{gr}}^{1/n_f}}^{\gamma_{\mathcal{N}} \mathcal{N}^{1/n_f}} e^{-(\mathcal{N}' - \mu_{Gf})^2/2\delta_{Gf}^2} d\mathcal{N}' \tag{7.28}$$

where $\gamma_{\mathcal{N}} = N_0^{-1/n_f}$, $\mathcal{N}_{\text{gr}} = \Delta\sigma_{\text{gr}}^{n_f} N_0$ and $s_{\mathcal{N}} = N_0 \Delta s_{0f}^{n_f}$.

For structures of positive geometry, the fatigue lifetime of a structure is the shortest fatigue lifetime of its RVEs. Therefore, we could use the weakest-link model based on the assumption that the fatigue lifetimes of RVEs are independent random variables.

The cdf of fatigue lifetime under a prescribed cyclic load can be written as

$$P_f(\mathcal{N}_f) = 1 - \prod_{i=1}^{n_s} \left[1 - P_{1\mathcal{N}}(\mathcal{N}_f \langle s(\boldsymbol{x}_i)\rangle^{n_f})\right] \tag{7.29}$$

or
$$\ln(1 - P_f) = \sum_{i=1}^{n_s} \ln\left[1 - P_{1\mathcal{N}}(\mathcal{N}_f \langle s(\boldsymbol{x}_i)\rangle^{n_f})\right] \tag{7.30}$$

where $\mathcal{N}_f = \Delta\sigma_{n0}^{n_f} N_f$, N_f = fatigue lifetime of the structure, $\Delta\sigma_{n0}$ = amplitude of the applied nominal stress, and $s(x_i)$ = dimensionless principal stress field such that $\Delta\sigma_{n0} s(\boldsymbol{x}_i)$ = amplitude of maximum principal stress at the center of the ith RVE.

By comparing the expressions for the cdfs of creep lifetime (Eqs. 7.15, 7.16, and 7.17) and fatigue lifetime (Eqs. 7.27, 7.28, and 7.29), it is clear that the cdfs of fatigue lifetime and creep lifetime have the same functional form and, therefore, are subjected to a similar size effect. For small size, the cdf of fatigue lifetime can be described by a predominantly Gaussian distribution transformed by a power-law function, with its far-left tail being Weibullian. As the structure size increases, the Weibullian part grows and the upper part, which deviates from the Weibull distribution, can be approximately modeled as a chain of lognormal distributions. For sufficiently large structures, where only the tail part of the lifetime cdf of one RVE matters for the failure of the entire structure, the cdf of fatigue lifetime of large structures becomes Weibullian:

$$P_f(\mathcal{N}_f) = 1 - \exp\left[-\left(\int_V \langle s(\boldsymbol{x}_i)\rangle^{n_f \bar{m}_f} \frac{\mathrm{d}V(\boldsymbol{x})}{V_0}\right)\left(\frac{\mathcal{N}_f}{s_{\mathcal{N}}}\right)^{\bar{m}_f}\right] \tag{7.31}$$

Based on this result, we can define the equivalent number of RVEs for the Weibullian fatigue cdf for large-size structures:

$$N_{\mathrm{eq},\mathcal{N}_f} = \int_V \langle s(\boldsymbol{x}_i)\rangle^{n_f \bar{m}_f} \mathrm{d}V(\boldsymbol{x})/V_0 \tag{7.32}$$

7.3 Size Effect on Mean Structural Strength

Based on the probability distribution function of the structural strength, the mean structural strength can be calculated as $\bar{\sigma}_N = \int_0^1 \sigma_N(P_f)\,\mathrm{d}P_f$. Since, by definition, the nominal strength is always positive, we can evaluate the mean strength as

$$\bar{\sigma}_N = \int_0^\infty \prod_{i=1}^{n_s} [1 - P_{1\sigma}(\sigma_N \langle s(\boldsymbol{x}_i)\rangle)]\,\mathrm{d}\sigma_N \tag{7.33}$$

where $P_{1\sigma}$ = cdf of the monotonic strength of one RVE, which is given by Eqs. 5.84a and 5.84b. By considering geometrically similar structures of different sizes, we obtain the size effect on the mean structural strength. Figure 7.4 presents the mean size effect curve together with the size-dependent probability density function of the structural strength, varying from the Gauss–Weibull grafted distribution at the small size to the Weibull distribution at the large size. It can be seen that, at the large-size limit, the

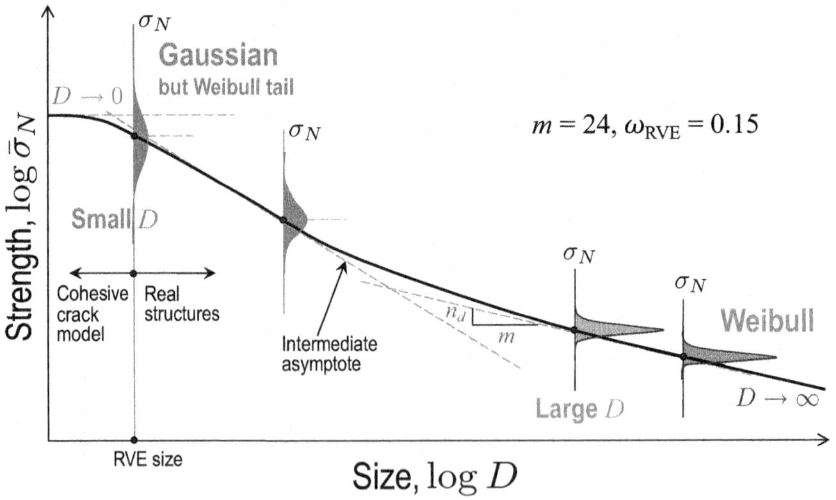

Figure 7.4. Size effect on the mean structural strength (type 1).

size effect curve approaches a straight line in the logarithmic plot. This agrees well with the classical power-law size effect of Weibull statistics because the cdf of strength approaches the Weibull distribution (Eq. 7.3) as the structure size increases. For small-size structures, the size effect curve deviates significantly from the straight line. This is due to the fact that the RVE size (or equivalently the size of material inhomogeneities) is not negligible compared to the structure size and therefore the finite weakest-link model predicts a non-Weibullian cdf of strength.

Based on Eqs. 5.84a and 5.84b, it is clear that the strength cdf of one RVE can be fully determined by four statistical parameters: ω_0 (coefficient of variation, CoV, of one RVE), P_{gr} (grafting probability), m (Weibull modulus), and s_0 (scale parameter of Weibull tail). Then the size effect on the mean strength must be characterized by these four parameters. Figure 7.5 presents the calculated mean size effect on structural strength

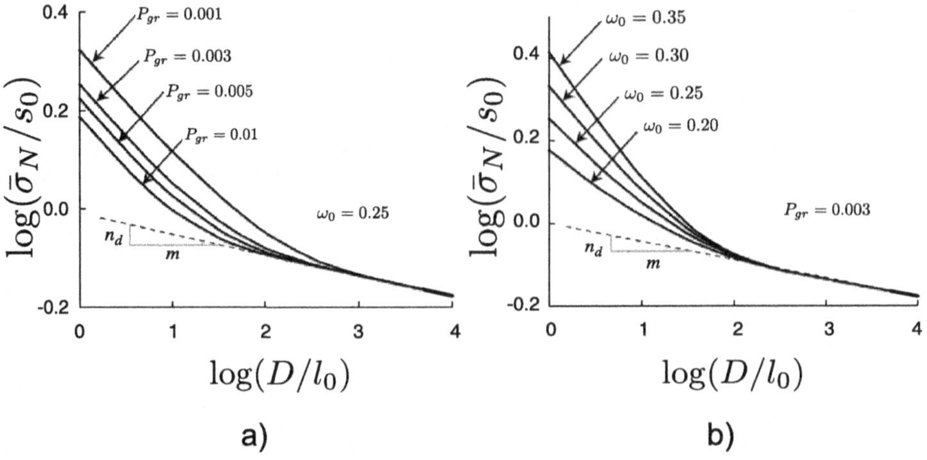

Figure 7.5. Mean size effect curves for (a) different grafting probabilities and (b) different CoVs.

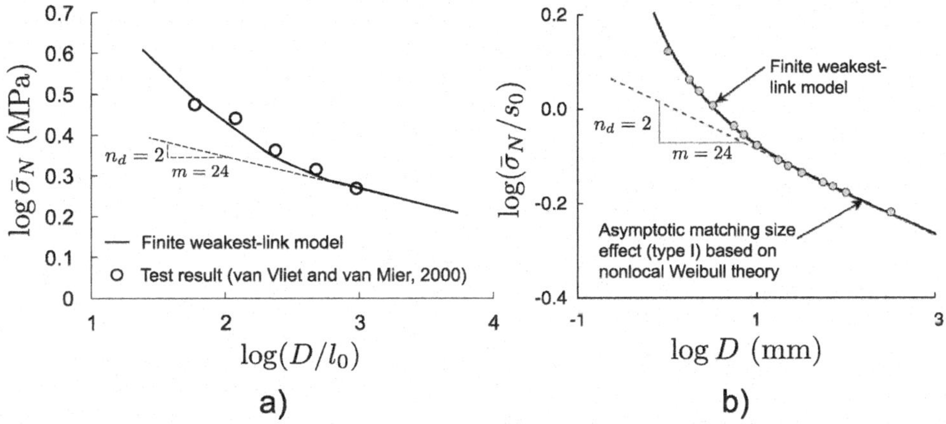

Figure 7.6. a) Optimum fix of mean size effect curve for uniaxial tensile concrete specimens and b) Optimum fix of finite-weakest link model by the type 1 size effect law.

for different grafting probabilities P_{gr} and different CoVs of one RVE, ω_0 (Bažant & Pang 2007).

Figure 7.6a shows the optimum fitting of the measured mean size effect on the strength of uniaxial tensile concrete specimens by the finite weakest-link model. The experiment involved a set of 2D geometrically similar tensile specimens with a size range of $1:16$ (van Vliet & van Mier 2000). The maximum aggregate size is about 8 mm. By considering the RVE size equal to three times the maximum aggregate size, it turns out that the largest specimen contains about 1600 RVEs. This indicates that, for the largest specimen size, the mean size effect curve would approach, but not fully reach, the Weibull asymptote, as shown in Figure 7.6a. Therefore, this measured size effect curve has to be explained by the finite weakest-link model, and it is seen that the present model agrees well with the test result. If the entire measured size effect curve were fitted by the classical Weibull size effect, one would obtain an erroneous value of the Weibull modulus ($m = 12$), which is much lower than the actual Weibull modulus of concrete ($m = 24$).

Figure 7.6b shows that the mean size effect calculated by the present theory matches very well the prediction by the nonlocal Weibull theory (Section 3.9) (Bažant & Novák 2000*b*, 2000*c*), which is an older theory (Bažant & Xi 1991) capable of describing the mean statistical size effect and the failure statistics of quasibrittle structures in the central range of the cdf, but not in the tail.

Recent studies have also shown that, for the small and intermediate size ranges, the size effect curve calculated by the finite weakest-link model also matches well with the deterministic calculation by the nonlocal damage model, crack band model, and cohesive crack model. This is because the mean size effect behavior for small-size and intermediate-size structures is caused mainly by the stress redistribution mechanism, which can be well predicted by the nonlinear deterministic calculation.

At the same time, this mechanism can also be captured by the finite weakest-link model, in which the statistical multiscale transition model used to formulate the cdf of RVE strength (as presented in Chapter 5) consists of statistical bundles and chains that statistically represent the damage localization and load redistribution mechanisms at different scales (only approximately, though, since only the elastic stresses are used). Nevertheless, when the structure size becomes too small, so small that a substantial stress redistribution would occur in one RVE, the statistical representation of stress and strain redistribution in the RVE by the chain and bundle models becomes insufficient (Le, Eliáš, & Bažant 2012). Therefore, the size effect calculated by the weakest-link model for structures of very small sizes starts to deviate from the deterministic calculation. For large-size structures, the zone of stress redistribution becomes negligible compared to the structure size and the size effect is caused mainly by the randomness of material strength leading to the Weibull size effect, which cannot be captured by the deterministic calculations.

With the grafted strength distribution of one RVE (Eqs. 5.84a and 5.84b), it is impossible to obtain an analytical expression for the mean strength $\bar{\sigma}_N$. But an approximate analytical formula has been obtained by asymptotic matching (Bažant & Novák 2000*a*; Bažant 2004*b*, 2005). It agrees well with the experiments on concrete (Rocco 1995) and on fiber composite (Bažant et al. 2006), and reads

$$\bar{\sigma}_N = \left[\frac{N_a}{D} + \left(\frac{N_b}{D}\right)^{n_d\psi/m}\right]^{1/\psi} \tag{7.34}$$

where N_a, N_b, ψ, and m are constants to be determined by asymptotic properties of the size effect curve, and n_d = number of spatial dimensions in which the structure is scaled, i.e., $n_d = 1, 2$. It is clear that Eq. 7.34 converges to the large-size asymptote $(N_b/D)^{n_d/m}$, which is a power law represented by a straight line of slope $-n_d/m$ in the size effect plot of $\log \sigma_N$ versus $\log D$.

The Weibull asymptote of strength cdf at the large-size limit indicates that the exponent, m, must be equal to the Weibull modulus of strength distribution, which represents the slope of the left tail of the strength histogram plotted in Weibull scale. The other three parameters, N_a, N_b, and ψ, can be determined by solving three simultaneous equations expressing three asymptotic conditions, $[\bar{\sigma}_N]_{D\to l_m}$, $[\mathrm{d}\bar{\sigma}_N/\mathrm{d}D]_{D\to l_m}$, and $[\bar{\sigma}_N D^{n_d/m}]_{D\to\infty}$, where l_m = minimum structure size for which the finite weakest-link model is applicable.

In contrast to the finite weakest-link model, some researchers proposed that the deviations from the classical Weibull statistics can be explained by size and shape dependence of the Weibull modulus, m. For example, Batdorf (1982) derived a failure envelope of unidirectional fiber composites relating the failure stress to the critical number of fiber breaks. Based on this envelope, he finds the curve of the size effect on failure stress to consist, on the log–log scale, of a sequence of linear segments of decreasing slopes, m, supposed to imply a size-dependent Weibull modulus. This interpretation is questionable because the fact that the Weibull modulus is obtained from the mean size effect curve implies the hypothesis that Weibull statistics applies. However, we have shown that the Weibull statistics applies only if the size of structure is sufficiently large.

The present finite weakest-link theory implies that Weibull modulus, m, is an intrinsic material property, determined by the multiscale transition from the nanoscale to the RVE

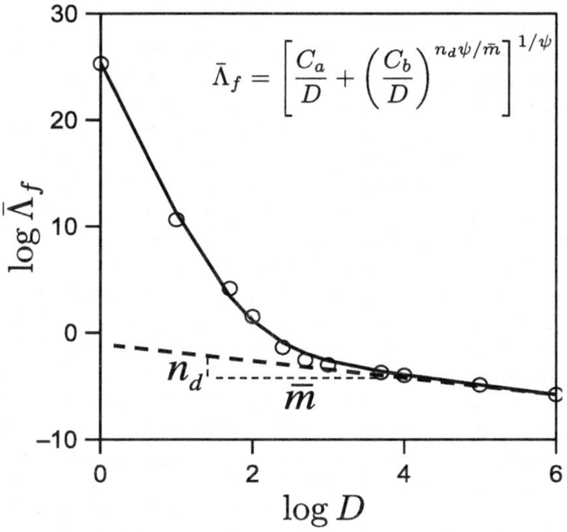

Figure 7.7. Size effect on mean structural lifetime.

scale. As discussed in Chapter 5, m is equal to the minimum of the sum of the power exponents of the Weibull tails among all the possible cuts separating the hierarchical model into two halves (e.g., Fig. 5.10).

On the macroscale, if the structure follows the weakest-link model, then the exponent of the power-law tail, which is equal to the Weibull modulus, remains unchanged.

7.4 Size Effects on Mean Structural Lifetimes and Stress-Life Curves

Similar to the aforementioned analysis of scaling of mean structural strength, we can also calculate the size effect on the creep and fatigue lifetimes. To this end, we consider geometrically similar specimens of different sizes and write

$$\text{Mean creep lifetime:} \quad \bar{\Lambda}_f = \bar{\lambda}\sigma_{n0}^{n_c} = \int_0^\infty \prod_{i=1}^{n_s} \{1 - P_{1\Lambda}[\langle s(\boldsymbol{x}_i)\rangle^{n_c}\Lambda]\}\mathrm{d}\Lambda \tag{7.35}$$

$$\text{Mean fatigue lifetime:} \quad \bar{\mathcal{N}}_f = \bar{N}_f \Delta\sigma_{n0}^{n_f} = \int_0^\infty \prod_{i=1}^{n_s} \{1 - P_{1\mathcal{N}}[\langle s(\boldsymbol{x}_i)\rangle^{n_f}\mathcal{N}_f]\}\mathrm{d}\mathcal{N}_f \tag{7.36}$$

Figure 7.7 shows a typical size effect on the mean creep lifetime calculated from the present model. From Figures 7.6 and 7.7, it is clear that the size effect curve of creep lifetime shares the same characteristics as the size effect curve of mean strength. Again, the size effect on the creep lifetime significantly deviates from the Weibull statistics for structures of small and intermediate sizes, due to the finite number of RVEs. Since the cdfs of creep and fatigue lifetimes have the same form, and since the power-law exponents of crack growth law under creep and fatigue loading are of similar order, it is expected that the size effect on the mean fatigue lifetime would be very similar to the size effect curve shown in Figure 7.7.

Similar to the size effect on the mean strength, it is impossible to obtain an analytical formula for the size effect on the mean lifetime. Nevertheless, it has been shown that the size effects on mean creep and fatigue lifetimes can reasonably be approximated by equations of a similar form as Eq. 7.34:

$$\bar{\Lambda}_f = \bar{\lambda}\sigma_{n0}^{n_c} = \left[\frac{C_a}{D} + \left(\frac{C_b}{D}\right)^{n_d\varphi/\bar{m}}\right]^{1/\varphi} \tag{7.37}$$

$$\bar{\mathcal{N}}_f = \bar{N}_f \Delta\sigma_{n0}^{n_f} = \left[\frac{C_p}{D} + \left(\frac{C_q}{D}\right)^{n_d\eta/\bar{m}_f}\right]^{1/\eta} \tag{7.38}$$

where $\bar{m}$, $\bar{m}_f$ = Weibull moduli of creep and fatigue lifetime cdfs. The constants C_a, C_b, and φ can be derived from three asymptotic conditions: $[\bar{\Lambda}_f]_{D\to l_m}$, $[\mathrm{d}\bar{\Lambda}_f/\mathrm{d}D]_{D\to l_m}$, and $[\bar{\Lambda}_f D^{(n_d/\bar{m})}]_{D\to\infty}$. Similarly, the constants C_p, C_q, and η ensue by matching the asymptotic conditions: $[\bar{\mathcal{N}}_f]_{D\to l_m}$, $[\mathrm{d}\bar{\mathcal{N}}_f/\mathrm{d}D]_{D\to l_m}$, and $[\bar{\mathcal{N}}_f D^{n_d/\bar{m}_f}]_{D\to\infty}$. It should be noted that C_a, C_b, C_p, and C_q depend on the structural geometry.

It is obvious that the size effect on the mean structural lifetime is much stronger than that on the mean strength. This is physically plausible. Consider two geometrically similar beams, with a size ratio 1:8. Let the nominal strength of the small beam be ς. Owing to the size effect on the mean strength, the nominal strength of the large beam could be as low as $\varsigma/2$ (Bažant 2005). If one applied nominal stress $\varsigma/2$ to both beams, the large beam would fail within standard laboratory testing period (i.e., about 5 minutes) while the small beam would be expected to survive at that load for decades.

Meanwhile, Eqs. 7.37 and 7.38 clearly indicate an inverse power-law form of the stress-life curves for a given structural geometry, under both creep and fatigue loading and various sizes. Figure 7.8a and b shows the recent experimental results on the stress-life curve of 99.9% Al_2O_3 and tetragonal zirconia polycrystal (TZP) under constant loads (Kawakubo 1995). Equation 7.37 can be used to predict the structural lifetime at low stress based on the structural lifetime at high stress. For concrete, n_c was estimated to be about 24 (Bažant & Prat 1988). Let f_t' denote the tensile strength of concrete obtained in a test lasting about 5 minutes. For applied stress $0.2f_t'$, Eq. 7.37 predicts the lifetime of the specimen to be about 5.7×10^{11} years, which is 40 times longer than the age of the universe (1.379×10^{10} years); for $0.5f_t$, Eq. 7.37 predicts about 160 years (thus the objection that there is no threshold stress is moot). The size effect on creep stress-life curve is demonstrated in Fig. 7.8c, which indicates that the stress-life curve shifts to the left horizontally as the structure size increases.

The inverse power law of the stress-life curve for fatigue loading agrees well with the well-known Basquin law (Basquin 1910) (commonly known as the "S–N" curve), which is supported by numerous test data on quasibrittle materials such as ceramics (Kawakubo 1995; Lee, Park, & Hahn 1995; Sakai & Hoshide 1995) and cortical bones (Turner, Wang, & Burr 2001). Similar to the size effect on the stress-life curve for creep loading, Eq. 7.38 implies that, on a bilogarithmic plot, the S–N curve must shift horizontally to the left as the structure size increases.

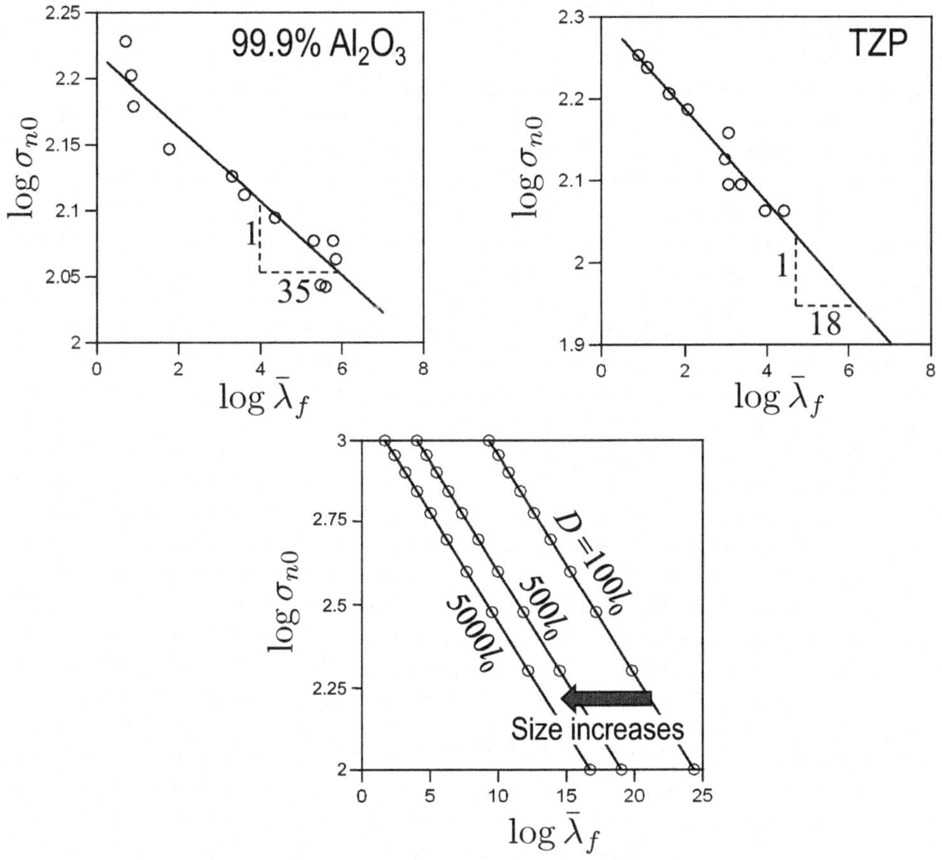

Figure 7.8. Experimental measured stress-life curves of engineering ceramics.

Figure 7.9 shows the experimentally measured S-N curves of sintered SiC under three point bending and uniaxial tension, fitted by Eq. 7.38 (Okabe & Hirata 1995). Owing to the difference in stress distribution, for the same specimen, the equivalent size of the specimen under three point bending is much smaller than it is under uniaxial tension. As seen in Fig. 7.9, there is a marked size effect on the S-N curve. For the same applied nominal stress amplitude, the lifetime of the specimen under three point bending is about six orders of magnitude longer than the lifetime of the specimen under uniaxial tension.

It should nevertheless be emphasized that the stress-life curves are here the consequence of the power laws for creep and fatigue crack growth. As discussed in Chapter 6, these power laws can be justified physically by assuming self-similarity of the function that relates the number microcracks to the applied stress (for creep loading) or applied stress amplitude (for fatigue loading). Nonetheless, for very small or very large constant stress and stress amplitudes, self-similarity may not be valid, and one may expect the stress-life curves to deviate from the power-law form (Suresh 1998).

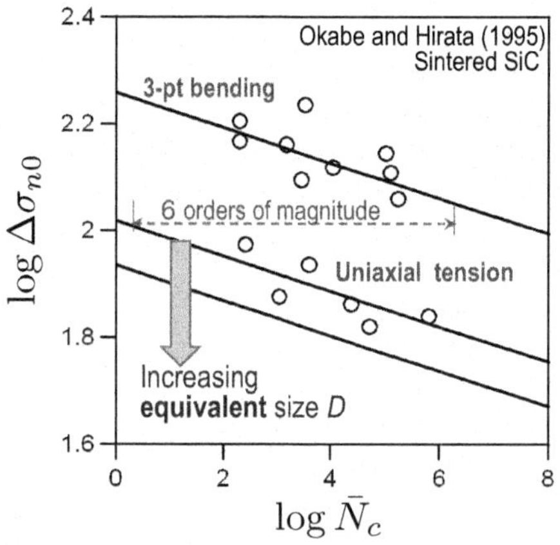

Figure 7.9. Size effect on the S-N curve.

Another noteworthy point is that the slopes of the stress-life curves are independent of structure size. The underlying reason is that these slopes are determined by the exponent of the creep and fatigue crack growth laws at the level of one RVE, which agrees well with the test data shown as Fig. 7.9. Unfortunately, almost no experimental data seem to exist on the scaling of the stress-life curve for creep loading.

7.5 Effect of Temperature on Strength and Lifetime Distributions

Since the present theory is based on thermally activated nanoscale processes, it is justified to use the Arrhenius factor for the temperature dependence of strength and lifetime distributions. This dependence naturally arises from the temperature dependence of the crack growth rate on the nanoscale (Eqs. 6.3 and 6.5), which is transferred over the scales without any change to the crack growth rate on the macroscale. The Arrhenius type of temperature effect on the growth rate of macrocracks in concrete has been shown to lead to the correct temperature dependence of the fracture energy of concrete (Bažant & Prat 1988).

Generally, many activation energy barriers exist on the surface of the free energy potential of a nanoelement (Krausz & Krausz 1988). Various extraneous factors, such as temperature range and corrosive agents, influence the dominant activation energy barrier, Q_0. This barrier, Q_0, may be assumed to dominate over the other smaller barriers, and to be approximately independent of temperature, T, within a certain range. This simplification allows avoiding the details of the transition rate theory, with its changing activation energy barriers.

Consider one RVE at two different temperatures T_1 and T_2, and assume that this temperature difference causes no change in Q_0 as well as no chemical reactions that would change the material properties. Based on Eq. 7.5, the strengths of the same RVE at these two temperatures are related as

$$\sigma_2 = \sigma_1 \exp\left[\frac{Q_0}{(n_c+1)k}\left(\frac{1}{T_2} - \frac{1}{T_1}\right)\right] \tag{7.39}$$

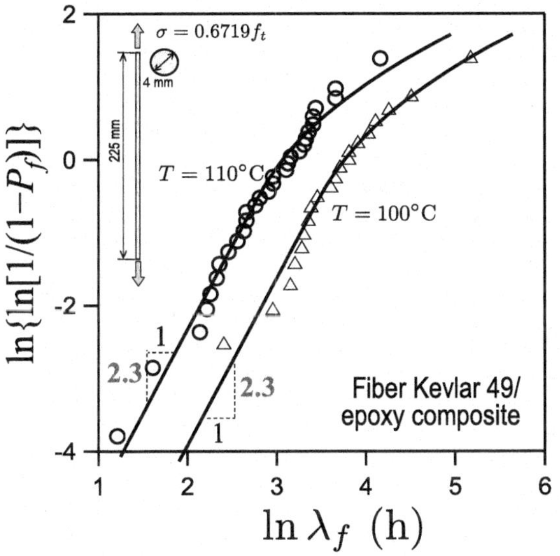

Figure 7.10. Optimum fits of lifetime distributions of Kevlar-49 fiber composites at elevated temperatures.

Equation 7.39 makes it possible to predict the cdf of structure strength $P_{f,T_2}(\sigma_N)$ at temperature T_2 if the cdf $P_{f,T_1}(\sigma_N)$ at temperature T_1 is known:

$$P_{f,T_2}(\sigma_N) = P_{f,T_1}\left(C_1^{-1}\sigma_N\right) \tag{7.40}$$

$$\text{where} \quad C_1 = \exp\left[\frac{Q_0}{(n_c+1)k}\left(\frac{1}{T_2} - \frac{1}{T_1}\right)\right] \tag{7.41}$$

Similar analysis also applies to the creep lifetime distribution. Based on Eq. 7.7, the cdf of structural lifetime $P_{f,T_2}(\Lambda_f)$ at temperature T_2 can be deduced from the lifetime cdf $P_{f,T_1}(\Lambda_f)$ at temperature T_1:

$$P_{f,T_2}(\Lambda_f) = P_{f,T_1}\left(C_2^{-1}\Lambda_f\right) \tag{7.42}$$

$$\text{where} \quad C_2 = \exp\left[\frac{Q_0}{k}\left(\frac{1}{T_2} - \frac{1}{T_1}\right)\right] \tag{7.43}$$

In the Weibull scale, the strength cdf or lifetime cdf at temperature T_2 can simply be obtained through a horizontal shift of the strength or lifetime cdf at temperature T_1 by the distance of $\ln C_1$ or $\ln C_2$, respectively.

Chiao, Sherry, and Hetherington (1977) investigated the lifetime histograms of organic fiber (Kevlar 49) composites at elevated temperatures (100°C and 110°C). In the experiment, bar specimens were subjected to a constant uniform tensile stress equal to 67% of the mean tensile strength. Here the lifetime histogram at 100°C is first fitted by the present finite weakest-link model, and the lifetime histogram at 110°C is then obtained by extrapolating the calibrated lifetime cdf for 100°C, based on Eq. 7.42.

Figure 7.10 shows that the horizontal shift predicted by the present theory yields excellent fits of both histograms. These fits show that the dominant activation energy barrier of this organic fiber composite in the temperature range 100 to 110°C is $Q_0 =$ 0.79 eV (and $Q_0 = 9225$ K).

The temperature dependence of strength and lifetime cdfs thus provides an effective way to determine the dominant Q_0. Equation 7.39 indicates not only the effect of temperature on the random strength of one RVE, but also the temperature dependence of the mean strength of structure. If the mean structural strengths, $\bar{\sigma}_1$ and $\bar{\sigma}_2$, are measured at two different temperatures T_1 and T_2, and if the temperature difference is not so large as to cause a change of Q_0, then

$$Q_0 = (n_c + 1)k \ln\left(\frac{\bar{\sigma}_2}{\bar{\sigma}_1}\right)\left(\frac{1}{T_2} - \frac{1}{T_1}\right)^{-1} \tag{7.44}$$

where the exponent n_c of the power law for creep crack growth rate can be obtained by the standard test measuring the crack growth velocity.

Alternatively, if the mean structural lifetimes $\bar{\lambda}_1$ and $\bar{\lambda}_2$ are measured at two different temperatures and at the same constant loading, then the dominant activation energy barrier can be determined without knowing n_c, i.e.,

$$Q_0 = k \ln\left(\frac{\bar{\lambda}_2}{\bar{\lambda}_1}\right)\left(\frac{1}{T_2} - \frac{1}{T_1}\right)^{-1} \tag{7.45}$$

If test data are available for both the mean structural strengths and the lifetimes at two different temperatures, then, based on Eqs. 7.44 and 7.45, one can determine both the dominant activation energy barrier Q_0 and the exponent n_c of the power law for creep crack growth rate.

Here we expect the temperature dependence of the fatigue lifetime to be the same as that of the creep lifetime. For a certain range of temperature changes (without changing the dominant activation energy barrier), the cdfs of fatigue lifetimes at two different temperatures can be written as

$$P_{f,T_2}(\mathcal{N}_f) = P_{f,T_1}(C_2^{-1}\mathcal{N}_f) \tag{7.46}$$

and, for the same loading history, the mean fatigue lifetimes at these two temperatures are simply related by

$$\bar{N}_f \mid_{T_2} = C_2 \bar{N}_f \mid_{T_1} \tag{7.47}$$

8 Computation of Probability Distributions of Structural Strength and Lifetime

Chapters 5 to 7 demonstrated that the grafted Gauss–Weibull distribution of the strength of one material representative volume element (RVE) leads to intricate size effects on the strength and lifetime distributions of quasibrittle structures. According to the finite weakest-link model of failure statistics of the entire structure, such a grafted distribution also causes a size effect on the shape of the probability distributions of both the structural strength and lifetime. Compared to the classical Weibull statistics, the grafted distribution of RVE strength and lifetime complicates the computation of the cumulative distribution functions (cdfs) of structural strength and lifetime, making closed-form solutions impossible. For computer simulations, the finite weakest-link model requires the structure to be subdivided into elements equal to the RVE size. However, for structures of general geometries, the subdivision is usually nonunique and subjective. This may lead to an inconsistent estimation of the failure probability.

This chapter presents three general methods for calculating the strength and lifetime distributions of quasibrittle structures. These methods are used to validate the finite weakest-link theory through optimum fitting of strength and lifetime histograms of structures that are made of quasibrittle materials, such as fiber composite or engineering and dental ceramics.

8.1 Nonlocal Boundary Layer Model for Strength and Lifetime Distributions

The original nonlocal models in continuum mechanics (Eringen 1966, 1972; Kröner 1966, 1967; Eringen & Edelen 1972) captured the effect of material heterogeneity on elastic and hardening plastic behaviors. For strain localization in strain-softening materials and generally for deterministic modeling of quasibrittle fracture, the nonlocal continuum concept was proposed by Bažant et al. (1984) and by Pijaudier-Cabot & Bažant (1987), and then improved in Bažant & Pijaudier-Cabot (1988); see also Bažant & Jirásek (2002). Later the nonlocal continuum concept was generalized to probabilistic modeling (Bažant & Novák 2000*b*, 2000*c*). In such a generalization, the failure probability at a particular material point depends not only on the local stress but also on the stress in its neighborhood of a size approximately equal to the RVE size l_0. In deterministic calculations, the nonlocal concept is necessary for regularizing the boundary value problem with strain-softening distributed damage and ensuring the convergence of fracture energy dissipation. This prevents spurious localization converging to failure at zero energy dissipation, and avoids spurious mesh sensitivity. In stochastic calculations, nonlocal averaging provides a convenient way to introduce a spatial correlation (Breysse & Fokwa 1992; Bažant & Novák 2000*b*).

We first demonstrate the nonlocal boundary layer model for strength distribution. Its extension to lifetime distributions is then straightforward. To avoid the subjective discrete subdivision of structures into RVEs, we may replace the finite sum in the logarithmic form of Eq. 7.2 with a continuous integral. With the nonlocal concept, the weakest-link model can then be written as (Bažant & Novák 2000*b*, 2000*c*):

$$\ln(1-P_f) = \int_V \ln\{1 - P_{1\sigma}[\overline{\sigma}(\boldsymbol{x})]\}\frac{\mathrm{d}V(\boldsymbol{x})}{V_0} \tag{8.1}$$

where $V_0 = l_0^{n_d}$; n_d = number of spatial dimensions in which the structure is to be scaled; l_0 = RVE size and $\overline{\sigma}(\boldsymbol{x})$ is the nonlocal stress, which can be calculated as $\overline{\sigma} = E\overline{\epsilon}$; and E is the elastic modulus (Bažant & Novák 2000*b*). The nonlocal strain $\overline{\epsilon}$ can be expressed as

$$\overline{\epsilon}(\boldsymbol{x}) = \frac{1}{\overline{\alpha}(\boldsymbol{x})}\int_V \alpha(\boldsymbol{x}'-\boldsymbol{x})\epsilon(\boldsymbol{x}')\,\mathrm{d}\boldsymbol{x}' \tag{8.2}$$

$$\text{and:}\quad \overline{\alpha}(\boldsymbol{x}) = \int_V \alpha(\boldsymbol{x}'-\boldsymbol{x})\,\mathrm{d}\boldsymbol{x}' \tag{8.3}$$

There are many choices for an appropriate weighting function $\alpha(\boldsymbol{x}'-\boldsymbol{x})$, such as Gaussian function, polynomials, and so on. Although the results do not strongly depend on the choice (Bažant & Novák 2000*b*), from the perspective of numerical convergence it is better to have a smooth weighting function (Bažant & Cedolin 1991). Here a smooth fourth-order polynomial weighting function is considered:

$$\alpha(\boldsymbol{x}'-\boldsymbol{x}) = \left(1 - \frac{||\boldsymbol{x}'-\boldsymbol{x}||^2}{\rho^2 l_0^2}\right)^2 \qquad (||\boldsymbol{x}'-\boldsymbol{x}|| \le \rho l_0) \tag{8.4}$$

$$\alpha(\boldsymbol{x}'-\boldsymbol{x}) = 0 \qquad (||\boldsymbol{x}'-\boldsymbol{x}|| \ge \rho l_0) \tag{8.5}$$

where ρ is a coefficient selected so that the volume under $\alpha(\boldsymbol{x}'-\boldsymbol{x})$ would be the same as that under $\alpha = 1$ for $||\boldsymbol{x}'-\boldsymbol{x}|| \le l_0/2$ and $\alpha = 0$ for $||\boldsymbol{x}'-\boldsymbol{x}|| \ge l_0/2$. Thus ρ is equal to $15/16$ for one dimension, $\sqrt{3/4}$ for two dimensions, and $(105/192)^{1/3}$ for three dimensions (Bažant & Cedolin 1991). For points whose nonlocal influence zone is completely contained within the structural domain, this nonlocal concept can be applied without any change. However, a major question, which still persists for all deterministic nonlocal models, is how to treat the weighting function $\alpha(\boldsymbol{x}'-\boldsymbol{x})$ when it protrudes through the structure boundary.

To get around this problem, a nonlocal boundary layer (NBL) model was proposed (Bažant, Le, & Hoover 2010). In this model, the structure is divided into two domains: a boundary layer V_b and an interior domain V_I; see Fig. 8.1. The boundary layer, which completely surrounds the interior domain, has a thickness equal to the RVE size l_0. In the interior domain, we can use the conventional nonlocal model to evaluate the failure probability of each local material point (Bažant & Jirásek 2002; Bažant & Novák

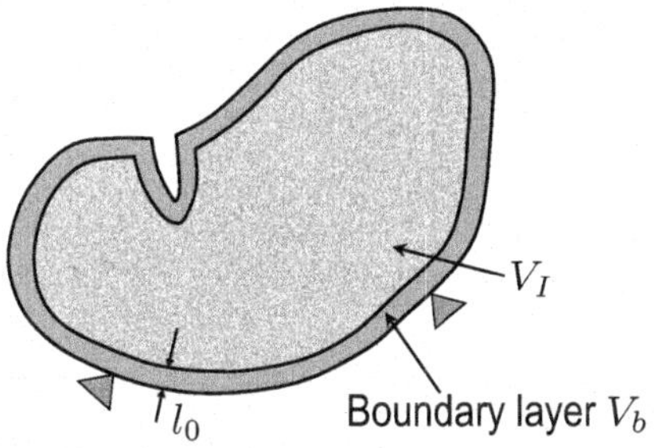

Figure 8.1. Concept of nonlocal boundary layer model.

2000*b*). Since the boundary layer thickness is equal to l_0, the nonlocal zone for the points of the interior domain will not protrude outside of the structure boundary.

Inside the boundary layer, the continuum stress used to determine the failure probability P_f is calculated from the deformation averaged over the thickness of the boundary layer l_0. Usually the (randomly scattered) strain profile across the boundary layer is statistically almost linear, and then the averaged strain can be taken as the continuum strain at the middle surface Ω_M of the boundary layer. Therefore, the failure probability, as determined by the weakest-link model, Eq. 8.1, can be rewritten as

$$\ln[1 - P_f(\sigma_N)] = l_0 \int_{\Omega_M} \ln\{1 - P_{1\sigma}[\sigma(\boldsymbol{x}_M)]\} \frac{d\Omega(\boldsymbol{x}_M)}{V_0} + \int_{V_I} \ln\{1 - P_{1\sigma}[\overline{\sigma}(\boldsymbol{x})]\} \frac{dV(\boldsymbol{x})}{V_0} \tag{8.6}$$

When the structure size increases, the thickness of the boundary layer becomes negligible when compared to the structure size. Thus, the nonlocal stress at a point approaches the local stress, and Eq. 8.6 converges to Eq. 7.1.

Although the stress in the boundary layer is treated as local, it is actually a nonlocal stress because it corresponds to the average strain in the boundary layer, which is a feature encompassing spatial correlation. The difference from the nonlocal stress at the points of the interior domain is that it is defined only for the boundary layer thickness as a whole and not for an arbitrary generic point across the thickness. Here it may be noted that the nonlocal treatment of the stress in the boundary layer has also been advocated for the deterministic nonlocal models, to introduce diminishing nonlocality as the structural boundary is approached (Krayani, Pijaudier-Cabot, & Dufour 2009).

Consider a simple example of a rectangular beam, with a span-to-depth ratio of 4, subjected to pure bending. To estimate the stress and strain fields, the engineering beam theory is used. The nominal strength of the beam can be defined as the modulus of rupture $\sigma_N = 6M_{\max}/bD^2$ where $M_{\max}$ is the maximum bending moment in the beam, D is the depth of the beam, and b is the thickness of the beam. Five methods are used to calculate the probability distribution of σ_N:

1. Following the weakest-link model, the beam is subdivided into a whole number of RVEs and the failure probability is calculated according to Eq. 7.1. The calculation is based solely on the underlying weakest-link model, and is considered as the benchmark solution.

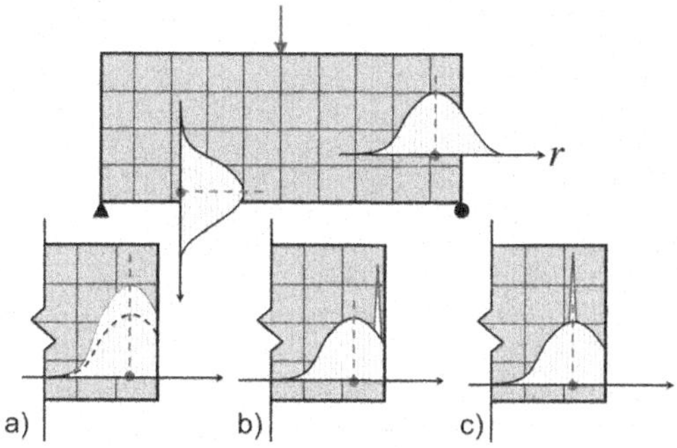

Figure 8.2. Different treatments of nonlocal weighting function at the structure boundary: a) uniform scaling; b) Dirac-delta function placed at boundary, and c) Dirac-delta function placed at the point of interest.

2. A nonlocal approach using Eqs. 8.1 to 8.3, in which the weighting function part protruding through the boundary is deleted and the remaining part is scaled so that its volume be equal to 1 (Fig. 8.2a).
3. The aforementioned NBL model, using Eqs. 8.2 to 8.6.
4. A nonlocal approach in which the protruding volume of the weighting function is lumped as a Dirac delta function at the structure boundary (Fig. 8.2b). Equation 8.2 can be rewritten to express this as

$$\bar{\epsilon}(\boldsymbol{x}) = \frac{1}{\bar{\alpha}_\infty}\left[\int_V \alpha(\boldsymbol{x}' - \boldsymbol{x})\epsilon(\boldsymbol{x}')\,\mathrm{d}\boldsymbol{x}' + \gamma(\boldsymbol{x})\int_\Gamma \epsilon(\boldsymbol{y})\,\mathrm{d}\boldsymbol{y}\right] \tag{8.7}$$

where Γ denotes the boundary of the structure through which the nonlocal domain protrudes, $\gamma(\boldsymbol{x}) = \bar{\alpha}_\infty(\boldsymbol{x}) - \bar{\alpha}(\boldsymbol{x})$, and $\bar{\alpha}_\infty(\boldsymbol{x})$ is the volume of the weighting function for the points far away from the boundary.
5. A nonlocal approach according to Borino et al. (2003), where the protruding volume of the weighting function is lumped at the central coordinate $\boldsymbol{x}$ of the weighting function (Fig. 8.2c). Using this concept, Eq. 8.2 can be rewritten as

$$\bar{\epsilon}(\boldsymbol{x}) = \left(1 - \frac{\bar{\alpha}(\boldsymbol{x})}{\bar{\alpha}_\infty}\right)\epsilon(\boldsymbol{x}) + \frac{1}{\bar{\alpha}_\infty}\int_V \alpha(\boldsymbol{x}' - \boldsymbol{x})\epsilon(\boldsymbol{x}')\,\mathrm{d}\boldsymbol{x}' \tag{8.8}$$

The second term in Eq. 8.8 is the same as Eq. 8.2, whereas the first term emphasizes only the local strain at point $\boldsymbol{x}$. For points $\boldsymbol{x}$ far away from the boundary, $\bar{\alpha}(\boldsymbol{x})$ approaches a constant value equal to $\bar{\alpha}_\infty$. Thus the local term contributes to Eq. 8.8 only if the weighting function protrudes the structural domain. When the point of interest $\boldsymbol{x}$ is far away from the boundary, the local term vanishes and Eq. 8.8 converges to Eq. 8.2.

The five aforementioned methods are now used to calculate the mean flexural strength of the sample beam. Figure 8.3 compares the mean size effect curves simulated by these methods. The NBL model delivers the closest agreement with the benchmark

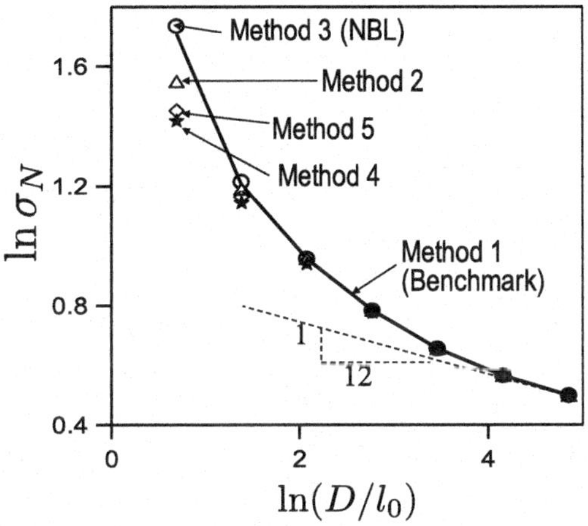

Figure 8.3. Mean size effect curves simulated by the weakest-link model and different nonlocal models.

solution for all beam sizes. The other three methods considerably underestimate the mean strength for the smaller size beams.

As discussed in Chapter 7, the nonlocal concepts, including their present form with the boundary layer, lose physical meaning for very small structures for which the RVE size or the boundary layer exceeds about one quarter of the cross-sectional dimension. In such cases, it is better to use the random discrete lattice-particle models. Note, however, that the geometrical randomness of discrete models is insufficient to simulate the size effect. To that end, one must include an autocorrelated random field of local strength, with the autocorrelation length equal to the RVE size (Carmeliet & de Borst 1995; Vořechovský & Sadílek 2008; Grassl & Bažant 2009).

As the structure size approaches infinity, the volume of the boundary layer V_b becomes negligible as compared to the total structure volume. In the other nonlocal models, the effects of the boundary treatment are also minimal for larger sizes because the nonlocal strain at a point approaches the local strain. Therefore, all models converge to the same value for larger sizes.

The preceding framework can readily be extended to the lifetime statistics. In this case, the NBL model can simply be written as

$$\text{Creep lifetime: } \ln[1 - P_f(\Lambda_f)] = \frac{l_0}{V_0} \int_{\Omega_M} \ln\{1 - P_{1\Lambda}[\langle s(\boldsymbol{x}_M)\rangle^{n_c} \Lambda_f]\}\, \mathrm{d}\Omega + \int_{V_I} \ln\{1 - P_{1\Lambda}\{\langle \bar{s}(\boldsymbol{x})\rangle^{n_c} \Lambda_f\}\} \frac{\mathrm{d}V(\boldsymbol{x})}{V_0} \tag{8.9}$$

$$\text{Fatigue lifetime: } \ln[1 - P_f(\mathcal{N}_f)] = \frac{l_0}{V_0} \int_{\Omega_M} \ln\{1 - P_{1\mathcal{N}_f}[\langle s(\boldsymbol{x}_M)\rangle^{n_f} \mathcal{N}_f]\}\, \mathrm{d}\Omega + \int_{V_I} \ln\{1 - P_{1\mathcal{N}_f}\{\langle \bar{s}(\boldsymbol{x})\rangle^{n_f} \mathcal{N}_f\}\} \frac{\mathrm{d}V(\boldsymbol{x})}{V_0} \tag{8.10}$$

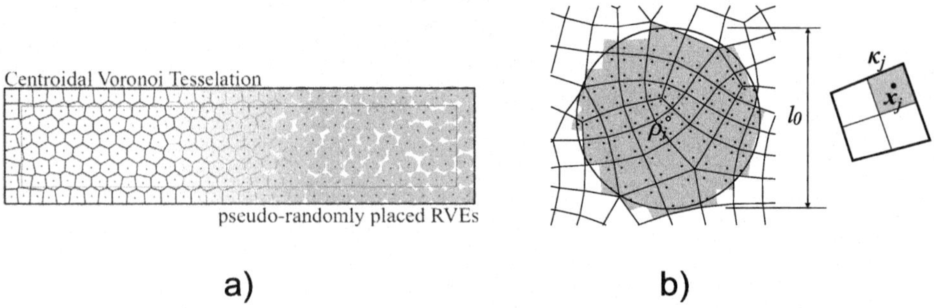

Figure 8.4. Pseudo-random placing of RVEs. (a) Example of centroidal Voronoi tessellation and imaginary locations of RVEs, and (b) stress averaging of the ith RVE volume.

8.2 Computation by Pseudo-random Placing of RVEs

Besides the aforementioned NBL model, another way to avoid the subjective division of structures into RVEs is through a general numerical scheme based on the pseudo-random placing of the RVEs, referred to as the Direct RVE Placing Method. This method directly corresponds to the original weakest-link model.

There are many possible ways of placing the RVEs in the structure, and each of them could yield a different strength cdf. Therefore, for statistically isotropic materials, it is preferable to make repeated pseudo-random choices of the RVE locations, and then average the results of all the realizations. The main advantage of pseudo-random placing of the RVEs is that it eliminates the directional bias, which inevitably arises with a regular placing.

For small-size structures where the RVE size is comparable to the structure size, one would need to generate many realizations of the random placing of RVEs. For large-size structures where the RVE size is negligible compared to the structure size, the RVE converges to a point in the structure and the random placing of RVEs would have a negligible effect on the resulting strength cdf. Therefore, fewer realizations are needed for large-size structures.

Here we employ the centroidal Voronoi tessellation to generate a set of random locations of n_r points, where $n_r = V/V_0$. Each of these points represents the center of one RVE. This tessellation produces a set of points corresponding to the centroids of the Voronoi cells, which satisfy two essential requirements:

1. The distance between any two adjacent points is approximately equal to the RVE size l_0.
2. The minimum distance of any point from the structure boundary is close to $l_0/2$ (Fig. 8.4a).

Once the centers of the RVEs are fixed, the average stress for each RVE can be obtained based on the stress field calculated by the standard finite element method. The RVE center $\boldsymbol{\rho}$ generally does not coincide with the integration point of the finite elements, $\boldsymbol{x}$. So the nonlocal stress for each RVE is calculated by averaging the stresses at the integration points that are enclosed within the RVE (see Fig. 8.4b), i.e.,

$$\bar{\sigma}(\boldsymbol{\rho}_i) = \frac{\sum_{j=1}^{n_r} \sigma(\boldsymbol{x}_j)\alpha(\boldsymbol{x}_j - \boldsymbol{\rho}_i)\kappa_j}{\sum_{j=1}^{n_s} \alpha(\boldsymbol{x}_j - \boldsymbol{\rho}_i)\kappa_j} \tag{8.11}$$

where $\alpha(\boldsymbol{x}_j - \boldsymbol{\rho}_i) =$ weighting function and $\kappa_j =$ the volume associated with integration point $\boldsymbol{x}_j$. To compute the average stress in one RVE, we consider only the integration points that lie within that RVE. Therefore, the function $\alpha(x)$ must have a cutoff that excludes all the integration points outside the RVE. It is clear that the weighting function $\alpha(x)$ could protrude the structural boundary for RVEs along the boundary. Here we choose a uniform weighting function for the RVE along the structural boundary:

$$\alpha_b(\boldsymbol{x}' - \boldsymbol{x}) = \begin{cases} 1/V_0 & (||\boldsymbol{x}' - \boldsymbol{x}|| \leq l_0/2) \\ 0 & (||\boldsymbol{x}' - \boldsymbol{x}|| \geq l_0/2) \end{cases} \tag{8.12}$$

By combining the weakest-link model with the centroidal Voronoi tessellation, the cdfs of structural strength and lifetime can be calculated as follows:

$$\text{Strength:} \qquad \ln(1 - P_f) = \sum_{i=1}^{n_r} \ln\{1 - P_{1\sigma}[\bar{\sigma}(\boldsymbol{\rho}_i)]\} \tag{8.13}$$

$$\text{Creep lifetime:} \qquad \ln(1 - P_f) = \sum_{i=1}^{n_r} \ln\{1 - P_{1\Lambda}[\langle \bar{s}(\boldsymbol{\rho}_i)\rangle^{n_c} \Lambda_f]\} \tag{8.14}$$

$$\text{Fatigue lifetime:} \qquad \ln(1 - P_f) = \sum_{i=1}^{n_r} \ln\{1 - P_{1\mathcal{N}}[\langle \bar{s}(\boldsymbol{\rho}_i)\rangle^{n_f} \mathcal{N}_f]\} \tag{8.15}$$

where $\bar{s}(\boldsymbol{\rho}_i) =$ a dimensionless stress field. Note that this approach resembles the Monte Carlo integration of the original weakest-link model with a fixed number n_r of integration points [e.g., Dimov (2008)], although the present method of placing the integration points is more complicated.

In the centroidal Voronoi tessellation, there is always a layer of RVEs placed along the structure boundary (Fig. 8.4a), which is consistent with the concept of the NBL model. One difference, though, is that the NBL model uses the stress at the middle surface of the layer while the current approach uses the average stress across the RVE. As a result, in the NBL model, switching between the cubical RVE shape, $V_0 = l_0^3$ (squares $V_0 = l_0^2$) and the spherical shape, $V_0 = 4\pi l_0^3/3$ (circles $V_0 = \pi l_0^2/4$), has no effect, while in the direct RVE placing method, the RVE shape would affect the number n_r of the RVEs and consequently their mutual distances as well as the minimal distance to the structure boundary.

Consider the case of using spherical RVEs. Since an assembly of non-overlapping spheres would occupy a volume larger than the structure volume V, the spheres must overlap each other to fit this assembly into the structure volume V. Thus the average mutual distance between the RVE centers must be slightly less than l_0. This feature also affects the distance between the centers of circumferential RVEs and the structure boundary, which would again be slightly less than $l_0/2$. Therefore, in general, the random RVE placing method would predict slightly higher failure probabilities than the NBL method does.

8.3 Approximate Closed-Form Expression for Strength and Lifetime Distributions

So far we presented two general numerical schemes that allow us to calculate the failure statistics of quasibrittle structures of any geometry. Nevertheless, it is desirable to obtain some closed-form solutions even if they are limited to certain special cases. In Chapter 7, we have shown that, as the structure size becomes very large, the strength and lifetime cdfs follow the classical two-parameter Weibull distribution with closed-form expressions, while the effects of structure size and geometry are accounted for by the so-called equivalent number of RVEs, N_{eq}. However, for small- and intermediate-size structures, N_{eq} would explicitly depend on the magnitude of the applied stress, which makes it impossible to obtain a closed-form expression. This section presents a simple method called the Taylor expansion method, which leads to approximate closed-form expressions for strength and lifetime cdfs for structures having some simple stress fields. We first present this method for the strength cdf and then extend it to the cdfs of creep and fatigue lifetime.

As discussed in Chapter 5, the cdf of strength of one RVE is separated into two parts: a Weibull tail (Eq. 5.84a) and a Gaussian core (Eq. 5.84b). Within the framework of the NBL model, we can likewise subdivide both the boundary layer Ω_M and the interior part V_I into two parts:

1. The Weibullian region, where the principal stress is less than the grafting stress, causing the failure probability to be governed by the Weibull tail, Eq. 5.84a.
2. The Gaussian region, where the principal stress is larger than the grafting stress, causing the failure probability to be governed by the Gaussian core, Eq. 5.84b.

Figure 8.5 shows such a division for beams under three point bending and pure bending. Based on such a subdivision, Eq. 8.6 may then be rewritten as follows (Le et al. 2012):

$$\ln(1-P_f) = \underbrace{\int_{V_{\sigma \mathrm{W}}} \ln[1-P_{\sigma \mathrm{W}}(\bar{\sigma}(\boldsymbol{x}))]\frac{\mathrm{d}V(\boldsymbol{x})}{V_0}}_{I_{V_{\sigma \mathrm{W}}}} + \underbrace{\int_{V_{\sigma \mathrm{G}}} \ln[1-P_{\sigma \mathrm{G}}(\bar{\sigma}(\boldsymbol{x}))]\frac{\mathrm{d}V(\boldsymbol{x})}{V_0}}_{I_{V_{\sigma \mathrm{G}}}}$$
$$+ \underbrace{l_0\int_{\Omega_{\sigma \mathrm{W}}} \ln[1-P_{\sigma \mathrm{W}}(\sigma(\boldsymbol{x}))]\frac{\mathrm{d}\Omega(\boldsymbol{x})}{V_0}}_{I_{\Omega_{\sigma \mathrm{W}}}} + \underbrace{l_0\int_{\Omega_{\sigma \mathrm{G}}} \ln[1-P_{\sigma \mathrm{G}}(\sigma(\boldsymbol{x}))]\frac{\mathrm{d}\Omega(\boldsymbol{x})}{V_0}}_{I_{\Omega_{\sigma \mathrm{G}}}} \quad (8.16)$$

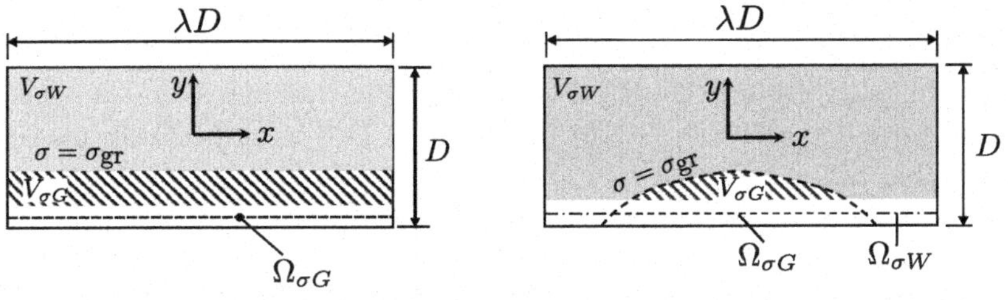

Figure 8.5. Division of the stress field into Weibull and Gaussian regions for (a) beam under pure bending and (b) beam under three point bending.

where $P_{\sigma \mathrm{W}}$ = Weibullian part of the strength cdf of one RVE (Eq. 5.84a), $P_{\sigma \mathrm{G}}$ = Gaussian part of the strength cdf of one RVE (Eq. 5.84b), $V_{\sigma \mathrm{W}}$ = Weibullian region of the interior part of structure $= \{\boldsymbol{x} \,|\, \boldsymbol{x} \in V_I \,\wedge\, \bar{\sigma}(\boldsymbol{x}) \le \sigma_{\mathrm{gr}}\}$; $V_{\sigma \mathrm{G}}$ = Gaussian region of the interior part of structure $= \{\boldsymbol{x} \,|\, \boldsymbol{x} \in V_I \,\wedge\, \bar{\sigma}(\boldsymbol{x}) > \sigma_{\mathrm{gr}}\}$; $\Omega_{\sigma \mathrm{W}}$ = Weibullian region of the boundary layer $= \{\boldsymbol{x} \,|\, \boldsymbol{x} \in \Omega_M \,\wedge\, \sigma(\boldsymbol{x}) \le \sigma_{\mathrm{gr}}\}$; and $\Omega_{\sigma \mathrm{G}}$ = Gaussian part of the boundary layer $= \{\boldsymbol{x} \,|\, \boldsymbol{x} \in \Omega_M \,\wedge\, \sigma(\boldsymbol{x}) > \sigma_{\mathrm{gr}}\}$. Clearly, we have $V_{\sigma \mathrm{W}} \cap V_{\sigma \mathrm{G}} = \emptyset$, $V_{\sigma \mathrm{W}} \cup V_{\sigma \mathrm{G}} = V_I$, $\Omega_{\sigma \mathrm{W}} \cap \Omega_{\sigma \mathrm{G}} = \emptyset$ and $\Omega_{\sigma \mathrm{W}} \cup \Omega_{\sigma \mathrm{G}} = \Omega_M$.

The integrals for the Weibullian region can easily be computed by using the concept of the equivalent number of RVEs (Section 2.4):

$$I_{V_{\sigma \mathrm{W}}} = -\left(\frac{\sigma_N}{s_0}\right)^m N^V_{\mathrm{eq},\sigma} \tag{8.17}$$

$$\text{where} \quad N^V_{\mathrm{eq},\sigma} = \int_{V_\mathrm{W}} \langle \bar{s}(\boldsymbol{x}) \rangle^m \frac{\mathrm{d}V(\boldsymbol{x})}{V_0} \tag{8.18}$$

$$I_{\Omega_{\sigma \mathrm{W}}} = -\left(\frac{\sigma_N}{s_0}\right)^m N^\Omega_{\mathrm{eq},\sigma} \tag{8.19}$$

$$\text{where} \quad N^\Omega_{\mathrm{eq},\sigma} = \int_{\Omega_\mathrm{W}} \langle s(\boldsymbol{x}) \rangle^m \frac{\mathrm{d}\Omega(\boldsymbol{x}) l_0}{V_0} \tag{8.20}$$

where $s(\boldsymbol{x})$ and $\bar{s}(\boldsymbol{x})$ are the dimensionless local and nonlocal stress fields for the Weibullian parts of the boundary layer and the interior part of structure, respectively. It should be pointed out that the size of Weibullian regions $V_{\sigma \mathrm{W}}$ and $\Omega_{\sigma \mathrm{W}}$ is explicitly dependent on σ_N.

In contrast to integrals $I_{V_{\sigma \mathrm{W}}}$ and $V_{\Omega_{\sigma \mathrm{W}}}$ for the Weibullian region, closed-form expressions for integrals $I_{V_{\sigma \mathrm{G}}}$ and $I_{\Omega_{\sigma \mathrm{G}}}$ over the Gaussian region could not be obtained. Here we propose to approximate the integrands of $I_{V_{\sigma \mathrm{G}}}$ and $I_{\Omega_{\sigma \mathrm{G}}}$ by means of the Taylor series expansion of the grafted Gauss–Weibull distribution of the RVE strength. The weakest-link model implies that the material elements with sufficiently small principal stress make negligible contributions to the failure of the entire structure. Thus, for the purpose of calculating the failure probability, we can neglect the part of structure that is subjected to relatively small stress. For example, consider the Weibull distribution with Weibull modulus $m = 24$. If the failure probability of the element with principal stress σ is denoted as p, then the failure probabilities of the elements with 0.8σ, 0.6σ, and 0.4σ are about $4.7 \times 10^{-3}p$, $4.7 \times 10^{-6}p$, and $2.8 \times 10^{-10}p$, respectively, the last two being negligible.

Therefore, when evaluating integrals $I_{V_{\sigma G}}$ and $I_{\Omega_{\sigma G}}$, we can limit our consideration to the elements in which the principal stress is within the range $\sigma \in [\mu\sigma_N, \sigma_{\max}]$, where $\sigma_{\max}$ = maximum principal stress used in the calculation and $\mu = \max(\sigma_{gr}/\sigma_N, 0.6\sigma_{\max}/\sigma_N)$. For such a limited stress range, we can approximate the integrands of $I_{V_{\sigma G}}$ and $I_{\Omega_{\sigma G}}$ by the Taylor series expansion with respect to the maximum principal stress $\sigma_{\max} = t\sigma_N$.

The number of terms that must be retained in the Taylor series expansion depends on σ_N, t, μ and the parameters of the grafted Gauss–Weibull cdf. Using too many terms would defeat the goal of attaining a simple analytical solution. Therefore, we truncate here the Taylor series expansion after its third derivative term. However, as will be seen later, the truncation after the third derivative is insufficient for an accurate approximation of the function $\ln[1 - P_{\sigma G}(\sigma)]$ for the range of $\sigma \in [\mu\sigma_N, t\sigma_N]$.

To further improve the approximation, we consider a linear combination of the Taylor series expansions of $\ln[1 - P_{\sigma G}(\sigma)]$ at $\sigma = \mu\sigma_N$ and at $\sigma = t\sigma_N$:

$$\ln[1 - P_{\sigma G}(\sigma)] = \omega(\sigma) \sum_{k=0}^{3} \frac{f^{(k)}(\mu\sigma_N)}{k!}(\sigma - \mu\sigma_N)^k + [1 - \omega(\sigma)] \sum_{k=0}^{3} \frac{f^{(k)}(t\sigma_N)}{k!}(\sigma - t\sigma_N)^k \tag{8.21}$$

where $f^{(k)}(\sigma) = \mathrm{d}^k \ln[1 - P_{\sigma G}(\sigma)]/\mathrm{d}\sigma^k$. To make the approximated function match the asymptotic properties of function $\ln[1 - P_{\sigma G}(\sigma)]$ at $\sigma = \mu\sigma_N$ and $\sigma = t\sigma_N$, it is clear that the function $\omega(\sigma)$ must decay from 1 at $\sigma = \mu\sigma_N$ to 0 at $\sigma = t\sigma_N$. We propose for this decay a quadratic function for $\omega(\sigma)$:

$$\omega(\sigma) = 1 - \left[\frac{\sigma - \mu\sigma_N}{t\sigma_N - \mu\sigma_N}\right]^2 \tag{8.22}$$

The performance of this approximation is evaluated for different values of $t\sigma_N$ corresponding to the failure probabilities $P_{1\sigma}(t\sigma_N) = 0.01$, 0.10, 0.50, and 0.99. Figure 8.6 presents the comparison among the exact solution of $\ln[1 - P_{\sigma G}(\sigma)]$, the three-term Taylor series expansion of $\ln[1 - P_{\sigma G}(\sigma)]$ at $\sigma = t\sigma_N$, the three-term Taylor series expansion of $\ln[1 - P_{\sigma G}(\sigma)]$ at $\sigma = \mu\sigma_N$, and the present approximation (Eqs. 8.21 and 8.22) for the stress range $\sigma \in [\mu\sigma_N, t\sigma_N]$. It can be seen that the three-term Taylor series expansion at either $\sigma = t\sigma_N$ or $\mu\sigma_N$ is unable to provide a close approximation while the proposed approximation function agrees well with the exact value of $\ln[1 - P_{\sigma G}(\sigma)]$.

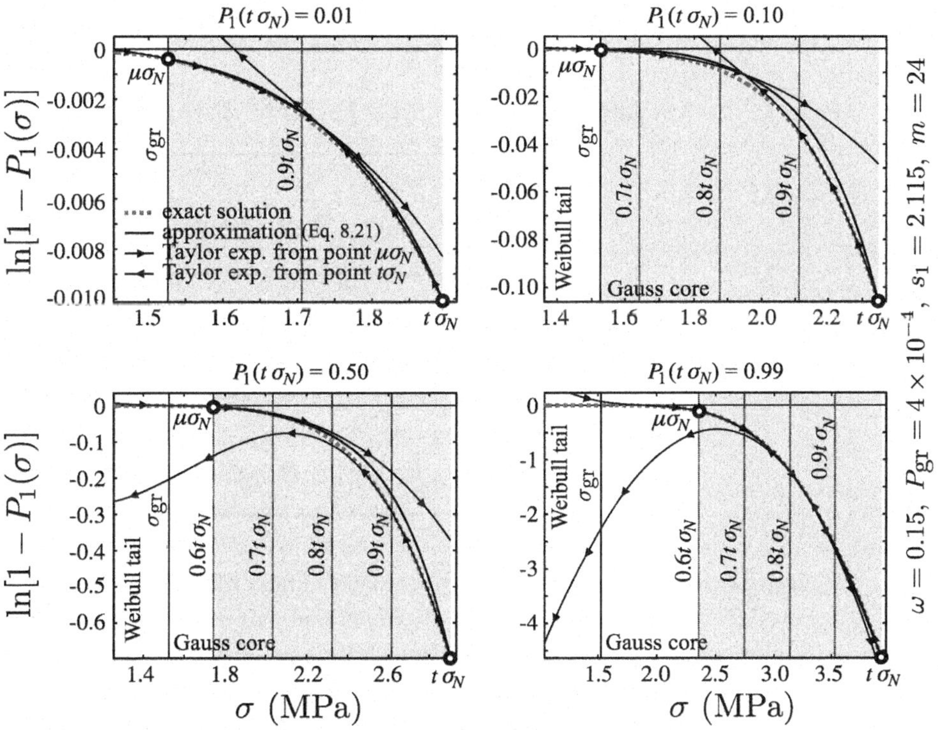

Figure 8.6. Approximation of $\ln[1 - P_{\sigma G}(\sigma)]$ by using Taylor expansion.

Consequently, integrals $I_{V_{\sigma G}}$ and $I_{\Omega_{\sigma G}}$ can be rewritten as

$$I_{V_{\sigma G}}(t,\mu) = \sum_{k=0}^{3} \frac{\sigma_N^k}{k!} f^{(k)}(\mu\sigma_N)\Delta_{V_{\sigma G},1}(k,t,\mu) + \sum_{k=0}^{3} \frac{\sigma_N^k}{k!} f^{(k)}(t\sigma_N)\Delta_{V_{\sigma G},2}(k,t,\mu) \tag{8.23}$$

$$I_{\Omega_{\sigma G}}(t,\mu) = \sum_{k=0}^{3} \frac{\sigma_N^k}{k!} f^{(k)}(\mu\sigma_N)\Delta_{\Omega_{\sigma G},1}(k,t,\mu) + \sum_{k=0}^{3} \frac{\sigma_N^k}{k!} f^{(k)}(t\sigma_N)\Delta_{\Omega_{\sigma G},2}(k,t,\mu) \tag{8.24}$$

where
$$\Delta_{V_{\sigma G},1}(k,t,\mu) = \frac{1}{V_0}\int_{V_{\sigma G}(t,\mu)} \left[1 - \left(\frac{\bar{s}(\boldsymbol{x}) - \mu}{t-\mu}\right)^2\right][\bar{s}(\boldsymbol{x}) - \mu]^k \,\mathrm{d}V \tag{8.25}$$

$$\Delta_{V_{\sigma G},2}(k,t,\mu) = \frac{1}{V_0(t-\mu)^2}\int_{V_{\sigma G}(t,\mu)} [\bar{s}(\boldsymbol{x}) - \mu]^2[\bar{s}(\boldsymbol{x}) - t]^k \,\mathrm{d}V \tag{8.26}$$

$$\Delta_{\Omega_{\sigma G},1}(k,t,\mu) = \frac{l_0}{V_0}\int_{\Omega_{\sigma G}(t,\mu)} \left[1 - \left(\frac{s(\boldsymbol{x}) - \mu}{t-\mu}\right)^2\right][s(\boldsymbol{x}) - \mu]^k \,\mathrm{d}\Omega \tag{8.27}$$

$$\Delta_{\Omega_{\sigma G},2}(k,t,\mu) = \frac{l_0}{V_0(t-\mu)^2}\int_{\Omega_{\sigma G}(t,\mu)} [s(\boldsymbol{x}) - \mu]^2[s(\boldsymbol{x}) - t]^k \,\mathrm{d}\Omega \tag{8.28}$$

The foregoing analysis can be extended directly to the lifetime cdfs. For calculating the cdfs of structural creep and fatigue lifetime, we can rewrite Eq. 8.16 as

$$\begin{aligned}
\ln(1-P_f) = & \underbrace{\int_{V_{XW}} \ln[1 - P_{XW}(\bar{s}(\boldsymbol{x})^{n_x} X)] \frac{dV(\boldsymbol{x})}{V_0}}_{I_{V_{XW}}} \\
& + \underbrace{\int_{V_{XG}} \ln[1 - P_{XG}(\bar{s}(\boldsymbol{x})^{n_x} X))] \frac{dV(\boldsymbol{x})}{V_0}}_{I_{V_{XG}}} \\
& + \underbrace{l_0 \int_{\Omega_{XW}} \ln[1 - P_{XW}(s^{n_x}(\boldsymbol{x}) X)] \frac{d\Omega(\boldsymbol{x})}{V_0}}_{I_{\Omega_{XW}}} \\
& + \underbrace{l_0 \int_{\Omega_{XG}} \ln[1 - P_{XG}(s^{n_x}(\boldsymbol{x}) X)] \frac{d\Omega(\boldsymbol{x})}{V_0}}_{I_{\Omega_{XG}}}
\end{aligned} \tag{8.29}$$

where $X = \Lambda, \mathcal{N}$; $n_x = n_f, n_c$ for creep and fatigue lifetime (see Eqs. 7.15–7.16, and Eqs. 7.27–7.28) respectively; P_{XW} = Weibullian part of the lifetime cdfs of one RVE, P_{XG} = core of the lifetime cdfs of one RVE (i.e., Gaussian distribution transformed by a power-law function), V_{XW} = Weibullian region of the interior part of structure $= \{\boldsymbol{x} \,|\, \boldsymbol{x} \in V_I \,\wedge\, \bar{s}(\boldsymbol{x})^{n_x} X \le X_{\text{gr}}\}$; V_{XG} = non-Weibullian region of the interior part of the structure $= \{\boldsymbol{x} \,|\, \boldsymbol{x} \in V_I \,\wedge\, \bar{s}(\boldsymbol{x})^{n_x} X > X_{\text{gr}}\}$; Ω_{XW} = Weibullian region of the boundary layer $= \{\boldsymbol{x} \,|\, \boldsymbol{x} \in \Omega_M \,\wedge\, s^{n_x}(\boldsymbol{x}) X \le X_{\text{gr}}\}$; and Ω_{XG} = non-Weibullian part of the boundary layer $= \{\boldsymbol{x} \,|\, \boldsymbol{x} \in \Omega_M \,\wedge\, s^{n_x}(\boldsymbol{x}) X > X_{\text{gr}}\}$.

The integrals over the Weibullian regions can easily be expressed in a closed form:

$$I_{V_{XW}} = -\left(\frac{X}{s_X}\right)^{m_x} N^V_{\text{eq},X} \tag{8.30}$$

$$\text{where} \quad N^V_{\text{eq},\sigma} = \int_{V_{XW}} \langle \bar{s}(\boldsymbol{x}) \rangle^{n_x m_x} \frac{dV(\boldsymbol{x})}{V_0} \tag{8.31}$$

$$I_{\Omega_{XW}} = -\left(\frac{X}{s_X}\right)^{m_x} N^\Omega_{\text{eq},X} \tag{8.32}$$

$$\text{where} \quad N^\Omega_{\text{eq},X} = \int_{\Omega_{XW}} \langle s(\boldsymbol{x}) \rangle^{n_x m_x} \frac{l_0 d\Omega(\boldsymbol{x})}{V_0} \tag{8.33}$$

where $m_X = \bar{m}$ and $\bar{m}_f$ = Weibull moduli of cdfs of creep and fatigue lifetime, respectively.

Similar to the Gaussian region of the strength cdf, the integrals for the non-Weibullian regions of the lifetime cdfs can be approximated through the Taylor series expansion of function P_{XG}. It is clear that the contribution of each RVE to the failure of the structure is determined by the dimensionless stress $s(\boldsymbol{x})$. To determine the level of $s(\boldsymbol{x})$ at which the contribution of the RVE can be neglected, we note that for cdfs of both creep and fatigue lifetime the non-Weibullian part can be described by a Gaussian distribution transformed by a power-law function, i.e., by transformed variables $s^{\frac{n_c}{n_c+1}}(\boldsymbol{x})\Lambda^{1/(n_c+1)}$

and $s(\boldsymbol{x})\mathcal{N}_f^{1/n_f}$ follow a Gaussian distribution, which is similar to the core of the strength cdf. Therefore, it is expected that the range of $s(\boldsymbol{x})$ in which the RVE could effectively contribute to the foregoing calculation of structural lifetime would be the same as that for the calculation of structural strength, i.e., $s^{n_x}(\boldsymbol{x}) \in [\mu_X, t_X]$, where $t_X = \max s^{n_x}(\boldsymbol{x}_i)$ and $\mu_X = \max(X_{\mathrm{gr}}/X, 0.6^{n_x} t_X)$. Equation 8.21 can be rewritten as

$$\ln\{1 - P_{X\mathrm{G}}[s^{n_x}(\boldsymbol{x})X]\} = \omega_X[s^{n_x}(\boldsymbol{x})X] \sum_{k=0}^{3} \frac{g_X^{(k)}(\mu_X X)X^k}{k!}[s^{n_x}(\boldsymbol{x}) - \mu_X]^k$$
$$+ \{1 - \omega_X[s^{n_x}(\boldsymbol{x})]\} \sum_{k=0}^{3} \frac{g_X^{(k)}(t_X X)X^k}{k!}[s^{n_x}(\boldsymbol{x}) - t_X]^k \quad (8.34)$$

where $g_X^{(k)}(x) = \mathrm{d}^k \ln[1 - P_{X\mathrm{G}}(x)]/\mathrm{d}x^k$ and the weighting function ω_X follows the same form as $\omega(\sigma)$:

$$\omega_X[s^{n_x}(\boldsymbol{x})] = 1 - \left[\frac{s^{n_x}(\boldsymbol{x}) - \mu_X}{t_X - \mu_X}\right]^2 \quad (8.35)$$

The integrals for the non-Weibullian region can then easily be calculated:

$$I_{V_{X\mathrm{G}}}(t, \mu) = \sum_{k=0}^{3} \frac{X^k}{k!} g_X^{(k)}(\mu X)\Delta_{V_{X\mathrm{G}},1}(k, t, \mu) + \sum_{k=0}^{3} \frac{X^k}{k!} g_X^{(k)}(tX)\Delta_{V_{X\mathrm{G}},2}(k, t, \mu) \quad (8.36)$$

$$I_{\Omega_{X\mathrm{G}}}(t, \mu) = \sum_{k=0}^{3} \frac{X^k}{k!} g_X^{(k)}(\mu X)\Delta_{\Omega_{X\mathrm{G}},1}(k, t, \mu) + \sum_{k=0}^{3} \frac{X^k}{k!} g_X^{(k)}(tX)\Delta_{\Omega_{X\mathrm{G}},2}(k, t, \mu) \quad (8.37)$$

where

$$\Delta_{V_{X\mathrm{G}},1}(k, t, \mu) = \frac{1}{V_0}\int_{V_{X\mathrm{G}}(t,\mu)} \left[1 - \left(\frac{\bar{s}^{n_x}(\boldsymbol{x}) - \mu}{t - \mu}\right)^2\right][\bar{s}^{n_x}(\boldsymbol{x}) - \mu]^k \,\mathrm{d}V \quad (8.38)$$

$$\Delta_{V_{X\mathrm{G}},2}(k, t, \mu) = \frac{1}{V_0(t-\mu)^2}\int_{V_{X\mathrm{G}}(t,\mu)} [\bar{s}^{n_x}(\boldsymbol{x}) - \mu]^2[\bar{s}^{n_x}(\boldsymbol{x}) - t]^k \,\mathrm{d}V \quad (8.39)$$

$$\Delta_{\Omega_{X\mathrm{G}},1}(k, t, \mu) = \frac{l_0}{V_0}\int_{\Omega_{X\mathrm{G}}(t,\mu)} \left[1 - \left(\frac{s^{n_x}(\boldsymbol{x}) - \mu}{t - \mu}\right)^2\right][s^{n_x}(\boldsymbol{x}) - \mu]^k \,\mathrm{d}\Omega \quad (8.40)$$

$$\Delta_{\Omega_{X\mathrm{G}},2}(k, t, \mu) = \frac{l_0}{V_0(t-\mu)^2}\int_{\Omega_{X\mathrm{G}}(t,\mu)} [s^{n_x}(\boldsymbol{x}) - \mu]^2[s^{n_x}(\boldsymbol{x}) - t]^k \,\mathrm{d}\Omega \quad (8.41)$$

For structures with some simple stress fields, such as the linear and bilinear stress profiles, $\Delta_{V_{\sigma\mathrm{G},i}}(k, t, \mu)$, $\Delta_{\Omega_{\sigma\mathrm{G},i}}(k, t, \mu)$, $\Delta_{V_{X\mathrm{G},i}}(k, t, \mu)$, $\Delta_{\Omega_{X\mathrm{G},i}}(k, t, \mu)$ $(i = 1, 2)$ can be expressed as polynomials and integrated analytically. This is likely possible for many structures, for which the stress field near the point with the largest maximum principal stress can be approximated by a linear or bilinear function.

8.4 Analysis of Strength Statistics of Beams under Flexural Loading

The three point and four point bend tests have commonly been adopted to measure the strength histograms of quasibrittle structures (Weibull 1939; Munz & Fett 1999; Tinschert, Zwez, Marx, & Ausavice 2000; Lohbauer et al. 2002). For beams with a large span-to-depth ratio, the engineering beam theory may be used to calculate the maximum principal stress field. Accordingly, for three point bend tests, the stress field can be represented by a two-way linear stress gradient along both the beam depth and span. We now compare the aforementioned numerical NBL model, random RVE placing method, and Taylor series expansion method by applying them to the strength statistics of beams under three point and four point bending.

In four point bending tests in which the two loading points are far apart, the failure is governed by the middle portion between the loading points, which is essentially experiencing a constant bending moment, i.e., pure bending. The stress field of the middle portion of the beam can be represented by a one-way stress gradient along the beam depth. Nevertheless, if the two loading points are not far apart, the contribution of the end portion of beam between the loading point and the support, which has a two-way linear stress distribution, can easily be added.

For the case of pure bending, the stress varies linearly in the y-direction (normal to the beam axis). Let us consider a prismatic beam of depth D (y-direction) and length λD (x-direction). Two-dimensional analysis presumes that the failure occurs simultaneously along the whole prism thickness (in the z direction). The elastic stress in the body can be described by

$$\sigma(\boldsymbol{x}) = \sigma_x(x, y) = -\frac{2y}{D}\sigma_N \tag{8.42}$$

Since the stress varies linearly, the nonlocal principal stress field coincides with the local principal stress field. The maximum stress in the boundary layer is $\sigma_{\max}^{\Omega} = (1 - l_0/D)\sigma_N$ and the maximum stress in the interior part is $\sigma_{\max}^{V} = (1 - 2l_0/D)\sigma_N$. Within the framework of the aforementioned Taylor series expansion method, the boundary between the Weibullian and Gaussian regions is given by the straight line $y = -\sigma_{\mathrm{gr}}D/(2\sigma_N)$ (Fig. 8.5a). For the Taylor series expansion method, $N_{\mathrm{eq},\sigma}^{V}$, $N_{\mathrm{eq},\sigma}^{\Omega}$, $\Delta_{V_{\sigma G},1}(k,t,\mu)$, $\Delta_{V_{\sigma G},2}(k,t,\mu)$, $\Delta_{\Omega_{\sigma G},1}(k,t,\mu)$, $\Delta_{\Omega_{\sigma G},2}(k,t,\mu)$ can be calculated analytically [for detailed expressions, see Le et al. (2012)]. By substituting these expressions into Eqs. 8.16, 8.23, and 8.24, we can obtain the corresponding strength cdf.

For beams under three point bending, the stress field $\sigma(x, y)$ can be assumed to be linear in two directions (Fig. 8.5b). The stress field can be written as

$$\sigma = \frac{2y}{D}\left(\frac{2|x|}{\lambda D} - 1\right)\sigma_N \tag{8.43}$$

Similar to the case of pure bending, the quantities that are required for the Taylor series expansion method and the resulting strength cdf can be obtained analytically (Le et al. 2012).

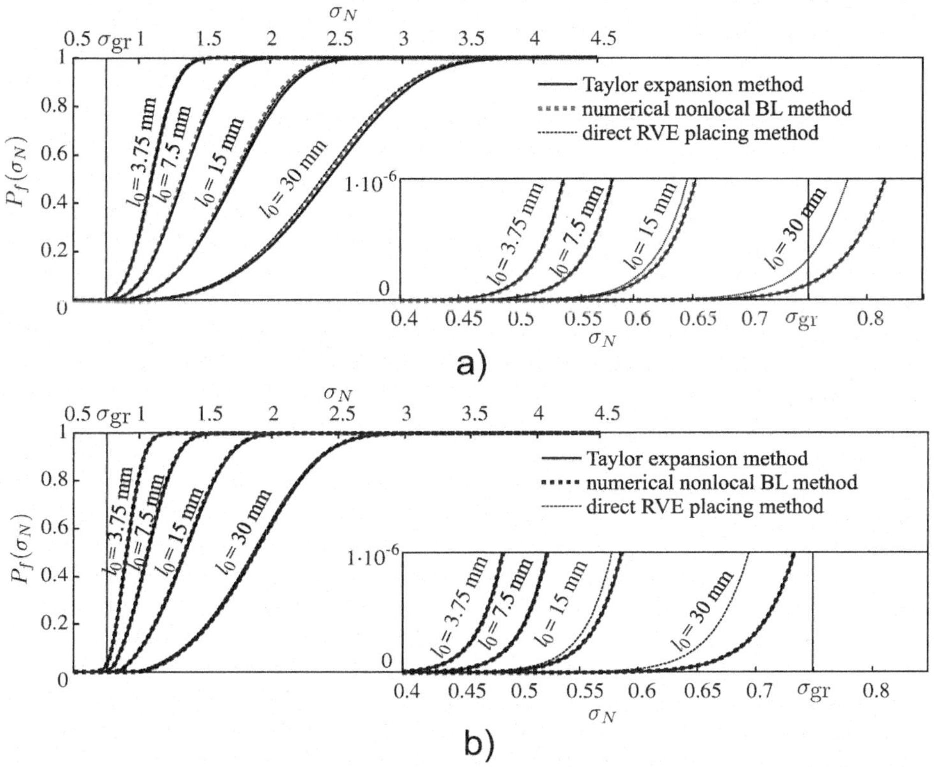

Figure 8.7. Comparison of strength cdfs of four geometrically similar beams under (a) three point bending, and (b) four point bending.

Note that, for the interior region, we can use the local stress to approximate the nonlocal stress, which is virtually exact for most of the structure except for the mid-span. Such an approximation will not cause any error in the calculation of the cdf of both the small-size and large-size beams because (1) for small-size beams, the failure probability is governed entirely by the boundary layer, and (2) for large-size beams, the nonlocal stress field converges to the local stress field. As will be demonstrated later, the error in calculating the cdf of strength for intermediate-size beams is insignificant.

Figure 8.7a and b presents a comparison of the calculated strength cdfs of beams under three point and four point bending by using the Taylor series expansion method, the direct RVE placing method, and the NBL method. For the four point bend beam, the two loading points are placed at a quarter length from the supports. We consider a linear one-way stress distribution for the middle portion of the beam and a linear two-way stress distribution for the two end portions of the beam. In the calculation, we fix the beam dimension, $D = 0.1$ m and $\lambda = 4$, and consider four different RVE sizes $l_0 = 3.75$ mm, 7.5 mm, 15 mm, and 30 mm. The statistical parameters used in the calculations are as follows: $s_1 = 1$ MPa, $m = 24$, $\mu_G = 2.0956$ MPa, and $\delta_G = 0.5361$ MPa. For the direct RVE placing method, 10, 50, 100, and 100 simulations are performed for beams with RVE size $l_0 = 3.75$ mm, 7.5 mm, 15 mm, and 30 mm, respectively, and the calculated cdfs are averaged for each size.

As seen from Figure 8.7a and b, the strength cdfs calculated by the Taylor series expansion method and the NBL method agree with each other for all beam sizes. The direct RVE placing method predicts a higher failure probability because the average distance to the boundary is slightly smaller than $l_0/2$, which implies that the stresses in the boundary RVEs are higher than the stresses in the center of the boundary layer.

The aforementioned calculations demonstrate that the Taylor series expansion method can accurately calculate the failure probability. Note that the present formulation is generally applicable to the strength cdf of structures whose critical stress region, i.e., $\sigma \in [\mu\sigma_N, t\sigma_N]$, has approximately a linear one-way or two-way stress profile.

8.5 Optimum Fits of Strength and Lifetime Histograms

For decades, tremendous efforts have been devoted to the experimental investigation of strength and lifetime distributions of structures made of quasibrittle materials such as Portland cement mortar (Weibull 1939), fiber composites (Chiao et al. 1977; Wanger, Phoenix, & Schwartz 1984; Wagner, Schwartz, & Phoenix 1986; Wagner 1989) and industrial or dental ceramics (Stanley & Inanc 1985; Fett & Munz 1991; Hoshide 1995; Okabe & Hirata 1995; Salem, Nemeth, Powers, & Choi 1996; Munz & Fett 1999; Tinschert et al. 2000; Lohbauer et al. 2002; Dos Santos et al. 2003; Bigleya et al. 2007; Studarta, Filser, Kochera, & Gauckler 2007; Studarta et al. 2007). The two-parameter Weibull distribution has been widely used to fit these histograms. However, systematic deviations have been observed – first for the strength of Portland cement mortar (Weibull 1939), for which the sizes large enough to render brittle response are unattainable, and recently for the strength of coarse-grained or toughened ceramics on the normal testing scale and fine-grained ceramics on the microscale (Le, Ballarini, & Zhu 2015). Similar deviations were also observed for histograms of both creep and fatigue lifetime. Based on the finite weakest-link theory of structural strength and lifetime, it is clear that such deviations are caused by the fact that the RVE is not negligible compared to the structure size.

We now demonstrate the optimum fitting of the experimentally measured strength and lifetime histograms of quasibrittle structures by the finite weakest-link theory. Since most of the histogram tests of strength and lifetime are carried out on structures with simple geometries, such as uniaxial tension bars, three point and four point bend beams, which have the simple uniform, linear, and bilinear stress profiles. Therefore, we can use the aforementioned Taylor expansion method to perform the optimum fitting.

8.5.1 Optimum Fits of Strength Histograms

Figure 8.8a–f shows the plots of the strength histograms of various industrial ceramics with the optimum fits by the finite weakest-link theory, where the details of these histogram tests are described as follows:

(a) The strength distribution of porcelain was first investigated by Weibull (1939). He conducted three point bend tests of 102 cylindrical unglazed porcelain rods with diameter 18.6 mm and length 100 mm.

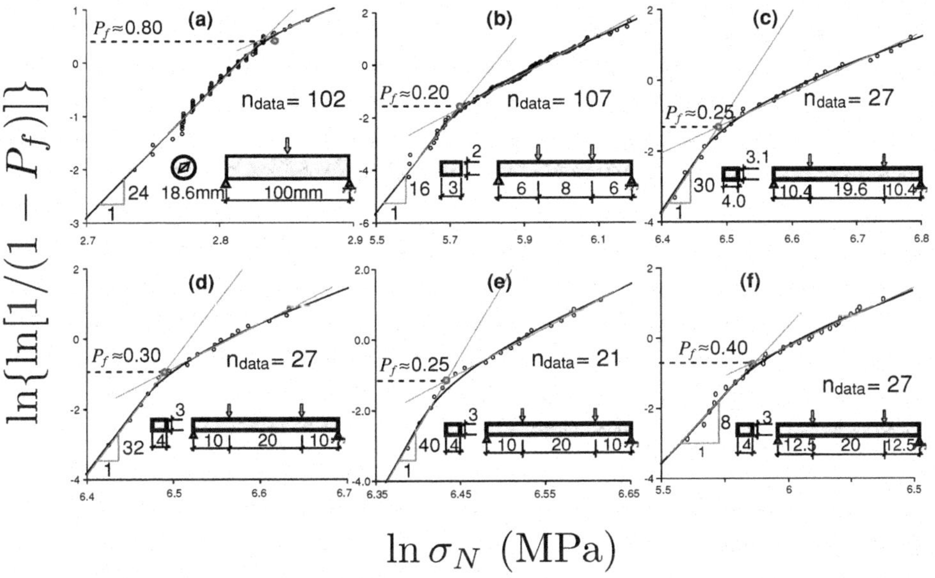

Figure 8.8. Optimum fits of strength distribution of ceramics by the finite weakest-link model. (a) Porcelain, (b) sintered α-SiC, (c) sintered Si_3N_4, (d) sintered Si_3N_4–Al_2O_3–Y_2O_3, (e) sintered Si_3N_4–Al_2O_3–CTR_2O_3, and (f) alumina/glass composite.

(b) Silicon carbide (SiC) ceramics, which are characterized by very high hardness and wear resistance and are widely used in automotive components, heat exchanger tubes, mechanical seals, and so on, were studied by Salem et al. (1996). They tested, under four point bend loading, 108 prisms of commercially available sintered alpha silicon carbide (Carborundum of Hexoloy, α-SiC), with dimensions $2 \times 3 \times 25$ mm; 36 of their prisms were produced with 0° grinding angle without annealing, the next 36 the same but with high-temperature annealing, and 36 more with a 90° grinding angle and with high-temperature annealing. For all the three groups of 36, they found almost the same strength distribution (Salem et al. 1996).

(c–e) Silicon nitride ceramics (Si_3N_4) (developed in the 1960s and 1970s in a search for dense, high-strength and high-toughness materials, and used, e.g., for reciprocating engine components and turbochargers, bearings, and metal cutting or shaping tools) were investigated by several researchers (Okabe & Hirata 1995; Gross 1996; Dos Santos et al. 2003). Gross (1996) conducted four point bending tests of 27 prismatic beams of sintered silicon nitride (SNW-1000), of dimensions $3.1 \times 4 \times 40.4$ mm. Dos Santos et al. (2003) tested two types of silicon nitride ceramics with performance improving additives: one was a sintering additive (Si_3N_4–Al_2O_3–Y_2O_3), and another was an aluminum additive (Al_2O_3/Y_2O_3) (which elevates the bending strength and fracture toughness). Dos Santos et al. (2003) further tested 21 four point bend beams, with dimensions $3 \times 4 \times 45$ mm, made of silicon nitride sintered with rare earth oxide additive (Si_3N_4–Al_2O_3–CTR_2O_3), which yields similar improvements but at lower costs.

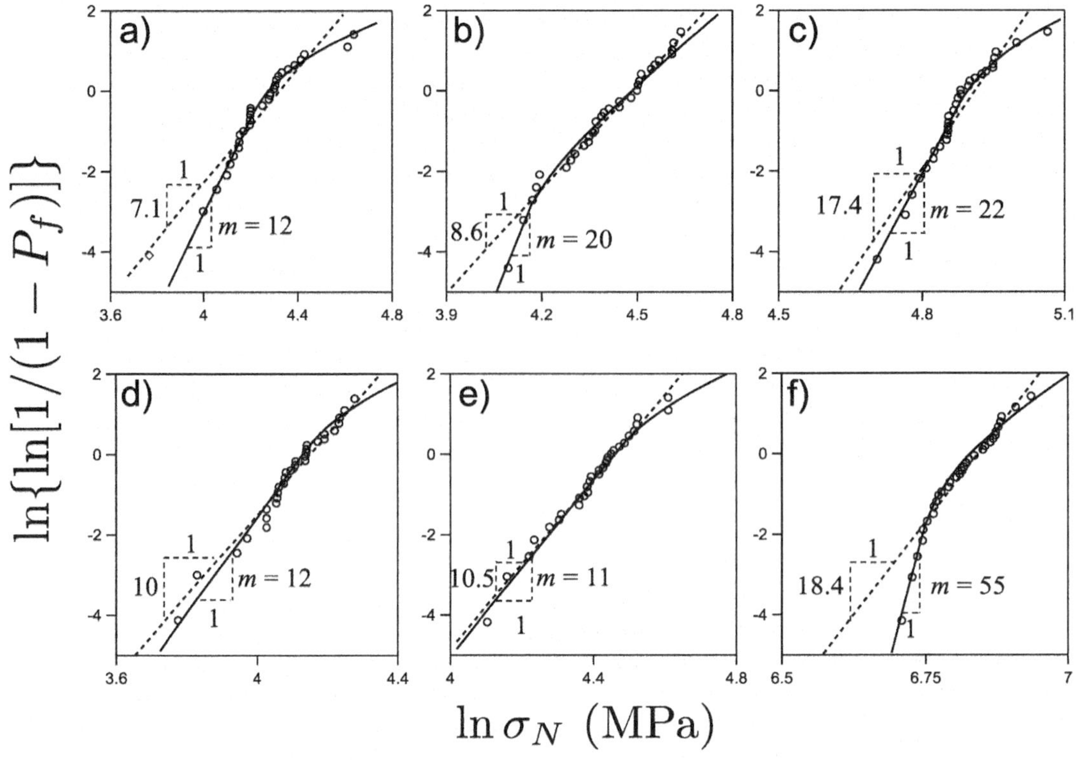

Figure 8.9. Optimum fits of strength distribution of dental ceramics by the finite weakest–link model. (a) Dicor (tetrasilicic fluoromica glass-ceramic), (b) IPS Empress (IE & leucite-reinforced porcelain), (c) Vitadur Alpha Core (alumina-reinforced feldspathic porcelain), (d) Vitadur Alpha Dentin (feldspathic porcelain), (e) Vita VMK 68 (feldspathic porcelain), and (f) Zirconia-TZP (partially stabilized zirconia ceramic).

(f) Lanthanum-glass–infiltrated alumina glass ceramics [which are attractive for restorative dentistry owing to their aesthetics and biocompatibility and offer high strength and fracture toughness (Ironside & Swain 1998)] were studied by Lohbauer et al. (2002). They tested alumina–glass composites dry-pressed and presintered (α-Al_2O_3) with a medium grain size. All alumina were CAD/CAM machined into prisms, with dimension $3 \times 4 \times 45$ mm before infiltration with 25 wt% (by weight) of glass of the following weight percentages (Lohbauer et al. 2002): 39–41% La_2O_3, 16–17% SiO_2, 15–18% Al_2O_3, 15% B_2O_3, 5% TiO_2, and 4% CeO_2. Twenty-seven four point bend specimens were tested, under dry conditions.

Figure 8.9a–f presents the optimum fitting of the strength histogram of dental ceramics, tested by Tinschert et al. (2000). The specimens were made of six dental restorative ceramics that are commonly used in crown, veneer, and inlay construction. The six ceramics materials are (a) Dicor (tetrasilicic fluoromica glass-ceramic), (b) IPS Empress (IE & leucite-reinforced porcelain), (c) Vitadur Alpha Core (alumina-reinforced

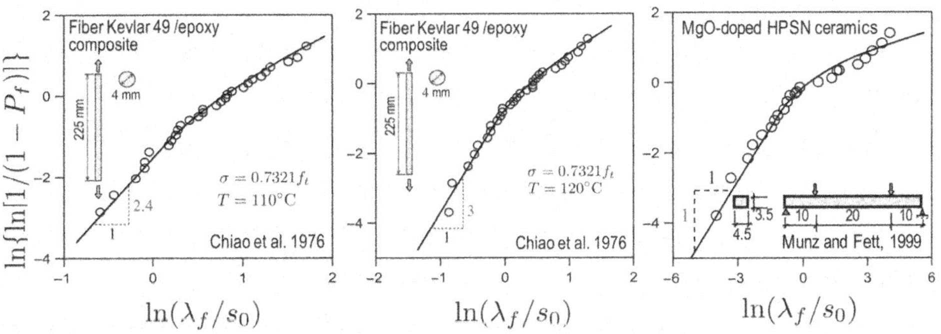

Figure 8.10. Optimum fits of creep lifetime distribution of composites and engineering ceramics by the finite weakest-link model.

feldspathic porcelain), (d) Vitadur Alpha Dentin (feldspathic porcelain), (e) Vita VMK 68 (feldspathic porcelain), and (f) zirconia-TZP (partially stabilized zirconia ceramic). For each material, 30 four point bend beam specimens with size $1.5 \times 3 \times 30$ mm were tested.

Despite rather low numbers of the specimens tested, one can discern from Figures 8.8a–f and 8.9a–f that the strength histograms plotted in Weibull scale are not straight lines, as required by the two-parameter Weibull distribution. As shown by the solid curves in Figures 8.8 and 8.9, these histograms can be fitted as closely as the scatter permits by the finite weakest-link theory, in which the Weibull scale histogram consists of two parts separated by a relatively abrupt kink. The left tail is a Weibull straight line, and the core deviates from it to the right as σ_N increases. Obviously, these two parts cannot be fitted by the two-parameter Weibull model simultaneously.

Based on the present framework, it is clear that the core part of histogram results from a chain of elements with the Gaussian part of the RVE strength. Smooth though the kink is, it is so abrupt that it may be approximated by a point transition at which both parts are grafted with a continuous cdf slope. The height of the grafting point characterizes the degree of brittleness of structure, which depends on the ratio of the structure size to the material inhomogeneity size.

For the fitting of strength histograms, we first estimate the RVE size l_0, which is approximately equal to two to three times the size of material inhomogenieties. The Weibull modulus can easily be deduced from the lower straight portion of the strength histogram in the Weibull scale. By knowing l_0, we can calculate the Weibull scale parameter s_0. Based on the analytical expression developed in the previous section, the Gaussian part of the strength cdf of RVE can be determined by the optimum fit of the upper curved portion of the strength histogram by the Levenberg–Marquardt algorithm.

8.5.2 Optimum Fits of Histograms of Creep Lifetime

Figure 8.10 presents the lifetime histograms of various fiber composites and engineering ceramics fitted by the finite weakest-link theory and the two-parameter Weibull distributions. Chiao et al. (1977) investigated the lifetime histograms of Kevlar-49 fiber composites. A total of 30 prismatic bar specimens were loaded in tension at elevated temperatures (100°C to 120°C) under a constant uniaxial tensile stress, which is about 70% of the mean short-time strength. Munz and Fett (1999) investigated the lifetime histograms of MgO-doped HPSN (hot-pressed silicon nitride) loaded at a temperature of 1100°C. Standard four point bend tests were used and the applied stress was about 50% of the mean short-time strength.

Similar to the strength histogram, the lifetime histogram is also seen to exhibit two segments in the Weibull scale. The lower part, which has for one RVE the same reach as the tail of strength cdf (i.e., cca $P_f < 0.001$), is a straight line. The upper part is a curve deviating to the right. The finite weakest-link theory is found to give optimum fits for these histograms. Compared to the strength cdf, the fitting of creep lifetime requires one more parameter, i.e., the power exponent of creep crack growth law n_c, which is typically in the range of 10 to 30.

The value of n_c can be determined in two ways: (1) perform standard tests of subcritical crack growth under constant loading and obtain n_c from the bilogarithmic plot of the crack growth rate versus the applied stress intensity factor; or (2) measure the Weibull moduli of strength and creep lifetime cdfs, and then use Eq. 7.21 to calculate n_c.

For the aforementioned optimum fitting of lifetime histograms of Kevlar-49, we estimate $n_c \approx 15$ based on Eq. 7.21 because it is known that the Weibull modulus of strength cdf of Kevlar-49 is about 40 and the Weibull modulus of creep lifetime cdf is about 2 to 3. For MgO-doped hot pressed silicon nitride (MgO-doped HPSN), Munz and Fett (1999) estimated n_c to be approximately equal to 25. Furthermore, it should be noted that, in the histograms fitted, we use the actual lifetime λ_f, which can easily be calculated from the transformed variable Λ_f (i.e., $\lambda_f = \Lambda_f/\sigma_0^{n_c}$, where σ_0 = applied nominal stress).

A more complete validation of the finite weakest-link theory is obtained by simultaneous fitting of the strength and creep lifetime cdfs of the same batch of specimens. Fett & Munz (1991) studied the strength and lifetime histograms of four point bend beams made of 99.6% Al_2O_3. Thirty specimens were used for each histogram test. Figure 8.11 shows the experimentally observed strength and lifetime histograms with the optimum fits by the present theory. Based on the lower straight portion of the strength and lifetime cdfs, the Weibull moduli for strength and lifetime distributions are estimated to be about 30 and 1.1, respectively. Based on Eq. 7.21, the exponent n_c of the power law for creep crack growth in this material (99.6% Al_2O_3) is found to be about 26. With one and the same set of parameters, the present theory is seen to give excellent fits of both the strength and lifetime histograms.

The fitting of these histograms reveals that the grafting probabilities of the strength and creep lifetime cdfs are about the same. This agrees well with the present theory, in which the grafting probability can be calculated as $P_{gr} = 1 - [1 - P_{gr,1}]^{N_{eq}}$. We note that the grafting probabilities $P_{gr,1}$ of strength and lifetime cdfs for one RVE are the same, i.e., $P_{1\sigma}(\sigma_{gr}) = P_{1\Lambda}(\Lambda_{gr})$, and the equivalent numbers N_{eq} of RVEs for strength and lifetime distributions are also about the same. Therefore, P_{gr} for the strength and creep lifetime cdfs must be about the same.

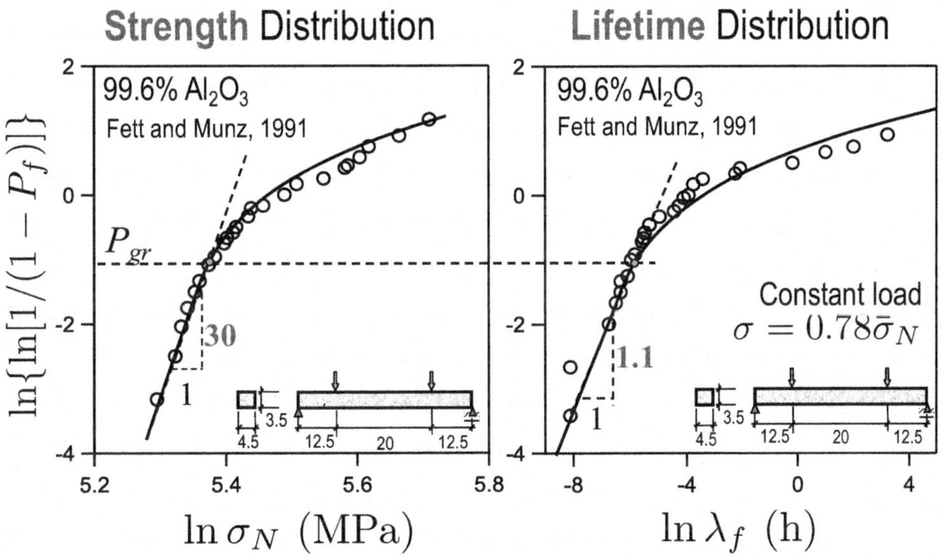

Figure 8.11. Optimum fits of strength and lifetime distributions of 99.6% Al_2O_3.

8.5.3 Optimum Fits of Histograms of Fatigue Lifetime

Figure 8.12 presents the optimum fits of fatigue lifetime histograms of engineering and dental ceramics by both the two-parameter Weibull distribution and the finite weakest-link theory. The experiments are summarized as follows.

(a–e) Structural alumina ceramics (99% Al_2O_3): Round bar specimens were tested under fully reversed cyclic load produced by a rotating bending machine (Sakai & Fujitani 1989; Sakai & Hoshide 1995). Five stress levels were used, where the maximum bending stress is about 42%, 49%, 56%, 63%, and 70% of the monotonic bending strength of the specimen. For each stress level, 20 identical specimens were tested.

(f–g) Dental ceramic composites: Glass infiltrated Al_2O_3–ZrO_2 with feldspathic glass (Inc-VM7) (Fig. 8.12f) and yttria-stabilized ZrO_2 with feldspathic glass (TZP-CerS) (Fig. 8.12g). For each material, 30 specimens with size 4 mm × 5 mm × 50 mm were tested under fully reversed cyclic bending (Studarta, Filser, Kochera, & Gauckler 2007; Studarta et al. 2007). The maximum bending stress used in the fatigue test is about 50% of the monotonic bending strength.

(h and i) Equine cortical bones: Specimens of different sizes, with dimensions 10.5 mm × 10 mm × 3 mm (Fig. 8.12h) and 21 mm × 10 mm × 3 mm (Fig. 8.12i), were extracted from 14 race horses with ages between 2 and 6 years. For each size, six specimens were subjected to direct tensile cyclic loads (Bigleya et al. 2007). Since these specimens came from different horses, the statistical parameters obtained from the histogram fitting represent average values for the tested group of horses.

Again, in the Weibull scale, we observe that the histogram of fatigue lifetime consists of two segments. The Weibull modulus is determined by the slope of the lower part of the histogram, which can be much larger than the Weibull modulus obtained by using

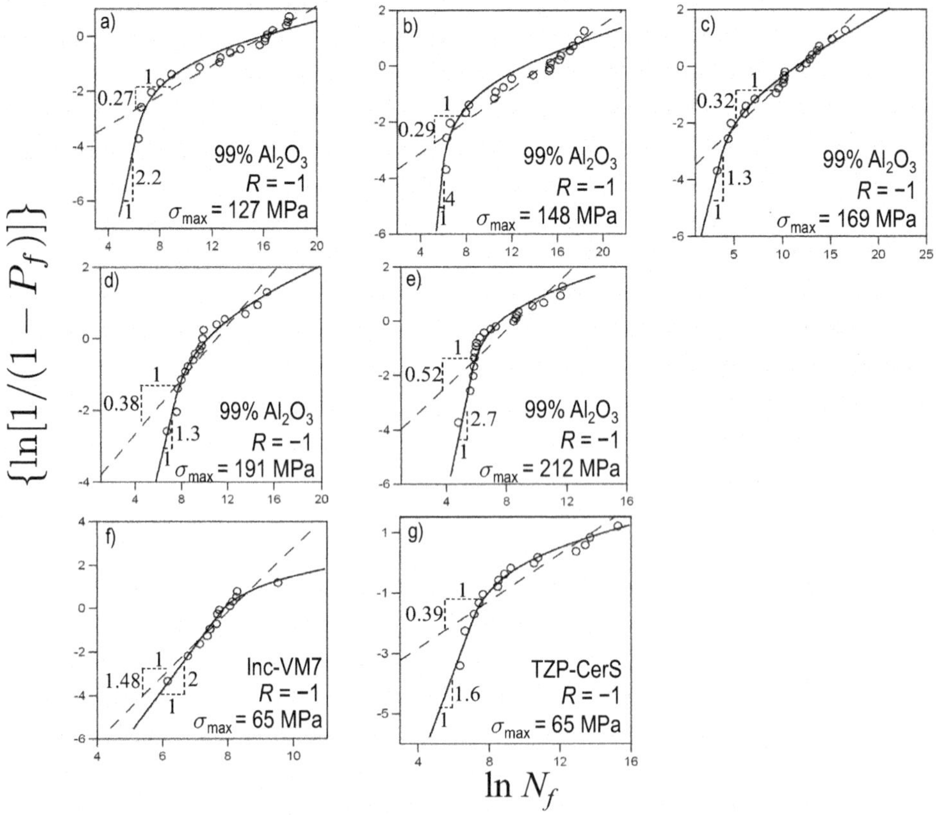

Figure 8.12. Optimum fits of fatigue lifetime distributions.

the Weibull distribution to fit the entire histogram. This makes a large difference in design lifetime for very low failure probabilities such as 10^{-6}.

Similar to the optimum fitting of the histograms of creep lifetime, we need to calibrate the exponent of the Paris law, n_f. Though, in principle, n_f can be obtained through either the test of fatigue crack growth or the ratio between the Weibull moduli of fatigue strength and fatigue lifetime cdfs, the fatigue strength cdf is impractical to measure because we would not know the fatigue strength of a structure for a prescribed loading duration. Therefore, n_f has to be determined from the tests of fatigue crack growth rate. For 99% Al_2O_3, it is known that the exponent of the Paris law is about 27 (Fett & Munz 1993). Studarta et al. (2007*a*, 2007*b*) reported that the values of n_f for Inc-VM7 and TZP-CerS are 29 and 25, respectively. The lifetime cdf is expressed in terms of the actual fatigue loading cycles N_f, which are calculated as $N_f = \mathcal{N}_f/\Delta\sigma_0^{n_f}$.

9 Indirect Determination of Strength Statistics of Quasibrittle Structures

As discussed in Chapter 8, one direct consequence of the finite weakest-link model is that the probability distribution of structural strength depends strongly on the structure size. This size dependence leads to a size effect on the mean structural strength, which agrees well with the predictions by the other widely accepted mechanical models such as nonlocal model and cohesive crack model. It is naturally expected that, based on the finite weakest-link model, the probability distribution of structural strength can be back calculated from the mean size effect curve.

In contrast to conventional histogram testing, where hundreds of identical specimens are needed, this indirect method requires only the mean strength tests on geometrically similar specimens of several sizes, which would greatly reduce the experimental cost. Another advantage, a major one, is that the mean strength values obtained by linear regression for size effect have a coefficient of variation that is much smaller than that of individual test data (it decreases roughly as the $-1/2$ power of the number of test data).

This chapter presents a mathematical formulation of this indirect method and verifies it through a comprehensive experimental analysis of the strength histograms and mean strengths of the asphalt mixture specimens of different sizes at a low temperature. Based on the experimental results, the applicability of the three-parameter Weibull distribution is also discussed.

9.1 Relation between Mean Size Effect Curve and Probability Distribution of RVE Strength

Based on the finite weakest-link model, it is clear that the calculation of the cumulative distribution function (cdf) of structural strength hinges on the knowledge of the strength cdf $P_{1\sigma}(\sigma)$ of one representative volume element (RVE). In Chapter 5, we showed that the atomistic fracture mechanics and the statistical bundle and chain models yield only the functional form of $P_{1\sigma}(\sigma)$, but not the values of the associated statistical parameters (Bažant & Pang 2007; Bažant et al. 2009; Le et al. 2011). Therefore, we must calibrate these parameters through macroscale experiments. Here we show that these statistical parameters can be related to the size effect curve of mean structural strength.

Recall that the strength distribution of one RVE can be approximated by a grafted Gauss–Weibull distribution (Eqs. 5.84a and 5.84b), i.e.,

$$P_{1\sigma}(\sigma) = 1 - e^{-\langle\sigma/s_0\rangle^m} \approx \langle\sigma/s_0\rangle^m \qquad (\sigma \leq \sigma_{\text{gr}}) \tag{9.1a}$$

$$P_{1\sigma}(\sigma) = P_{\text{gr}} + \frac{r_f}{\delta_G\sqrt{2\pi}}\int_{\sigma_{\text{gr}}}^{\sigma} e^{-(\sigma'-\mu_G)^2/2\delta_G^2}\,\mathrm{d}\sigma \qquad (\sigma > \sigma_{\text{gr}}) \tag{9.1b}$$

Although six statistical parameters are used to define the cdf of RVE strength (i.e., m, s_0, μ_G, δ_G, r_f, and σ_{gr}), due to the normalization requirement of the entire cdf and to the continuity of the probability density function (pdf) at σ_{gr}, only four of them are actually independent.

On the other hand, we have shown that the mean size effect curve can be approximately described by Eq. 7.34:

$$\bar{\sigma}_N = \left[\frac{N_a}{D} + \left(\frac{N_b}{D}\right)^{n_d\psi/m}\right]^{1/\psi} \tag{9.2}$$

where m = Weibull modulus, n_d = number of spatial dimensions in which the structure is to be scaled, and N_a, N_b, ψ = constants. In principle, the constants N_a, N_b, ψ, and m can be determined based on the experimentally measured size effect on the mean structural strength. The Weibull modulus m is one of the statistical parameters of $P_{1\sigma}(\sigma)$, and we can use N_a, N_b, and ψ to determine the other three statistical parameters of $P_{1\sigma}(\sigma)$, i.e., Weibull scale parameter s_0, the mean μ and the standard deviation δ_G of the Gaussian part.

First consider the large-size asymptote. Since $\psi n_d/m \ll 1$, Eq. 9.2 indicates that the mean strength at the large-size limit is governed by the term N_b/D, i.e.,

$$\bar{\sigma}_N = (N_b/D)^{n_d/m} \tag{9.3}$$

At the large-size limit, the weakest-link model implies that the power-law tail of strength cdf of one RVE governs the strength cdf of the entire structure. By rewriting the joint probability equation (Eq. 7.1) in the logarithmic form and considering the approximation $\ln(1-x) = -x$ for $x \to 0$, we have

$$\ln\left[1 - P_f(\sigma_N)\right] = -\sum_{i=1}^{n_s}\left(\frac{\sigma_N\langle s_i\rangle}{s_0}\right)^m \tag{9.4}$$

$$\text{or} \qquad P_f(\sigma_N) = 1 - \exp\left\{-\int_V \langle s(\boldsymbol{x})\rangle^m \frac{\mathrm{d}V(\boldsymbol{x})}{l_0^{n_d}}\left(\frac{\sigma_N}{s_0}\right)^m\right\} \tag{9.5}$$

where n_s = number of RVEs in the structure, l_0 = RVE size, and $s(\boldsymbol{x})$ = dimensionless stress field. The mean strength of large-size structures can easily be obtained as (Bažant & Pang 2007; Bažant et al. 2009; Le et al. 2011):

$$\bar{\sigma}_N = s_0\Gamma(1 + 1/m)\left(\frac{D_0}{D}\right)^{n_d/m} \tag{9.6}$$

where $D_0 = \left[\int_V \langle s(\boldsymbol{\xi})\rangle^m \mathrm{d}V(\boldsymbol{\xi})\right]^{-1/n_d} l_0$, $\boldsymbol{\xi}$ = dimensionless normalized coordinate vector $= \boldsymbol{x}/D$, and $\Gamma(x)$ = Eulerian gamma function. By comparing Eqs. 9.3 and 9.6, we have

$$s_0 = \frac{1}{\Gamma\left(1+\frac{1}{m}\right)}\left(\frac{N_b}{D_0}\right)^{n_d/m} \tag{9.7}$$

Equation 9.7 indicates that the Weibull scale parameter s_0 can be determined from the large-size limit of the mean structural strength. Note that, to be able to use Eq. 9.7, the RVE size l_0 must be known first. As discussed in Chapter 7, the RVE size is about two to three times the size of the material inhomogeneities (Bažant & Pang 2007), which is typically known from either the information on the material composition or the digital image examination of the material.

The mean and standard deviation of the Gaussian part of the strength cdf of one RVE can be obtained from the small-size asymptote of the size effect curve. The smallest characteristic size of the structure, l_m, that makes physical sense, can be determined by comparing the mean strength predicted by the weakest-link model to the mean strength obtained by the nonlinear continuum models such as nonlocal damage model and cohesive crack model (Le et al. 2012). For beams subjected to bending, which are commonly used in laboratory testing, the minimum depth of the beam for the weakest-link model to be applicable is about 4 RVEs (Le et al. 2012).

Based on Eq. 9.2, we have the following asymptotic conditions at the small-size limit:

$$D \to l_m: \quad \bar{\sigma}_N = \left[\frac{N_a}{l_m} + \left(\frac{N_b}{l_m}\right)^{n_d\psi/m}\right]^{1/\psi} \tag{9.8}$$

$$D \to l_m: \quad \frac{\mathrm{d}\bar{\sigma}_N}{\mathrm{d}D} = -\frac{1}{\psi}\left[\frac{N_a}{l_m^2} + \frac{n_d\psi}{ml_m}\left(\frac{N_b}{l_m}\right)^{\psi n_d/m}\right]\left[\frac{N_a}{l_m} + \left(\frac{N_b}{l_m}\right)^{\psi n_d/m}\right]^{1/\psi-1} \tag{9.9}$$

At the small-size limit, the finite weakest-link model indicates that the Weibull part of the strength cdf would terminate at a very low failure probability, which is on the order of 10^{-3}. Therefore, the mean structural strength is governed mainly by the Gaussian part of the grafted distribution of RVE strength. We can then replace the grafted distribution of one RVE by a Gaussian distribution with the mean μ_G and standard deviation δ_G for the purpose of calculating the mean strength at the small-size limit:

$$\bar{\sigma}_N \approx \int_0^\infty \prod_{i=1}^{n_m}\left[1 - \Psi\left(\frac{\eta_i\sigma_N - \mu_G}{\delta_G}\right)\right]\mathrm{d}\sigma_N \tag{9.10}$$

where $\Psi(x)$ = standard Gaussian (normal) cdf$= (2\pi)^{-1/2}\int_{-\infty}^x e^{-x^2/2}\,\mathrm{d}x$, $\eta_i\sigma_N$ = maximum principal stress at the center of ith RVE, and n_m = number of RVEs in the structure at the small-size limit. Based on Eqs. 9.8 and 9.10, we obtain a single equation relating μ_G and δ_G:

$$\int_0^\infty \prod_{i=1}^{n_m}\left[1 - \Psi\left(\frac{\eta_i\sigma_N - \mu_G}{\delta_G}\right)\right]\mathrm{d}\sigma_N = \left[\frac{N_a}{l_m} + \left(\frac{N_b}{l_m}\right)^{\psi n/m}\right]^{1/\psi} \tag{9.11}$$

The second equation for μ_G and δ_G can be obtained from the slope of the size effect curve at the small-size limit. In general, the slope of the mean size effect curve can be calculated as

$$\frac{\mathrm{d}\bar{\sigma}_N}{\mathrm{d}D} = \frac{n_d}{D}\int_0^\infty [1 - P_f(\sigma_N)]\ln[1 - P_f(\sigma_N)]\mathrm{d}\sigma_N \tag{9.12}$$

Based on the aforementioned approximation of the grafted cdf of RVE strength by the Gaussian cdf at the small-size limit, Eq. 9.12 can simply be rewritten as

$$\frac{\mathrm{d}\bar{\sigma}_N}{\mathrm{d}D} = \frac{n_d}{D}\int_0^\infty \prod_{i=1}^{n_m}\left[1 - \Psi\left(\frac{\eta_i\sigma_N - \mu_G}{\delta_G}\right)\right]\sum_{i=1}^{n_m}\ln\left[1 - \Psi\left(\frac{\eta_i\sigma_N - \mu_G}{\delta_G}\right)\right]\mathrm{d}\sigma_N \tag{9.13}$$

Comparing Eq. 9.13 with Eq. 9.9 furnishes the second equation relating μ_G and δ_G:

$$\frac{n_d}{D}\int_0^\infty \prod_{i=1}^{n_m}\left[1 - \Psi\left(\frac{\eta_i\sigma_N - \mu_G}{\delta_G}\right)\right]\sum_{i=1}^{n_m}\ln\left[1 - \Psi\left(\frac{\eta_i\sigma_N - \mu_G}{\delta_G}\right)\right]\mathrm{d}\sigma_N = -\frac{1}{\psi}\left[\frac{N_a}{l_m^2} + \frac{\psi n_d}{m l_m}\left(\frac{N_b}{l_m}\right)^{\psi n_d/m}\right]\left[\frac{N_a}{l_m} + \left(\frac{N_b}{l_m}\right)^{\psi n_d/m}\right]^{1/\psi - 1} \tag{9.14}$$

By solving Eqs. 9.11 and 9.14, we can obtain μ_G and δ_G. Together with calibrated s_0 (Eq. 9.7) and m, the strength cdf of one RVE can be determined from the mean size effect curve. In conjunction with the finite weakest-link model (Eq. 7.1), we can then predict the strength distribution of structures of any size and geometry.

9.2 Experimental Verification

9.2.1 Description of Experiments

To verify the aforementioned method of determining the strength distribution, a comprehensive set of experiments has recently been performed on asphalt mixtures at a low temperature including the size effects on both the mean strength and the strength distribution (Le, Cannone Falchetto, & Marasteanu 2013; Cannone Falchetto, Le, Turos, & Marasteanu 2014). The mixture was prepared with an asphalt binder (7.4% by weight) from performance grade PG 64-34 (American Association of State Highway and Transportation Officials 2010) and with a blend of aggregates consisting of taconite aggregates (55% MIN TAC tailings and 10% ISPAT tailings) and pit sand (35%). The nominal aggregate size was 1.22 mm.

Ideally the experiments should involve a large size range. However, the sizes of the testing machine and the climate chamber imposed a limit on the size of the three point bend specimens. To overcome this difficulty, it was noted that structures of different geometries have different stress fields, which could result in large differences in the failure probability. Consequently, we could equate the mean nominal strengths of structures of different geometries and of similar structures of different sizes.

Table 9.1 Test specimens

Specimen ID	Type of test	Specimen dimension ($L \times D \times W$)
A	Three point bend	$100 \times 16.7 \times 40$ mm
B	Three point bend	$173 \times 28.9 \times 40$ mm
C	Three point bend	$300 \times 50 \times 40$ mm
D	Direct tension	$255 \times 55 \times 55$ mm

For instance, considering a Weibull strength cdf, a concrete beam in direct tension is statistically equivalent to a beam of a depth 50 times larger under pure bending (Bažant & Planas 1998). Therefore, to reach a large size range, we can perform the mean strength test on direct tension specimens. The size of the specimen is purposely chosen to be much larger than the size of the RVE. This ensures that the strength distribution of the direct tension specimen follows the Weibull distribution so that we can use the classical Weibull statistics to calculate the equivalent size of the three point bend beam.

The test specimens were prepared from 26 asphalt mixture slabs with a size 380 mm $\times$ 200 mm. The slabs had the targeted air void content of 7% and were made by using a linear kneading compactor. Two slab thicknesses, 50 mm and 75 mm, were used to optimize the preparation of the asphalt mixture. Asphalt beams of three sizes (size ratio: $1 : \sqrt{3} : 3$) with a 1:6 depth to span ratio were cut from these slabs. The width of beam b was kept at 40 mm. In the meantime, asphalt mixture prisms of a single size were also prepared for the direct tension test. Detailed information on the specimens is given in Table 9.1.

For three point bend specimens, the characteristic size D of the structure is chosen to be the beam depth (Fig. 9.1a) whereas, for direct tension specimens, the width of the prism is chosen as the characteristic size (Fig. 9.1b). For specimen sets A and D, the test involved about 10 replicates and the mean strength of each was recorded. For specimen sets B and C, 30 replicates were tested for each set and the corresponding strength histograms were obtained.

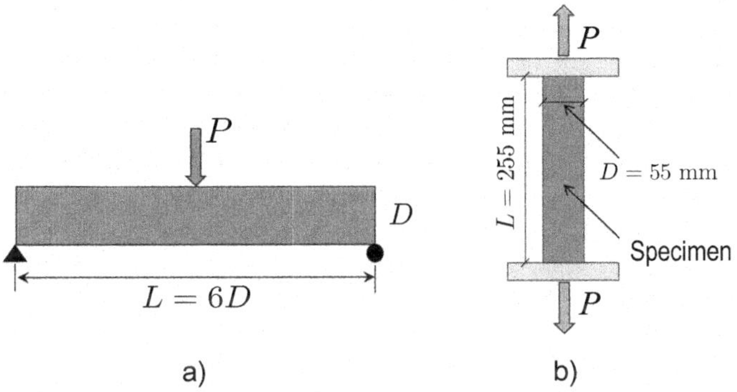

Figure 9.1. Schematics of three point bend and direct tension specimens.

The dimensions of the specimens were selected based on the amount of available material, the size limitations imposed by the current slab preparation method, as well as the dimension of the climatic chamber used to condition and test the specimens. The test temperature was chosen to be −24°C, which is within a few degrees of the glass transition temperature of the asphalt mixture. The effects of viscoelastic behavior were minimal given the short duration of the strength tests (Li & Marasteanu 2006; Turos, Cannone Falchetto, Tebaldi, & Marasteanu 2012).

All the tests were performed using an MTS servo-hydraulic testing system equipped with a 100 KN load cell. Liquid nitrogen was used to reach and maintain the test temperature in the environmental chamber, with an accuracy of ±1°C. Before testing, the specimens were conditioned for 3 hours at the test temperature. For the direct tension test, the specimens were glued to two metal plates by using an epoxy compound. Since we were interested only in the peak load, both direct tension and three point bend tests were performed in a load-control mode. Different loading rates were chosen for different specimen sizes and geometries so that all the specimens reached the peak load in about 5 minutes.

9.2.2 Analysis of Test Results

We define the nominal strengths for the three point bend specimen σ_N^B and the direct tension specimen σ_N^T as

$$\sigma_N^B = \frac{3P_{\max}L}{2bD^2} = \frac{9P_{\max}}{bD} \tag{9.15}$$

$$\sigma_N^T = \frac{P_{\max}}{bD} \tag{9.16}$$

For specimen sets B and C (Table 9.1), we calculate the corresponding probability distribution of nominal strength by ranking the measured strength values in an ascending order, $i = 1, \dots, n$, where i is the rank and n is the total number of test specimens. By using the midpoint position method (Rinne 2009), the strength cdf can be calculated as

$$\Pr\left(\sigma_N^B \le \sigma_i\right) = \frac{i - 0.5}{n} \tag{9.17}$$

Figure 9.2 plots the resulting strength histograms in the Weibull scale. It is seen that the strength histogram consists of two segments separated by a kink point, which is a characteristic shared with the observed strength histograms of many other quasibrittle materials such as engineering and dental ceramics shown in Chapter 8 (Figs. 8.8 and 8.9).

Figure 9.2 clearly shows that the strength distribution strongly depends on the structure size. As the structure size increases, the lower Weibull portion becomes more dominant. It is also seen that the slopes of the Weibull portions of these two histograms are the same, which confirms that the Weibull modulus does not vary with the structure size. By fitting the Weibull portion, we find the Weibull modulus m for this asphalt mixture to be 26.

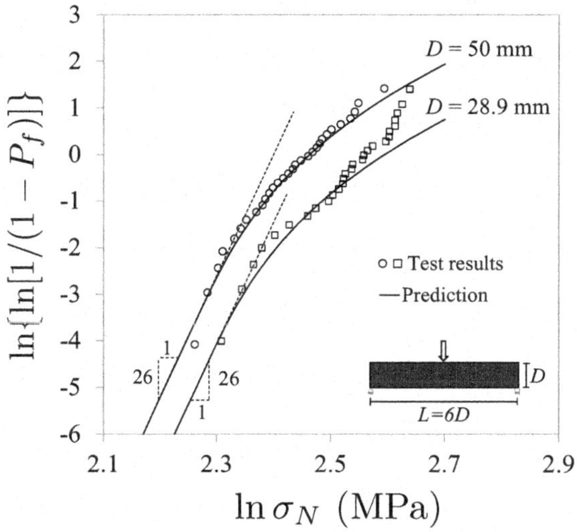

Figure 9.2. Measured and predicted strength distributions of asphalt mixture beams.

To plot the complete mean size effect curve, we first need to convert the direct tension specimen to its equivalent three point bend specimen. We note that the direct tension specimen has about 5.25×10^4 RVEs (the volume of the specimen is about 756 cm^3 and the volume of one RVE is about 0.0144 cm^3). The large number of RVEs implies that the strength distribution of the direct tension specimen must follow the two-parameter Weibull distribution. Its corresponding mean strength $\bar{\sigma}_N^T$ can be written as

$$\bar{\sigma}_N^T = N_T^{-1/m} s_0 \Gamma(1 + 1/m) \tag{9.18}$$

where N_T = the number of RVEs in the direct tension specimen. The strength distribution of the equivalent three point bend beam can be written as

$$P_f(\sigma_N^B) = 1 - \exp\left[-N_{\text{eq},B}(\sigma_N^B/s_0)^m\right] \tag{9.19}$$

where $N_{\text{eq},B}$ = equivalent number of RVEs of the beam, which differs from the actual number of RVEs in the beam due to the nonuniform stress distribution. $N_{\text{eq},B}$ can be calculated as

$$N_{\text{eq},B} = \frac{b}{l_0^3} \int \left[\sigma(x,y)/\sigma_N^B\right]^m \mathrm{d}x\mathrm{d}y \tag{9.20}$$

Using the engineering beam theory for three point bend beams, we can easily obtain $N_{\text{eq},B} = 0.5(m+1)^{-2}(V/V_0)$, where V = the volume of the beam. The mean strength of the beam is then given by

$$\bar{\sigma}_N^B = N_{\text{eq},B}^{-1/m} s_0 \Gamma(1 + 1/m) \tag{9.21}$$

By equating $\bar{\sigma}_N^T$ to $\bar{\sigma}_N^B$, we can calculate the equivalent characteristic size (i.e., depth) of the three point bend beam, which has a nominal strength equal to $\bar{\sigma}_N^T$:

$$D_{\text{eq}} = \sqrt{\frac{2(m+1)^2 N_T V_0}{6b}} = 2143 \text{ mm} \tag{9.22}$$

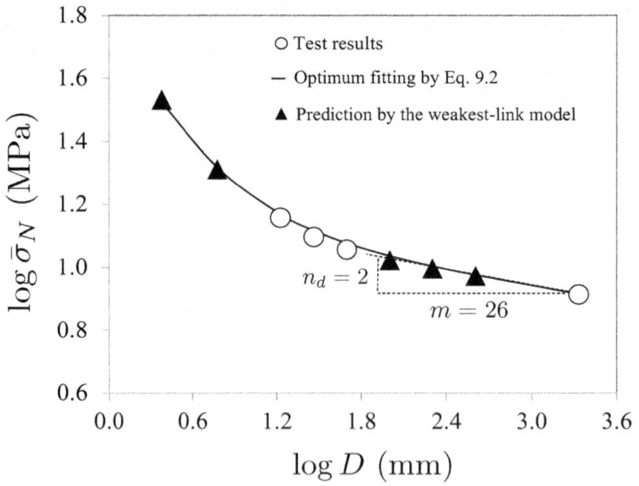

Figure 9.3. Size effect on mean structural strength of asphalt mixture specimens.

It can be seen that, by simply adjusting the loading configuration, we can achieve a large equivalent three point bend specimen, which allows us to obtain the size effect curve for a much wider size range. Meanwhile, based on the measured mean strength of direct tension specimens, Eq. 9.18 directly yields the value of the Weibull scale parameter $s_0 = 12.68$ MPa. For beam specimens with 2D scaling (i.e., $n_d = 2$), Eq. 9.7 becomes

$$N_b = \frac{(m+1)l_0 s_0^{m/2}}{\sqrt{3}} \left[\Gamma\left(1+\frac{1}{m}\right)\right]^{m/2} \tag{9.23}$$

Figure 9.3 plots the mean size effect curve of structural strength of the three point bend beams. The data point for the largest beam is obtained by the aforementioned size conversion of the direct tension specimens. The measured size effect curve has a trend similar to those observed in other quasibrittle structures (Bažant 2005; Bažant & Novák 2000*a*). However, it is clear that we do not have a sufficient number of points to determine the large-size asymptote of the size effect curve. As mentioned earlier, instead of using the size effect curve, the Weibull modulus m must be determined from the lower portion of the strength histogram, and N_b from the Weibullian mean of the strength of direct tension specimens.

The only parameters left to be determined for Eq. 9.2 are N_a and ψ. It is known that ψ typically ranges from 0.5 to 2 (Bažant 2005). In this study, we limit the range of ψ to be between 0.5 and 2.5 for the fitting of the mean size curve. Based on the nonlinear fitting, we obtain the following values: $N_a = 49.49$ mm· MPa$^{0.99}$ and $\psi = 1.01$. It is interesting to note that the fitted value of ψ agrees well with the analytically derived size effect on the modulus of rupture of three point bend beams (Bažant & Li 1995*b*; Bažant 1997, 2004*b*, 2005).

After finding these constants, we can now use Eqs. 9.11 and 9.14 to determine the remaining two statistical parameters for the strength cdf of one RVE, μ_G, and δ_G. Figure 9.4 shows the geometry of the beam at the small-size limit, i.e., $D = 4l_0$. Based on the elastic stress field calculated by the engineering beam theory, there are two layers of

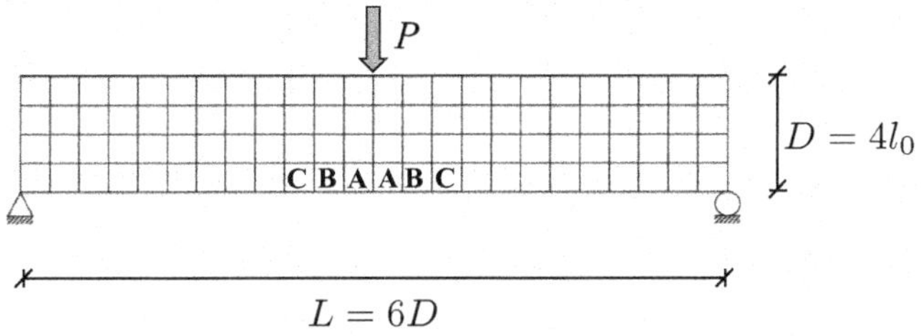

Figure 9.4. Geometry of three point bend beam at the small-size limit.

RVEs in tension. Along the beam depth, the stress at the center of the bottom-layer RVEs is twice larger than the stress at the center of the upper-layer RVEs. Along the beam span, the RVE stress decays from the mid-span. The average stress in the two RVEs at the mid-span (RVEs A in Fig. 9.4) is about 9.5% higher than the average stress in the RVEs adjacent to them (RVEs B in Fig. 9.4), and is about 26% higher than the average stress in the next adjacent RVEs (RVEs C in Fig. 9.4). Given the large value of the Weibull modulus, it is reasonable to include only the bottom-layer RVEs A and B for the calculation. Therefore, we can rewrite Eqs. 9.11 and 9.14 as

$$\int_0^\infty \prod_{i=1}^{2}\left[1-\Psi\left(\frac{\eta_i\sigma_N-\mu_G}{\delta_G}\right)\right]^2 d\sigma_N = \left[\frac{N_a}{4l_0}+\left(\frac{N_b}{4l_0}\right)^{\psi n_d/m}\right]^{1/\psi} \tag{9.24}$$

$$\begin{aligned}\frac{1}{l_0}\int_0^\infty \prod_{i=1}^{2}\left[1-\Psi\left(\frac{\eta_i\sigma_N-\mu_G}{\delta_G}\right)\right]^2 \sum_{i=1}^{2}\ln\left\{\left[1-\Psi\left(\frac{\eta_i\sigma_N-\mu_G}{\delta_G}\right)\right]\right\} d\sigma_N = \\ -\frac{1}{\psi}\left[\frac{N_a}{16l_0^2}+\frac{\psi n_d}{4l_0 m}\left(\frac{N_b}{4l_0}\right)^{\psi n_d/m}\right]\left[\frac{N_a}{4l_0}+\left(\frac{N_b}{4l_0}\right)^{\psi n_d/m}\right]^{1/\psi-1}\end{aligned} \tag{9.25}$$

We can numerically solve μ_G and δ_G from Eqs. 9.24 and 9.25, which yields $\mu_G = 45.24$ MPa, and $\delta_G = 14.82$ MPa. With the finite weakest-link model (Eq. 7.1), we can then predict the strength distributions of beams with sizes $D = 28.9$ mm and $D = 50$ mm.

Figure 9.2 shows a comparison between the predicted and measured strength distributions. Obviously, they agree well with each other for both beam sizes, which verifies the present method for determining the strength statistics from the mean strength size effect curve. Now we use the calibrated finite weakest-link model to calculate the strength distribution and the mean strength of beams with other different sizes. We see from Figure 9.3 that the mean structural strength predicted from the weakest-link model lies on the size effect curve represented by Eq. 9.2.

9.3 Determination of Large-Size Asymptotic Properties of the Size Effect Curve

When the amount of available material is limited, we may perform the mean strength test on direct tension specimens of only one size, even though this is insufficient to obtain the large-size asymptote of the size effect curve and to determine the Weibull modulus m and constant N_b. The Weibull modulus m may be estimated from the measured strength histogram, and the constant N_b is subsequently determined from the Weibullian mean of the strength of direct tension specimens. This would be unnecessary if we performed the mean strength test on the direct tension specimens of two different sizes.

Assume that, besides the present direct tension tests on specimens with N_T RVEs whose mean strength is $\bar{\sigma}_N^T$, we know the mean strength of the direct tension specimens of another sufficiently large size (say that the specimen consists of $N_{T,2}$ RVEs and the mean strength is $\bar{\sigma}_{N,2}^T$). Based on the Weibull statistics (Eq. 9.18), the Weibull modulus m can simply be determined from the mean strengths of specimens of these two sizes:

$$m = \frac{\ln(N_{T,2}) - \ln(N_T)}{\ln(\bar{\sigma}_N^T) - \ln(\bar{\sigma}_{N,2}^T)} \tag{9.26}$$

Knowing m, we could then calculate the sizes of the three point bend beams equivalent to these two direct tension specimens (Eq. 9.20). These two points would form the large-size asymptote of the size effect curve, from which the constant N_b could be determined easily.

9.4 Comparison with the Histogram Testing Method

Previous studies have shown that, according to the finite weakest-link model, the strength distribution of one RVE could be determined by fitting the strength histogram. The detailed fitting algorithm can be found in Le & Bažant (2009). It is worthwhile to compare the proposed size effect method and the conventional histogram testing method.

One advantage of the proposed method over the histogram testing method is the number of specimens needed. The proposed method only requires testing of the mean strengths of geometrically similar specimens of several different sizes, where three to five specimens are needed for each structure size and the size effect test should involve about three to five different sizes. Therefore, the total number of specimens required for the proposed method is about 20 specimens, which is much fewer than the number of specimens required for conventional histogram testing.

Another advantage of the present method is that, in contrast to individual tests, the coefficient of variation of the random experimental error of the mean strength obtained by regression decreases (according to the Central Limit Theorem) as the $-1/2$ power of the number of tests. The individual tests entering the strength histogram have a much larger scatter, and so many more tests are needed [the reason why Weibull's (1939) histograms are smooth is that he plotted points obtained as averages of groups of about 50 specimens each].

It is evident that the present method requires a mean size effect curve with a large size range. The size effect curve in the small-size range is usually easy to obtain, whereas the large-size asymptote, which is crucial for the determination of the Weibull modulus m and constant N_b, can be a challenge for some types of specimens owing to the size

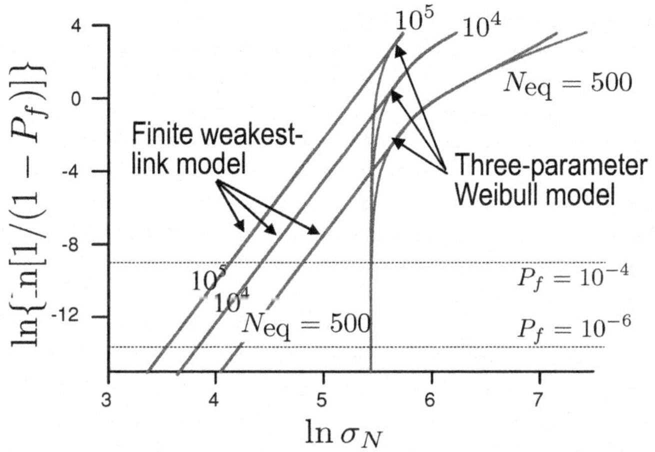

Figure 9.5. Strength distributions of geometrically similar specimens predicted by the finite weakest-link model and the three-parameter Weibull model.

constraint of the loading machine (e.g., three point bend beams considered here). This problem can be mitigated by varying both the size and geometry of the test specimens.

9.5 Problems with the Three-Parameter Weibull Distribution of Strength

It is evident that the observed strength histograms of quasibrittle structures (Figs. 8.8, 8.9, and 9.2) deviate significantly from the two-parameter Weibull distribution. As a remedy, the three-parameter Weibull distribution with a nonzero threshold has been widely accepted as an empirical probability distribution. It can be written as

$$P_f(\sigma_N) = 1 - \exp\left(-\int_V \langle \sigma_N s(\boldsymbol{x}) - \sigma_u \rangle^{m_w} / s_w^{m_w} \frac{\mathrm{d}V}{l_0^{n_d}}\right) \tag{9.27}$$

where σ_u = strength threshold, m_w = Weibull modulus of the three-parameter Weibull distribution, and s_w = scale parameter. Figure 9.5 shows a schematic plot of the strength cdfs of geometrically similar specimens predicted by the finite weakest-link model and the three-parameter Weibull model. It is clear that the three-parameter Weibull model would yield a much higher design strength for a low failure probability (e.g., $P_f = 10^{-6}$). Even though the three-parameter Weibull distribution may give, for a limited number of tests, a better fit of measured strength histograms compared to the two-parameter Weibull distribution, it is physically unsound and unsafe in extrapolation to $P_f = 10^{-6}$, for three reasons that follow.

9.5.1 Theoretical Argument

Based on Chapters 4 and 5, the zero strength threshold for the Weibull distribution ensues from (1) the classical transition rate theory, where the activation energy barriers for forward and backward jumps on the curve of the free energy potential of the

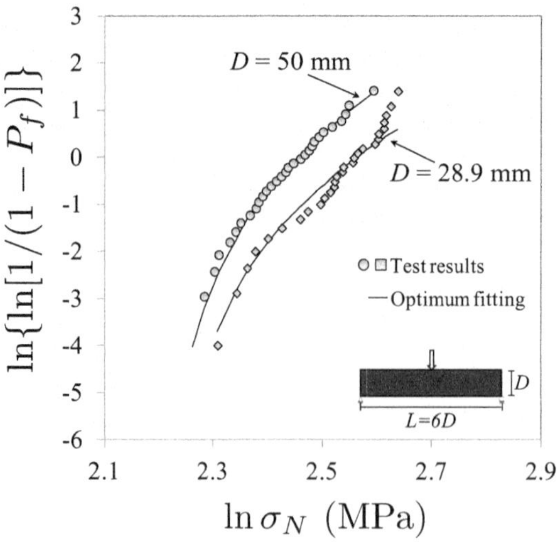

Figure 9.6. Optimum fits of strength distributions of asphalt mixture beams by the three-parameter Weibull model.

nanoscale structure as a function of atomic crack length must differ very little; and (2) the hypothesis that bridging up the ascending scales is equivalent to some combination of parallel and series couplings.

The finite threshold in strength cdf is inconsistent with the transition rate theory for stress-driven fracture. By reverse reasoning, it would require a finite threshold for the transition rate theory. But this is impossible. Besides, the zero threshold in the Gauss–Weibull distribution is not unrealistic because, for example, the P_f for stress equal to 10% of the mean strength is, despite zero threshold, of the order of 10^{-24} (which translates into a lifetime much longer than the age of the Earth).

Meanwhile, the random walk analysis in Chapter 4 (Section 4.3) indicates that the diffusion-driven failure of a nanostructure could happen at zero stress albeit the time-to-failure is extremely long (for more than the age of universe). Contrary to this result, the three-parameter Weibull model predicts that the structure will never fail if the applied stress is below the strength threshold.

9.5.2 Evidence from Histogram Testing

Figure 9.6 shows the optimum fitting of the measured strength histograms of asphalt mixture beams of series B and C by the three-parameter Weibull distribution. As seen, the fitting is as good as that by the finite weakest-link model. But are the histograms sufficiently broad to show the difference? They are not.

Significantly broader histograms, consisting of about 2868 data, have been obtained for Portland cement mortar, which is also a quasibrittle material and exhibits virtually the same fracture behavior as tough ceramics; see Weibull's tests (Weibull 1939) in Figure 9.7, presented in the Weibull scale with zero threshold, for three different ages

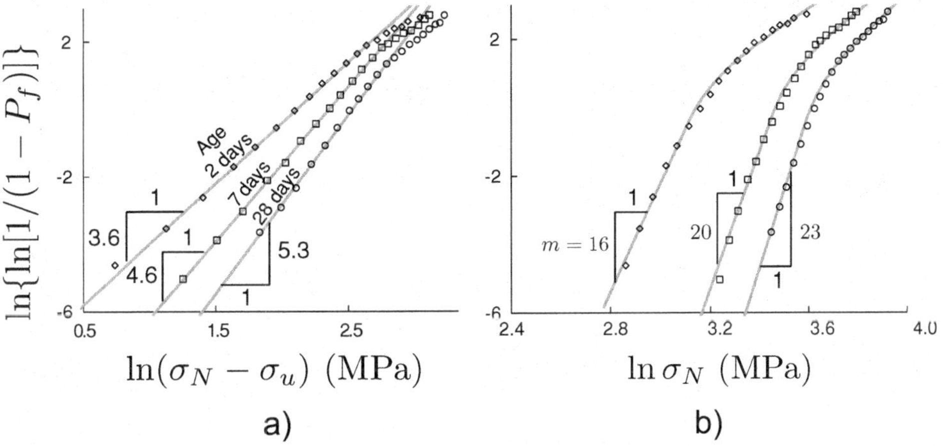

Figure 9.7. Optimum fits of strength distributions of Portland cement mortar by (a) the three-parameter Weibull model and (b) the finite weakest-link model.

of mortar, representing three different degrees of chemical hydration. In contrast to Figure 9.6, where each point represents just one test, each circle point in Figure 9.7 represents the mean strength of a group of about 50 specimens. This is what makes these histograms so smooth and unambiguous.

As seen in Figure 9.7, Weibull's broader histograms can again be fitted perfectly by the finite weakest-link model. However, the fit by Weibull distribution with nonzero threshold is less than perfect. Small but systematic deviations are seen at the high-probability end. So we must conclude that the overall experimental evidence, as it now exists, mildly favors the zero threshold. A clearer evidence would be desirable because predictions of both theories for very low failure probabilities such as $P_f = 10^{-6}$, which matter for practical design, are very different. But that would require a forbidding amount of testing.

It should also be pointed out that, in the present experiments, beam series B and C have a very limited size range, which is insufficient for checking the functional form of the strength cdf. For a wider size range, i.e., at least 1:4, we would be able to see the difference between the finite weakest-link model and the three-parameter Weibull distribution even with a relatively small number of specimens (say, about 30 specimens).

9.5.3 Mean Size Effect Analysis

It is clear that, to make the experimental evidence from histograms unambiguous, an enormous amount of testing would be needed. Therefore, we must seek and analyze experimental evidence for other predictions of the theory. Among these, the size effect is the most effective.

From the optimum fits of the strength histograms of beam series B and C by the three-parameter Weibull distribution (Fig. 9.6), we obtain the following values of the parameters: $\sigma_u = 8.8$ MPa, $m_w = 1.4$, and $s_w = 110$ MPa. Based on these parameters, we

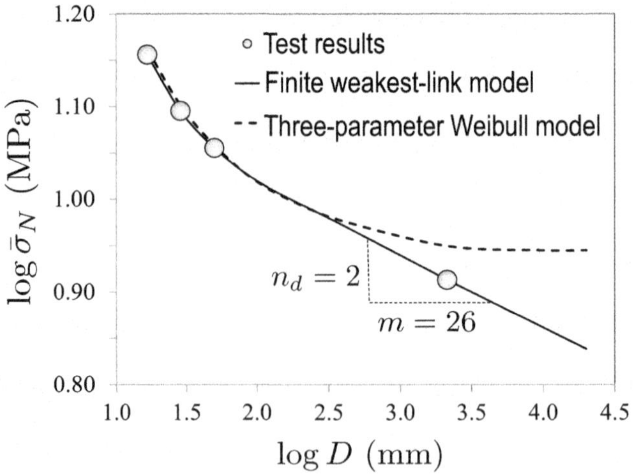

Figure 9.8. Comparison of mean size effect curves predicted by the finite weakest-link model and the three-parameter Weibull model.

can predict the mean structural strengths of beam series A and direct tension specimens: σ_N (beam series A) = 14.51 MPa, and σ_N (direct tension specimens) = 8.84 MPa. The measured mean strengths for these two specimen series are 14.31 MPa and 8.18 MPa, respectively. Clearly the three-parameter Weibull distribution fails to predict the mean strength of direct tension specimens, which corresponds to the large-size asymptote of the size effect curve. The present size effect test supports the finite weakest-link model as opposed to the three-parameter Weibull distribution.

To obtain a more complete picture, we plot the size effect curves predicted by the finite weakest-link model and the three-parameter Weibull distribution (Fig. 9.8). It can be seen that a significant deviation occurs at the large-size limit. Such a difference can be explained by considering a simple case where all the RVEs are subjected to the same stress. The three-parameter Weibull distribution yields the mean structural strength as

$$\bar{\sigma}_N = \int_0^\infty (1 - P_f)\,\mathrm{d}\sigma_N \tag{9.28}$$

$$= \int_0^\infty \left[\exp\left(-\left\langle\frac{\sigma_N - \sigma_u}{s_w}\right\rangle^{m_w}\right)\right]^{V/V_0} \mathrm{d}\sigma_N \tag{9.29}$$

$$= \sigma_u + s_w \Gamma\left(1 + 1/m_w\right)(V_0/V)^{1/m_w} \tag{9.30}$$

It is evident that the mean size effect curve predicted by the three-parameter Weibull distribution approaches the strength threshold at the large-size limit.

In contrast, the finite weakest-link model indicates that the large-size asymptote of the size effect curve must follow the classical Weibull size effect, i.e., $\sigma_N \propto D^{-n_d/m}$. For structures with nonuniform stress distribution, the three-parameter Weibull distribution also leads to a vanishing size effect at the large-size limit, which gives a constant nominal strength. Such a large-size asymptote of the size effect curve is inconsistent

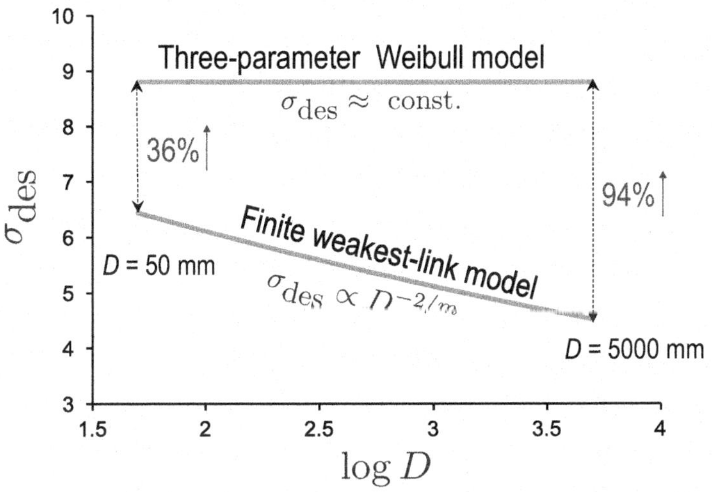

Figure 9.9. Design strengths corresponding to $p_f = 10^{-6}$ predicted by the three-parameter Weibull model and the finite weakest-link model.

with the experimental observations. The Weibull asymptote at the large-size limit is also clearly observed for the size effect on the modulus of rupture of concrete (Bažant 2005) presented in Chapter 3 (Fig. 3.15).

Therefore, we must conclude that the three-parameter Weibull distribution gives an incorrect mean strength behavior at the large-size limit. We can also invoke fine-grain ceramics or embrittled metal, in which the fracture process zone size is of the order of a micrometer or submicron. These materials behave as quasibrittle on the micrometer scale (Le et al. 2015). If we used the three-parameter Weibull distribution on that scale, instead of the Gauss–Weibull cdf, we would conclude from the weakest-link model that the statistical size effect on the macroscale (>0.01 m) would be nil, which is obviously wrong.

It is crucial to point out the consequence of using the three-parameter Weibull distribution for predicting the design strength σ_{des}, which corresponds to a low failure probability $P_f = 10^{-6}$. As shown in Figure 9.5, the three-parameter Weibull distribution would overestimate the design strength, and the degree of the overestimation gets larger as the specimen size increases. This is due to the fact that, for the finite weakest-link model, the strength cdf at such a low failure probability would follow a Weibull distribution. It is clear that the corresponding design strength would scale with the specimen size as $\sigma_{des} \propto D^{-n_d/m}$ (Section 2.3). However, the three-parameter Weibull distribution would predict that this design strength would converge to the strength threshold, as seen in Figure 9.5.

Following the aforementioned optimum fitting of strength histograms of asphalt mixtures by the finite weakest-link model and the three-parameter Weibull distribution, we plot the design strength σ_{des} predicted by these two models, as shown in Figure 9.9. It is seen that, for the specimens of the laboratory test scale, the three-parameter Weibull model overestimates the design strength by 36%, whereas for large-size specimens ($D = 5$ m), the three-parameter Weibull model would overestimate the design strength by 94%. Therefore, it is particularly unsafe to use the three-parameter Weibull distribution to predict the design strength of large-size structures for a low failure probability.

9.6 Alternative Proof of Strength Distribution of an RVE Based on Stability Postulate and Atomistic Analysis

Now that the theory of strength distribution and size effect has been thoroughly presented, it is appropriate to mention that the necessity of a grafted strength distribution can be proven by an alternative simple argument.

It is expected that the strength cdf satisfies the Gnedenko condition for extreme value statistics (Gnedenko 1943), which limits the choices to the three known extreme value distributions (Fréchet, Gumbel, and Weibull). Among these functions, only the Weibull cdf is admissible for strength, and since it has a power-law left tail, we obtain, independently of atomistic considerations and scale transitions, an *alternative proof* that the cdf tail of a single RVE must be a power law (or, more precisely, a function within the attraction domain of Weibull's extreme value distribution, which in practice means a power law). Indeed, if it were not, the infinite chain could not have the Weibull cdf (see Sec. 2.6).

This alternative proof, of course, does not imply anything for the cdf core of each RVE. This proof would, for instance, allow the entire cdf of an RVE to be Weibullian (as often erroneously assumed). In particular, it does not prove that the core must be Gaussian. This crucial feature can be proven, without atomistic analysis, by the fact that the strength of an RVE must be the sum of contributions from subscale elements within the RVE, which are forced to deform in overall conformity with the deformation of the RVE (and thus act as parallel coupling on the subscale of the RVE). On the other hand, this summation alone proves nothing about the tail (like the Central Limit Theorem, Sec. 2.7).

The foregoing arguments, however, do not suffice to prove that the threshold of strength distribution must be zero. The atomistic analysis, presented in Chapter 4, seems to be the only way to prove it *theoretically*.

The zero value of the strength threshold is a fundamental property, which has so far been ignored by many researchers and users although it makes enormous difference for structural safety (Sec. 9.5). This property is, in this book and the cited articles, amply verified experimentally – by *simultaneous* fitting (with the same material parameters) of strength histograms and mean size effect data.

Note (added in proof): It is interesting that the foregoing arguments do not suffice to determine the pdf of strength of imbricated lamellar materials such a nacre. In ongoing work at Northwestern University, these materials are statistically treated as a fishnet pulled diagonally. It turns out that the fishnet statistics is again transitional between the Gaussian and Weibull distributions but does lead to a grafted Gauss-Weibull distribution with a Weibull tail spreading into the core as the fishnet size increases. Rather the transition occurs gradually over the whole probability range.

10 Statistical Distribution and Size Effect on Residual Strength after Sustained Load

In many engineering applications, the risky overload would not occur right at the first load application. This chapter, based mostly on Salviato, Kirane, & Bažant (2014), deals with the residual strength of a structure at an accidental sudden overload that occurs later, after a period of sustained load high enough to weaken the structure. The residual strength is also important in the laboratory since, as will be shown, it is advantageous to replace lifetime tests by residual strength tests.

Practical examples of overload include an earthquake, a sudden excessive traffic load on a bridge, an excessive rise of water level behind a dam, accidental impact, or sudden overheating of an electronic component. Knowledge of the statistics of residual strength should allow improvements in the safety factors, taking into account the strength degradation of the structure depending on the load history and duration. It should also allow meaningful estimates of the remaining service life of structures for which maintenance design is a primary concern. This is of paramount importance from the perspective of cost reduction and safety, especially for modern large aircraft made of load-bearing quasibrittle composites (Lee, Ma, Thimm, & Verstraeten 2008).

The residual strength of different materials has been widely studied phenomenologically, but mainly for the case of cyclic loading. For example, Yang and Liu proposed a model for residual strength degradation and periodic proof tests for graphite-epoxy laminates under cyclic loading (Yang & Liu 1977; Yang 1978). A mechanistic attempt was provided by Halpin, Johnson, & Waddoups (1972), based on the kinetics of fracture growth. There have been some experimental studies, e.g., for concrete (Award & Hilsdorf 1972), but strictly deterministic. Unfortunately, no information exists in the literature on the residual strength in the statistical sense. In those few attempts that provided a statistical perspective (Kirchner & Walker 1971; Evans 1974; Hahn & Kim 1975; Thomas, Verrilli, & Calomino 2002; Duffy et al. 2003), the quasibrittleness of the structure was not considered.

In this chapter, following Salviato et al. (2014), we use the same framework as in Chapters 5 to 7 to extend the theory to the probabilistic distributions of residual strength after a period under sustained constant load. A relation between the initial strength and the residual strength for one representative volume element (RVE) is derived and then combined with the Gauss–Weibull distribution of strength to get the cumulative distribution function (cdf) of residual strength of RVE. The cdf of the residual strength of a structure of any size is determined within the framework of the finite weakest-link theory, and is validated by means of available test data on borosilicate and soda-lime

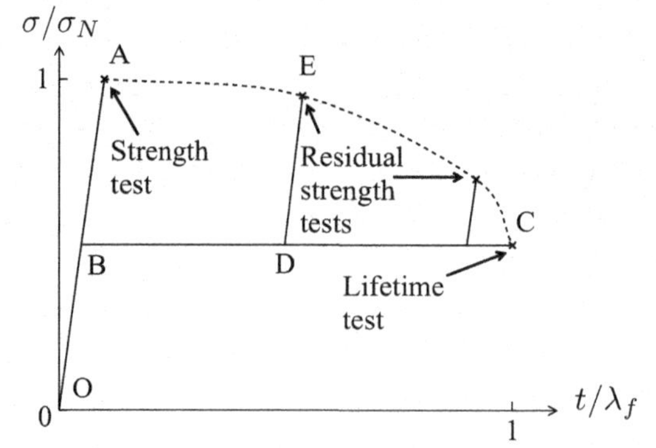

Figure 10.1. Schematic of loading histories of strength, lifetime, and residual strength tests.

silicate glasses. Finally, the size effect on the mean residual strength is examined and shown to be just as strong as the size effect on the initial strength.

10.1 Nanomechanics Based Relation between Monotonic Strength and Residual Strength of One RVE

As explained in Chapter 6, the subcritical crack growth under sustained load may be described by the Charles–Evans law (Charles 1958*a*, 1958*b*; Evans 1972) [which was justified in Bažant et al. (2009), Bažant & Le (2009), and Le et al. (2011) by atomistic arguments and scale bridging based on progressively refined energy arguments]. This law reads

$$\dot{a} = A\, e^{-Q_0/kT} K_R^{n_c} \tag{10.1}$$

where a = crack length, $\dot{a} = \mathrm{d}a/\mathrm{d}t$ (t = time); A = material constant, Q_0 = activation energy, k = the Boltzmann constant, and T = absolute temperature. The mode I stress intensity factor is denoted as K_R where the subscript R indicates the RVE level. So, $K_R = \sigma\sqrt{l_0}k_R(\alpha)$ where σ = nominal stress, which is defined as $\sigma = P/bl_0$, l_0 = RVE size, b = width of the RVE in the transverse direction, $\alpha = a/l_0$ = relative crack length, and k_R = dimensionless stress intensity factor. The foregoing equation thus becomes

$$\dot{a} = Ae^{-Q_0/kT}\sigma^{n_c} l_0^{n_c/2} k_R^{n_c}(\alpha) \tag{10.2}$$

In Chapter 7, the derivation of Eq. 7.11 led to a simple relation between the monotonic strength and the static fatigue lifetime of quasibrittle structures. We now apply the same approach to the residual strength test. Consider different load histories as illustrated in Fig. 10.1. The load history O–A corresponds to the (short-time monotonic) strength test and O–B–C to a static lifetime test. The history O–B–D–E corresponds to a residual strength test.

First consider load history O–A in which the load is increased linearly and rapidly until failure. After separating the variables from Eq. 10.2 we get

$$\frac{e^{Q_0/kT}}{A l_0^{(n_c-2)/2}} \frac{d\alpha}{k_R{}^{n_c}(\alpha)} = \sigma^{n_c}\, dt \tag{10.3}$$

Now let α_0 be the initial relative crack length and α_c the critical relative crack length. The loading rate is expressed as $r = \sigma/t$. Integration of the foregoing equation for history O–A, with appropriate limits on both sides, gives

$$\frac{1}{r}\int_0^{\sigma_N} \sigma^{n_c}\, d\sigma = e^{Q_0/kT} \int_{\alpha_0}^{\alpha_c} \frac{d\alpha}{A l_0^{(n_c-2)/2} k_R{}^{n_c}(\alpha)} \tag{10.4}$$

where the substitution $dt = d\sigma/r$ has been made. This yields

$$\sigma_N{}^{n_c+1} = r(n_c+1)e^{Q_0/kT} \int_{\alpha_0}^{\alpha_c} \frac{d\alpha}{A l_0^{(n_c-2)/2} k_R{}^{n_c}(\alpha)} \tag{10.5}$$

Next, we consider the lifetime test (history O–B–C), where the load is increased fast to σ_0 at a constant rate r and then held constant until failure. The time required for failure is the static lifetime denoted by λ_f. Integrating the same crack growth equation for this load history, we get

$$\sigma_0{}^{n_c+1} + r(n_c+1)\sigma_0{}^{n_c}(\lambda_f - t_0) = \sigma_N{}^{n_c+1} \tag{10.6}$$

Now consider the loading history O–B–D–E in which the applied stress is first raised rapidly from 0 to $\sigma_0 = t_0 r$, where r = loading rate. Then the applied stress is sustained for the period $t_R - t_0$, and at time t_R the applied stress is rapidly raised up to failure. The failure stress represents the residual strength σ_r. Integration over this history furnishes (Salviato et al. 2014)

$$\frac{1}{r}\int_0^{\sigma_0} \sigma^{n_c}\, d\sigma + \int_{t_0}^{t_R} \sigma_0{}^{n_c}\, dt + \frac{1}{r}\int_{\sigma_0}^{\sigma_r} \sigma^{n_c}\, d\sigma = e^{Q_0/kT} \int_{\alpha_0}^{\alpha_c} \frac{d\alpha}{A l_0^{(n_c-2)/2} k_R{}^{n_c}(\alpha)} \tag{10.7}$$

This yields

$$\sigma_r{}^{n_c+1} + r(n_c+1)\sigma_0^{n_c}(t_R - t_0) = r(n_c+1)e^{Q_0/kT} \int_{\alpha_0}^{\alpha_c} \frac{d\alpha}{A l_0^{(n_c-2)/2} k_R{}^{n_c}(\alpha)} \tag{10.8}$$

Now, substituting from Eq. 10.5, we get (Salviato et al. 2014),

$$\sigma_r{}^{n_c+1} + r(n_c+1)\sigma_0^{n_c}\left(t_R - \frac{\sigma_0}{r}\right) = \sigma_N{}^{n_c+1} \tag{10.9}$$

$$\text{or} \quad \sigma_r = \left[\sigma_N{}^{n_c+1} - \sigma_0^{n_c}(n_c+1)(r t_R - \sigma_0)\right]^{1/(n_c+1)} \tag{10.10}$$

Equation 10.10 describes the degradation of the residual strength as a function of two independent (deterministic) variables, applied load σ_0 and duration t_R of a sustained load application. This equation also represents a link between the short-time strength and the residual strength. The short-time strength test and the lifetime test can be identified as two limiting cases of this equation. For the strength test, $\sigma_0 = 0$ and $t_R = t_0$. Substituting these values, Eq. 10.10 reduces to $\sigma_r = \sigma_N$. Similarly, for the lifetime test, $t_R = \lambda_f$, and thus a comparison of Eq. 10.6 and Eq. 10.9 yields $\sigma_r = \sigma_0$.

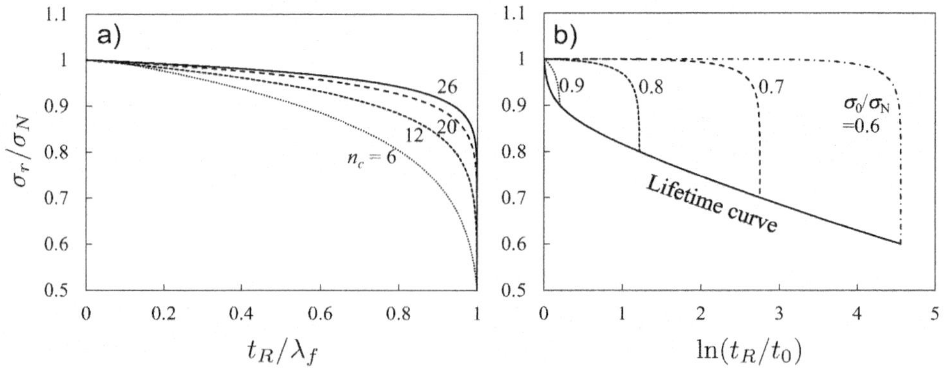

Figure 10.2. Predicted degradation curves for (a) various values of static crack growth exponent n_c and (b) various values of applied load.

10.2 Analysis of Residual Strength Degradation for One RVE

We may now proceed to analyze the effect of various parameters on Eq. 10.10 for the residual strength degradation. Figure 10.2a shows the degradation in strength of one RVE under static load for various values of n_c, and for $\sigma_0 = 0.5\sigma_N$. The time of load application normalized with respect to the lifetime is shown on the horizontal axis. For convenience, σ_N is assumed to be 1 MPa and the loading rate is taken as 0.5 MPa/s. As expected, the normalized residual strength decreases with the applied load until it reaches the lower limit σ_0/σ_N. In other words, the end condition that $\sigma_r = \sigma_0$ for $t_R = \lambda_f$ is satisfied.

As can be seen, the rate of strength degradation is initially negligible but then it progressively increases. Rapid strength degradation is seen in the end. This effect gets more and more pronounced for higher values of n_c. For, e.g., $n_c = 6$, the residual strength σ_r drops to 90% of the original strength σ_N after the hold time of $t_R = 0.543\lambda_f$, whereas for $n_c = 20$ and $n_c = 26$, this occurs after the hold time of $t_R = 0.873\lambda_f$ and $t_R = 0.929\lambda_f$, respectively. In view of this observation, the degradation curve can roughly be divided in *two regions*, one of relatively slow degradation and one of rapid degradation. The distinction between these two regions becomes more pronounced with high values of n_c.

Now consider the effect of applied load on the predicted degradation in strength. Figure 10.2b shows the normalized residual strength as a function of $\ln(t_R/t_0)$, for different values of applied load σ_0. Again, σ_N is assumed to be 1 MPa and the loading rate is taken as 0.5 MPa/s. A subcritical crack growth exponent $n_c = 26$ is assumed here, which is typical for ceramics such as alumina (Fett & Munz 1991; Kawakubo 1995).

For a given applied load, increasing the time of load application leads to a decrease of the residual strength until the lower limit σ_0 is reached, which occurs at the intersection with the lifetime curve. Again, it can be seen that the rate of degradation continuously increases. When $\sigma_0 = 0.6\sigma_N$, the residual strength is almost equal to the structural strength for about 90% of its lifetime. Then it drops rapidly toward the σ_0 value. As the applied load is increased, the lifetime is reduced and the curves become steeper and steeper. For $\sigma_0 = 0.9\sigma_N$, the residual strength drops to σ_0 within a small fraction of the lifetime.

To document the large effect of the applied load level on the lifetime, it may be noted that the ratio of the lifetimes of the last and the first curves (i.e., those for $\sigma_0 = 0.6\sigma_N$ and $\sigma_0 = 0.9\sigma_N$, respectively) is about 1.5×10^4. A similar enormous ratio was reported in Yang and Liu (1977), Yang (1978), and Diao, Mai, and Mai (1995) for strength degradation of composites under cyclic loading; in Nielsen (1996) for wood under static and cyclic loading; in Park and Lee (1997) for ceramics; and in Yavuz and Tessler (1993) for silicon carbide ceramics subjected to sustained loading at high temperatures.

This discussion reveals the usefulness of Eq. 10.10. For a given applied load, this equation can help determine the portion of lifetime for which the strength degradation is negligible. The only other parameter needed for this determination is the exponent n_c of the static crack growth law.

10.3 Probability Distribution of Residual Strength

10.3.1 Formulation of Statistics of Residual Strength for One RVE

In Chapter 7, we showed that the lifetime cdf of one RVE can be derived from the strength cdf and the relation between the monotonic strength and lifetime. Similarly, starting from the strength cdf, it is now possible to determine the cdf of the residual strength of one RVE by means of Eq. 10.10. Recall the grafted Gauss–Weibull cdf for RVE strength:

$$P_1(\sigma_N) = 1 - \exp\left(-\sigma_N^m/s_0^m\right) \qquad (\sigma_N < \sigma_{N,\mathrm{gr}}) \tag{10.11a}$$

$$P_1(\sigma_N) = P_{\mathrm{gr}} + \frac{r_f}{\sqrt{2\pi}\,\delta_G}\int_{\sigma_{N,\mathrm{gr}}}^{\sigma_N} e^{-(\sigma'-\mu_G)^2/2\delta_G^2}\,\mathrm{d}\sigma' \qquad (\sigma_N \geq \sigma_{N,\mathrm{gr}}) \tag{10.11b}$$

Now consider the residual strength σ_r' as a random variable. With Eq. 10.10, the cdfs of the residual strength and the monotonic strength can be related as

$$\Pr(\sigma_r' \leq \sigma_r) = \Pr\left[\sigma_N' \leq \left(\sigma_r^{n_c+1} + \sigma_A\right)^{1/(n_c+1)}\right] \tag{10.12}$$

where σ_N' = random monotonic strength of RVE, and $\sigma_A = \sigma_0^{n_c}(n_c+1)(rt_R - \sigma_0)$. Equation 10.12 yields (Salviato et al. 2014)

$$P_{1,R}(\sigma_r) = 1 - \exp\left[-\left(\langle\sigma_r^{n_c+1} + \sigma_A\rangle/s_R\right)^{\overline{m}}\right] \qquad (\sigma_0 \leq \sigma_r < \sigma_{r,\mathrm{gr}}) \tag{10.13a}$$

$$P_{1,R}(\sigma_r) = P_{\mathrm{gr}} + \frac{r_f}{\sqrt{2\pi}\,\delta_G}\int_{\sigma_{r,\mathrm{gr}}}^{(\sigma_R^{n_c+1}+\sigma_A)^{1/(n_c+1)}} e^{-(\sigma'-\mu_G)^2/2\delta_G^2}\,d\sigma' \qquad (\sigma_r \geq \sigma_{r,\mathrm{gr}} > \sigma_0) \tag{10.13b}$$

where $\sigma_{R,\mathrm{gr}} = (\sigma_{N,\mathrm{gr}}^{n_c+1} - \sigma_A)^{1/(n_c+1)}$, and $s_R = s_0^{n_c+1}$, $\overline{m} = m/(n_c+1)$; $P_{1,R}$ represents the probability of failure of one RVE under an overload, and $P_{1,R}(\sigma_0)$ represents the probability of failure of one RVE before the overload is applied. Note that only the part of the cdf where the residual strength is defined, i.e., where $\sigma_r \geq \sigma_0$, is considered.

Unlike the strength distribution, the residual strength cdf of one RVE does not have a pure Weibull tail. The threshold stress σ_A in the cdf stems from the fact that some specimens may fail already during the period of sustained preload, which excludes them from the statistics of the overload (Salviato et al. 2014). These are the specimens for which $\lambda_f < t_R$ or $\sigma_N < \sigma_0$. This is a crucial difference between the cdfs of strength and of residual strength. The threshold is described by σ_A, which is a function of the deterministic parameters σ_0 and t_R. Note also that when t_R or σ_0 becomes sufficiently small, the threshold σ_A becomes negligible because the number of specimens with $\sigma_N < \sigma_0$ or $\lambda_f < t_R$ tends to zero (Salviato et al. 2014). Then, the statistics of residual strength resembles the statistics of structural strength.

Similar to the cdfs of strength and lifetime, the cdf of residual strength is also a graft of two distributions. It was shown in Le et al. (2011) that although the cdf of strength is a Gauss–Weibull graft and the tail of the lifetime cdf is Weibullian, the rest of the lifetime cdf did not exactly follow the Gaussian distribution. Similarly, Eq. 10.13b here indicates that the majority part of the residual strength distribution of one RVE does not exactly follow the Gaussian distribution. Also, note that the Weibull moduli of the cdfs of lifetime and of the $(n_c + 1)$th power of the residual strength are the same.

10.3.2 Formulation of Residual Strength cdf of Geometrically Similar Structures of Different Sizes

As discussed in Chapter 1, quasibrittle structures failing at the initiation of a macrocrack can be statistically modeled as a chain of RVEs. The number of links of the chain is finite because the size of the fracture process zone (FPZ) is not negligible compared to the overall structure size. Therefore, the residual strength of this chain is equal to the smallest residual strength among its links. So, the weakest-link model can be used, in the same way as before, to compute the residual strength cdf of a structure of any size, i.e., a structure consisting of any number of RVEs. As in previous chapters, what matters is the equivalent number of RVEs, N_{eq}, obtained by weighting the RVEs according to the stress level [see Chapter 2 for the definition of N_{eq} (Bažant & Pang 2007; Le & Bažant 2011)].

Similar to the definition of nominal strength, the residual strength, i.e., the nominal applied stress at maximum load, is defined as $\sigma_0 = c_n P/bD$ or $c_n P/D^2$ for 2D- or 3D scaling, where $P =$ the applied load. Then, by applying the joint probability theorem to the survival probabilities, the residual strength distribution of the structure is obtained as

$$P_{f,R}(\sigma_r) = 1 - \prod_{i=1}^{N} \{1 - P_{1,R}[\langle \sigma_0 s(\mathbf{x}_i) \rangle, t_R, \sigma_r]\} \tag{10.14}$$

where $s(\mathbf{x}) =$ dimensionless stress field and $\mathbf{x} =$ position vector. Similar to the chain model for the cdf of structural strength, the residual strength of the ith RVE is here assumed to be governed by the maximum average principal stress $\sigma_0 s(\mathbf{x}_i)$ within the

RVE, which is valid provided that the other principal stresses in one RVE are fully statistically correlated. Similar to strength and lifetime, the probability of residual strength of structures can be effectively calculated with the nonlocal boundary layer model (Le et al. 2011, 2012), described in Chapter 8.

For small values of σ_A and sufficiently large structures (large N or N_{eq}), the weakest-link model shows the residual strength cdf to be determined by the far-left tail of the residual strength cdf of one RVE: $P_{1,R}(\sigma_r) = (\langle \sigma_r^{n_c+1} + \sigma_A \rangle / s_R)^{\overline{m}}$. Therefore, noting that $r = s(\mathbf{x})\, \mathrm{d}\sigma/\mathrm{d}t$, Eq. 10.14 can be rewritten as (Salviato et al. 2014)

$$P_{f,R}(\sigma_r) = 1 - \exp\left[- \int_V \langle s(\mathbf{x}) \rangle^m \frac{\mathrm{d}V(\mathbf{x})}{V_0} \left(\langle \sigma_r^{n_c+1} + \sigma_A \rangle / s_R \right)^{\overline{m}} \right] \tag{10.15}$$

where $V_0 = l_0^{n_d}$, n_d = number of spatial dimension in which the structure is scaled, and l_0 = size of one RVE. As for the strength, it is convenient to define $N_{\text{eq},\sigma_r} = \int_V \langle s(\mathbf{x}) \rangle^m \, \mathrm{d}V(\mathbf{x})/V_0$; here N_{eq,σ_r} represents the equivalent number of RVEs, which is the number of RVEs under uniform stress for which σ_r would give the same cdf of residual structure strength as does Eq. 10.14 for given σ_0 and t_R. Although, at the outset, N_{eq,σ_r} for residual strength and $N_{\text{eq},\sigma}$ for initial strength have different definitions, it turns out that their values are exactly the same.

The mean residual strength may be calculated as (Salviato et al. 2014)

$$\overline{\sigma}_R = \int_{P_{f,R}(\sigma_0)}^{1} \sigma_R \mathrm{d}P_{f,R} + \int_0^{P_f(\sigma_0)} \sigma_N \mathrm{d}P_f + \sigma_0 [P_{f,R}(\sigma_0) - P_f(\sigma_0)] \tag{10.16}$$

where $P_{f,R}(\sigma_r)$ and $P_f(\sigma_N)$ are the cdfs of the residual strength and of the initial strength, respectively. On the right-hand side, the first term takes into account the failures that occur during the overload, the second term the failures that occur initially before the sustained load is reached, and the third term the failures that occur during the sustained loading.

10.4 Comparison among Strength, Residual Strength, and Lifetime Distributions

Let us first investigate the qualitative behavior of the probability distribution of residual strength. In Chapter 8, we reported the optimum fits of strength and lifetime histograms on 99.6% Al_2O_3 tested by Fett and Munz (1991) (Fig. 8.11). Based on the fitting, the Weibull moduli for strength and lifetime, m and $\overline{m}$, are estimated to be about 30 and 1.1, respectively (Le et al. 2011). According to Eq. 7.21, the exponent n_c of the power-law for the creep crack growth is found to be about 26. It should be emphasized that, except for n_c, all the parameters of the lifetime distribution are determined by the strength histograms. Indeed, with the same set of parameters, the theory gives an excellent fit of the strength histograms and an excellent prediction of the lifetime distribution.

The same parameters are now used to predict the residual strength, which is an intermediate property between the initial strength and lifetime. Figure 10.3, based on Salviato et al. (2014), shows the Weibull scale plot of residual strength for different times t_R

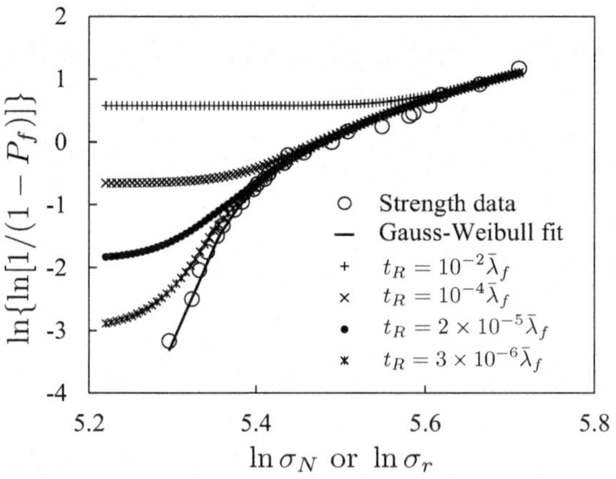

Figure 10.3. Predicted distribution of residual strength of 99.6% Al_2O_3 for various hold times.

of load application. The time is normalized by the mean lifetime, $\bar{\lambda}_f$, and the applied sustained load $\sigma_0 = 0.78\bar{\sigma}_N$ is chosen to correspond, according to histogram tests, to a lifetime well in excess of t_R. For reference, the strength data are shown in the same figure. Only the part of the strength cdf of the specimens that survive the sustained constant preload (i.e., those for which $\sigma_r \geq \sigma_0$) is considered.

The plot shows that, for high values of σ_r, the cdf of residual strength coincides with that of strength. This part of the distribution ensues from a chain of elements whose strength is sufficiently higher than the applied load so that degradation is almost nil. For lower values of σ_r, the distribution of residual strength diverges from the strength cdf and rapidly reaches the lowest probability $P_{f,R}(\sigma_0)$. This probability characterizes the elements of the chain that fail before overload.

For a given value of σ_0, the point at which the distribution begins to diverge shifts to the left for shorter t_R. For sufficiently short t_R, the divergence occurs below the grafting probability. Above the point of divergence, the residual strength distribution almost resembles the Gaussian and a part of the Weibullian distribution. In such a case, the grafting probability values of strength and residual strength are exactly the same. This is due to the fact that the equivalent numbers of RVEs for strength and residual strength have the same values. Because the grafting probabilities of a structure of any size can be calculated as $P_{gr} = 1 - [1 - P_{gr,1}]^{N_{eq}}$, and because the $P_{gr,1}$ values are the same (see Eqs. 10.11a and 10.13a), the grafting probability of the structure must also be the same. In the limiting case where $t_R \to t_0$, the two distributions are, of course, identical.

10.5 Experimental Validation

10.5.1 Optimum Fits of Strength and Residual Strength Histograms of Borosilicate Glass

The preceding section gave an example of predicting the cdf of residual strength based on Gauss–Weibull parameters obtained by fitting strength and lifetime histograms. Now,

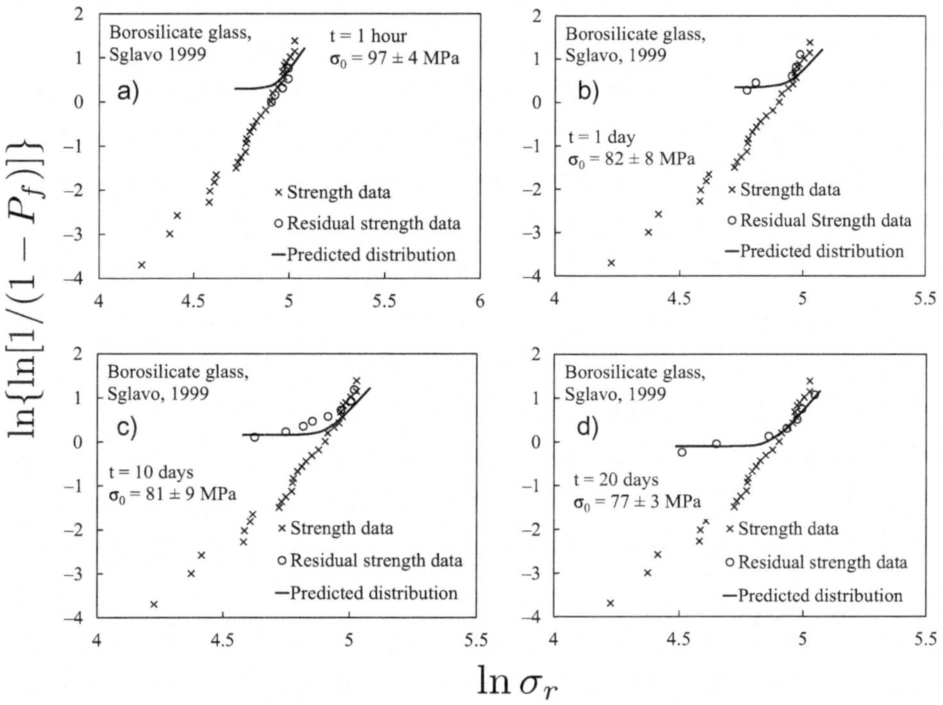

Figure 10.4. Optimum fits of residual strength histograms for borosilicate glass for different hold times: (a) 1 hour, (b) 1 day, (c) 10 days, and (d) 20 days.

following Salviato et al. (2014), these parameters are calculated by fitting the monotonic strength and residual strength histograms, using data for borosilicate glasses (Sglavo & Renzi 1999). Figure 10.4a–d shows the experimentally observed strength and residual strength histograms plotted in the Weibull scale. The reported data were obtained by conducting, in deionized water, four point bend tests of borosilicate glass rods with a nominal diameter of 3 mm and length of 100 mm. The loading rate was about 60 MPa/s, and tests were made for different sustained load durations.

Since the RVE size in glass is far smaller than the size of the specimens tested, the material is brittle rather than quasibrittle and thus the strength distribution is virtually undistinguishable from the Weibull distribution; see Figure 10.4a–d. The optimum fitting of strength and residual strength furnished Weibull modulus $m \approx 6$. For n_c, the value of 30 was estimated.

The fit, predicted by the statistical formulation and shown by the solid lines, is seen to be in good agreement with the experimental results (Salviato et al. 2014). Except for the 1-hour case, all the other plots show the deviation of the residual strength distribution from the strength distribution to reach the probability $P_{f,R}(\sigma_0)$. It should be emphasized that, despite the scatter and the low number of data, all the residual strength distributions are fitted with the same set of parameters.

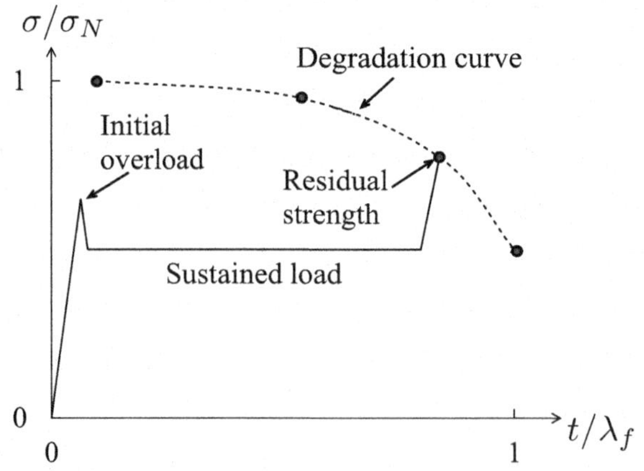

Figure 10.5. Load history in Hahn and Kim's tests.

10.5.2 Optimum Fits of Strength Histograms and Prediction of Lifetime and Mean Residual Strength for Unidirectional Glass/Epoxy Composites

The same methodology is now followed for the strength, lifetime, and residual strength data on unidirectional glass/epoxy composites, reported in Hahn and Kim (1975). Each specimen analyzed consisted of 8 unidirectional plies; 71 specimens were tested to obtain the strength and lifetime distributions. The same constant sustained load of $\sigma_0 =$ 758 MPa was applied in all the lifetime tests. Unlike the previously analyzed cases, an overload σ_p was applied initially. After that, the load was decreased to σ_0, then held constant up to time t_R, and then monotonically increased up to failure. None of the specimens broke before the final overload (Fig. 10.5).

Figure 10.6a shows, in the Weibull scale, the least-square fits of strength histograms by the grafted Gauss–Weibull distribution. Similar to the fits for alumina, this fit shows a kink in the curve where the Weibull distribution transitions to the Gaussian. By optimum least-square fitting, $m = 56$ and $n_c = 27$. Then, based on the parameter set derived from strength, the lifetime distribution is predicted. As can be seen from Fig. 10.6b, the prediction, reported in Salviato et al. (2014) and shown by the solid line, agrees quite well with the test data. Compared to the data scatter, the fitting errors are insignificant.

It should be noted that, prior to Salviato et al. (2014), a methodology to predict the cdf of lifetime from the cdf of strength was proposed in Hahn and Kim (1975). But its usefulness was limited because a purely Weibullian distribution of strength was assumed even though the composite used was surely quasibrittle.

Now that the distribution parameters have been identified, the theory is applied to predict the mean residual strength and compared to the experimental data. A comparison is made only for the mean since the number of available data does not suffice for the entire cdf. Furthermore, since the load history in the tests differed slightly from the one shown in Fig. 10.5, the following equation is used, instead of Eq. 10.10, to compute the strength degradation (Salviato et al. 2014):

$$\sigma_r = \left[{\sigma_N}^{n_c+1} - \sigma_0^{n_c}(n_c+1)rt_R' + 2\left(\sigma_0^{n_c+1} - \sigma_p^{n_c+1}\right)\right]^{1/(n_c+1)} \tag{10.17}$$

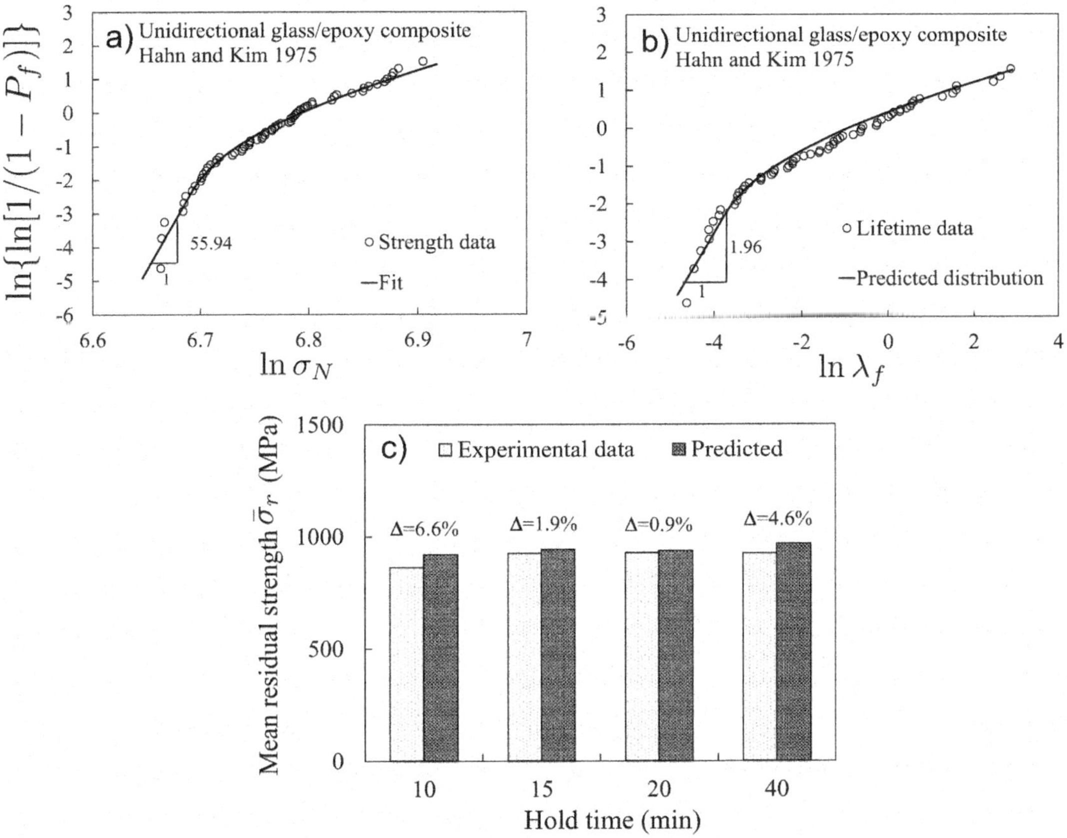

Figure 10.6. (a) Optimum Gauss–Weibull fit of strength histogram, (b) predicted lifetime distribution, and (c) comparison of predicted mean residual strength for unidirectional glass/epoxy composite.

Here σ_p = initial overload and t'_R = time of application of the sustained constant load measured after the initial overload, as reported in Hahn and Kim (1975). Equation 10.17 is derived from Eq. 10.1 similarly as Eq. 10.10 was.

The resulting cdf of residual strength is then used to compute the mean values. No failures before the application of final overload were reported, which is not surprising in view of the small number of specimens tested. Therefore, the means computed from these tests considered only the surviving specimens. The probability of failure of the specimens surviving at the start of overload, denoted as P'_f, can be determined as follows (Salviato et al. 2014):

$$P'_f = \frac{P_{f,R} - P_0}{1 - P_0} \tag{10.18}$$

where $P_0 = P_{f,R}(\sigma_0)$ = probability of failure before applying the final load. The results, taken from Salviato et al. (2014), are shown in Figure 10.6c for the different initial

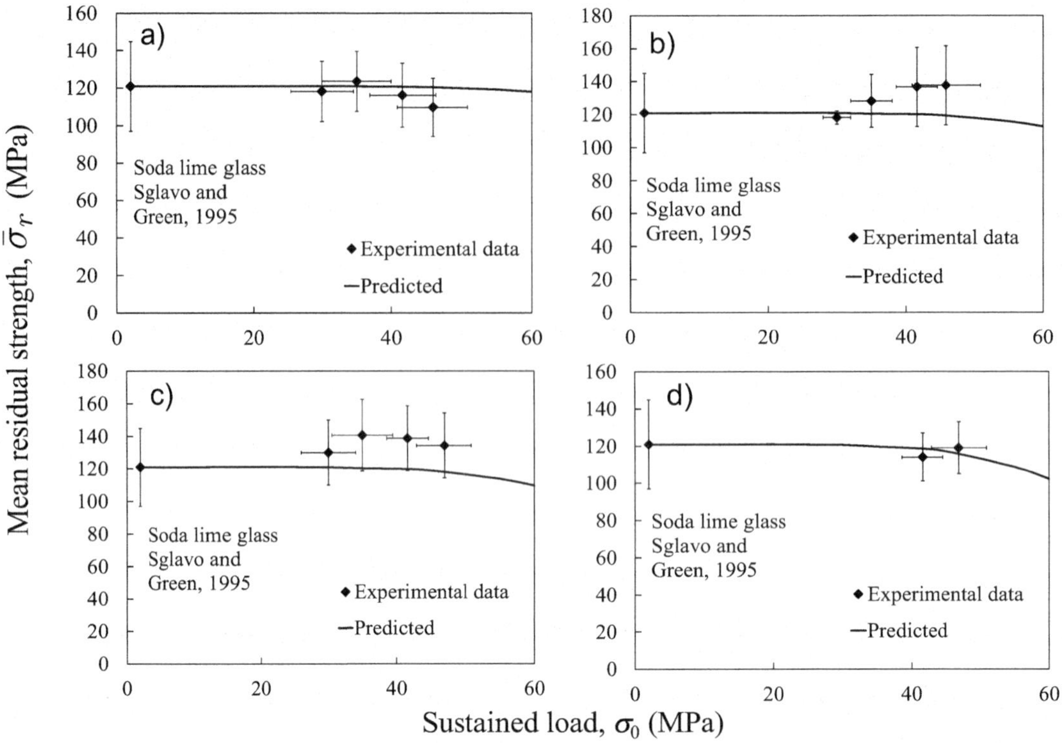

Figure 10.7. Comparison of predicted strength degradation curves with experimental values for soda lime silicate glass at (a) 1 hour, (b) 1 day, (c) 3 days, and (d) 20 days.

overloads and durations. The predictions agree with the experiments quite well, the difference always being less than 7%.

10.5.3 Prediction of Strength Degradation Curve for Soda-Lime Silicate Glasses

The experimental data for verifying the distributions of residual strength after a static preload are scarce. It is nevertheless possible to use other predictions for further verification of the theory.

Consider now the the measurements of mean residual strength of soda-lime glass, as reported in Sglavo & Green (1995), in which the strength histograms were determined by four point bend tests, in deionized water. The beams were soda-lime silicate glass rods. Their nominal diameter was 3.2 mm and length 90 mm. The loading rate was about 47 MPa/s. The mean residual strength was studied for different durations and different applied loads. The parameters of the cdf were in these tests determined by fitting the histograms of strength, which furnished $m = 5.7$. Exponent n_c was assumed as 24, since this is a typical value.

The mean residual strength is now computed according to Eq. 10.16 and is compared to the experimental data in Figure 10.7a–d, where it is shown as a function of σ_0 for

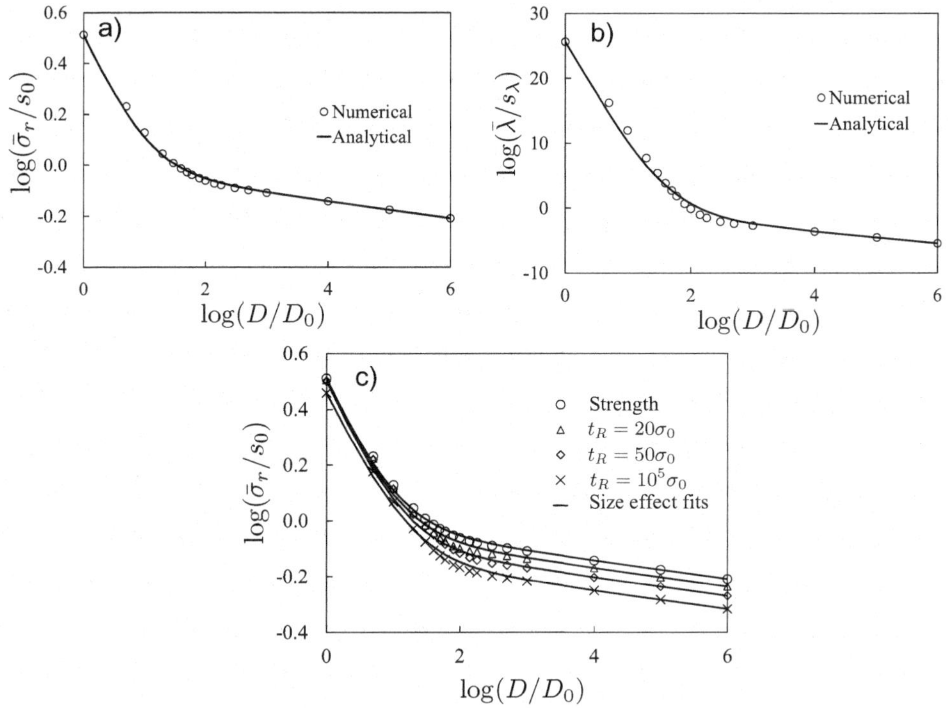

Figure 10.8. Calculated size effect curves on (a) the mean strength, (b) the mean residual strength at different hold times, and (c) the mean lifetime for 99.6% Al_2O_3.

different durations t_R of initial load rise (Salviato et al. 2014). Note that the mean degradation curve of strength agrees very well with the available experimental data. For a fixed duration, the strength is initially almost unaffected by the applied load, while later the effect of the applied load on the strength rapidly increases, lowering the mean residual strength of the structure significantly. Also, the effect of applied load on the strength degradation at higher values of t_R becomes more pronounced.

10.6 Comparison of Size Effects on Mean Strength, Residual Strength, and Lifetime

A more severe check on the theory would be to test the size effects on the mean lifetime and the mean residual strength. However, no such test data seem to be available in the literature. It is nevertheless interesting to predict these size effects from the theory. As discussed in Chapter 7, the mean structural strength and lifetime can be calculated as $\overline{\sigma}_N = \int_0^1 \sigma_N \mathrm{d}P_f$ and $\overline{\lambda} = \int_0^1 \lambda \mathrm{d}P_{f,\lambda}$, where P_f and $P_{f,\lambda}$ are the cdfs of the strength and of the lifetime respectively. When the specimens that have not survived the sustained preload are included, the residual strength can similarly be computed from Eq. 10.16.

Figure 10.8a and b shows the size effect on both the mean structural strength and lifetime of the 99.6% Al_2O_3, predicted on the basis of the strength and lifetime cdfs that were calibrated by Fett and Munz's (1991) histograms (Fig. 8.11). Note that, at the large-size limit, the curves of the size effect on both the mean strength and lifetime tend to straight lines in the logarithmic plot. This agrees well with the power-law size effect of Weibull statistics because the strength and lifetime cdfs approach the Weibull distribution as the structure size increases (Bažant & Pang 2006, 2007; Le & Bažant 2009; Le et al. 2011).

Figure 10.8c, based on Salviato et al. (2014), shows the size effect on the mean residual strength calculated using the parameter set determined in Section 10.4 from the strength and lifetime histograms. The applied load is $\sigma_0 = 0.78\overline{\sigma}_N$. As marked in the figure, different times of load application are considered, depending on the mean strength; i.e., $rt_R = \beta\overline{\sigma}_N$. Note that, for a given t_R, the mean residual strength shows a trend similar to those shown by the strength and lifetime at the large-size limit. In fact, the means tend to a straight line with the same slope as the mean strength.

The following explanation can be offered, with reference to Eq. 10.16. Since, in the large-size limit, the Weibull statistics applies, the mean strength scales according to $\overline{\sigma}_N = \overline{\sigma}^W_{N,0}(D_0/D)^{n_d/m}$ where $\overline{\sigma}^W_{N,0}$ is the intercept of the Weibull asymptote on the y-axis ($D = D_0$). Now, if the applied load is expressed as a fraction of the mean strength, i.e., $\sigma_0 = \alpha\overline{\sigma}_N$, then the probability of failure during the initial load application, $P_f(\sigma_0)$, does not depend on the size. In fact, for size D,

$$\begin{aligned} P_f(\sigma_0)|_D &= 1 - \exp\left[-\left(\frac{D}{D_0}\right)^{n_d}\left(\frac{\alpha\overline{\sigma}_N}{s_0}\right)^m\right] \\ &= 1 - \exp\left[-\left(\frac{\alpha\overline{\sigma}^W_{N,0}}{s_0}\right)^m\right] \\ &= P_f(\sigma_0)|_{D_0} \end{aligned} \tag{10.19}$$

The same holds true for the probability of failure during the application of the sustained load, $P_{f,R}(\sigma_0)$. Since $rt_R = \beta\overline{\sigma}_N$,

$$\sigma_A = \overline{\sigma}_N^{n_c+1}(n_c+1)(\beta - \alpha) = \chi\overline{\sigma}_N^{n_c+1} \tag{10.20}$$

Therefore,

$$\begin{aligned} P_{f,R}(\sigma_0)|_D &= 1 - \exp\left[-\left(\frac{D}{D_0}\right)^{n_d}(\alpha+\chi)^{m/n_c+1}\left(\frac{\overline{\sigma}_N}{s_0}\right)^m\right] \\ &= 1 - \exp\left[-(\alpha+\chi)^{m/(n_c+1)}\left(\frac{\overline{\sigma}^W_{N,0}}{s_0}\right)^m\right] \\ &= P_{f,R}(\sigma_0)|_{D_0} \end{aligned} \tag{10.21}$$

Now, for a given probability of failure, the strength and the residual strength scale, respectively, as

$$\sigma_N|_D = (D_0/D)^{n_d/m}\sigma_N|_{D_0} \tag{10.22}$$

$$\begin{aligned}\sigma_r|_D &= \left(\sigma_N^{n_c+1} - \beta\overline{\sigma}_N^{n_c+1}\right)^{1/(n_c+1)}|_D \\ &= (D_0/D)^{n_d/m}\left(\sigma_N^{n_c+1} - \beta\overline{\sigma}_N^{n_c+1}\right)^{1/(n_c+1)}|_{D_0} \\ &= (D_0/D)^{n_d/m}\sigma_r|_{D_0}\end{aligned} \tag{10.23}$$

Thus, at the large-size limit, all the terms in Eq. 10.16 scale with the same power law and the mean residual strength can now be expressed as (Salviato et al. 2014)

$$\overline{\sigma}_R = (D_0/D)^{n_d/m}\overline{\sigma}_{R,0}^W \tag{10.24}$$

where $\overline{\sigma}_{R,0}^W$ is the intercept of the Weibull asymptote at the y-axis ($D = D_0$). Thus it is shown that, at the large-size limit, the trend of mean residual strength is similar to the trend of strength except for the fact that the mean residual strength is lower, depending on the applied load and the time of load application. Upon increasing the applied load and the time of load application, the mean residual strength decreases until it reaches the value of the applied load. This happens when, for all the sizes, most specimens fail before the application of the overload.

Since the large-size asymptotic behavior of mean residual strength resembles that of mean strength and the shape of the size effect curve is similar, the size effect can reasonably be approximated by an equation similar to those proposed for the mean structural strength and lifetime (Eqs. 7.34 and 7.37):

$$\overline{\sigma}_R = \left[\frac{M_a}{D} + \left(\frac{M_b}{D}\right)^{\eta n_d/m}\right]^{1/\eta} \tag{10.25}$$

where m is the Weibull modulus of the cdf of strength and, similar to the size effect on the mean strength, M_a, M_b, and η can be derived by matching three asymptotic conditions: $[\overline{\sigma}_R]_{D\to l_0}$, $[d\overline{\sigma}_R/dD]_{D\to l_0}$, and $[\overline{\sigma}_R D^{\eta d/m}]_{D\to\infty}$. Figure 10.8c shows that Eq.10.25 provides a good approximation for all the different times of load application.

In deriving the foregoing result, the two ratios, i.e., the applied load ratio to strength and the hold time ratio to lifetime, were kept constant across the sizes. It is trivial to note, however, that if the absolute value of the applied load or the hold time, or both, are kept constant, the size effect will, of course, be much stronger. However, in this case, the mean residual strength does not resemble the strength curve and it cannot be described by Eq. 10.25.

As demonstrated in Chapter 9, an important practical merit of the theory of Bažant & Pang (2006, 2007) and Le et al. (2011, 2013), and of its present extension is that they provide a way to determine the strength, residual strength, and lifetime distributions without any histogram testing. In fact, if size effect tests of the mean strength and tests of the static crack growth rate are available, they suffice to calibrate the grafted distributions of both structural lifetime and residual strength. And once the set of parameters of the distribution are known, they can be used to obtain not only the cdf of strength but also the cdf of residual strength and the cdf of lifetime. Furthermore, the curve of mean size effect can be calibrated with much fewer tests because the mean has a much smaller coefficient of variation, particularly smaller by the factor $1/\sqrt{N_d}$ in which $N_d =$ number of individual data points.

Alternatively, the present theory can be used as an efficient way to predict the lifetime distribution by means of the tests of strength and of residual strength. For sufficiently high applied loads, the tests of mean residual strength are far less time consuming than the lifetime tests.

11 Size Effect on Reliability Indices and Safety Factors

Chapters 5–8 largely focused on the probability distribution of structural strength. For the actual reliability-based structural design, the structural strength is not the only random variable that needs to be considered. In general, the failure state of a structure can be described explicitly by a performance function, i.e., $Z = g(x_1, x_2, \ldots, x_n) < 0$, where $g(x_1, x_2, \ldots, x_n) =$ performance function and x_i $(i = 1, .., n) =$ random variables, which may represent random material properties, structural geometry, and applied loads. The failure probability of the structure can then be computed as

$$P_{FL} = \int_{g<0} \cdots \int f(x_1, x_2, \ldots, x_n)\, \mathrm{d}x_1\, \mathrm{d}x_2 \ldots \mathrm{d}x_n \tag{11.1}$$

where $f(x_1, x_2, \ldots, x_n)$ is the joint probability density function (pdf) of all x_i's. Direct integration of this joint pdf over multidimensional domain $g < 0$ can be cumbersome and therefore, various approximate methods, such as the first-order and second-order reliability methods, have been proposed to compute P_{FL} (Ang & Tang 1984; Haldar & Mahadevan 2000*a*; Lemaire 2009). With these methods, the failure probability can be calculated from the reliability indices, such as the Cornell index and Hasofer–Lind index. These indices can also be used to determine the safety factors for engineering design, in which the conventional deterministic structural analysis can be used to ensure a sufficiently small failure risk.

The concept of safety factors further leads to the development of load and resistance factors, which have been widely adopted in various structural design codes, e.g., the American Institute of Steel Construction (AISC) load resistance factor design (LRFD) (American Institute of Steel Construction 2011) and the American Concrete Institute (ACI) 318 design codes (American Concrete Institute Committee 318 2011).

In some complicated cases, the failure limit state cannot be expressed in an explicit mathematical form. The failure probability can be calculated by combining some advanced computational schemes (e.g., stochastic finite element method) with the reliability methods (e.g., perturbation method, response surface method, sensitivity-based method, and so on). Based on these calculations the relevant design parameters can be determined to achieve a tolerable failure risk (Haldar & Mahadevan 2000*b*; Lemaire 2009).

Similar to previous chapters, our primary interest is in the quasibrittle structures that fail under controlled loads (rather than displacements) once a macrocrack initiates from one material representative volume element (RVE). As usual, we consider a linear failure condition, i.e., $Z = \sigma_N - \sigma_L < 0$, where σ_N = nominal structural strength = $P_{\max}/bD$, σ_L = applied nominal stress = P/bD, $P_{\max}$ = maximum load capacity of the structure, and P = applied load. The load randomness can be estimated from the knowledge of load statistics and the statistics of structural strength σ_N is described by the finite weakest-link model, as discussed in Chapter 7. It should be emphasized that, although the failure condition is linear in terms of σ_N and σ_L, the finite weakest-link model and the Gauss–Weibull grafted distribution of RVE strength (Eqs. 5.84a and 5.84b) take into account the nonlinear stress redistribution and damage localization mechanisms.

Considering that σ_N and σ_L are statistically independent, we can calculate the structural failure probability as

$$P_{FL} = \iint_{x>y} f_L(x) f_R(y)\,\mathrm{d}x\,\mathrm{d}y = \int_0^\infty \left(\int_0^x f_R(y)\,\mathrm{d}y \right) f_L(x)\,\mathrm{d}x = \int_0^\infty P_R(x) f_L(x)\,\mathrm{d}x \tag{11.2}$$

where $f_L(x)$ and $f_R(x)$ are the probability density functions (pdfs) of the nominal stress σ_L due to loads and of the nominal strength σ_N, respectively, and $P_R(x)$ is the cumulative distribution function (cdf) of σ_N.

In Chapter 7, we demonstrated that $P_R(x)$ is strongly dependent on the structure size and geometry. Therefore, it is naturally expected that such size dependence would lead to a size effect on the reliability indices and safety factors of quasibrittle structures. In this chapter, which is largely based on Le (2015), we first discuss this size effect in the context of the mean-value first-order second-moment (MVFOSM) method and the advanced first-order second-moment (AFOSM) method. We then derive an approximate size effect equation on the central and nominal safety factors by using the asymptotic analysis. Finally, we analyze the ill-fated Malpasset dam to demonstrate the importance of including such a size effect in the reliability-based structural design.

11.1 Size Effect on the Cornell Reliability Index

It is clear that analytical integration of Eq. 11.2 for quasibrittle structures is not feasible. Therefore, a closed-form solution for the size effect on the failure probability P_{FL} is unavailable. Instead of using direct numerical integration for Eq. 11.2, we investigate the influence of structure size on the reliability indices of quasibrittle structures, which can further be used to compute the size effect on P_{FL}.

The Cornell reliability index is the most widely used approach for reliability-based analysis and design because of its simplicity (Cornell 1969; Ang & Tang 1984; Haldar & Mahadevan 2000*a*). The concept of the Cornell index represents the original formulation of the MVFOSM method. By considering a linear failure condition, the Cornell index can be calculated as

$$\beta = \frac{\mu_R - \mu_L}{\sqrt{\delta_R^2 + \delta_L^2}} \tag{11.3}$$

where μ_R, μ_L = mean values of the structural strength σ_N and the applied nominal stress σ_L, respectively, and δ_R, δ_L = standard deviations of σ_N and σ_L, respectively. If both σ_N and σ_L follow the Gaussian distributions, then the Cornell index can be directly used to compute the actual failure probability of the structure, i.e.,

$$P_{FL} = \Phi(-\beta) \tag{11.4}$$

where $\Phi(x)$ = standard Gaussian distribution function.

Based on the definition of the Cornell index, its dependence on the structure size naturally arises from the size effects on the mean and variance of σ_N, which can be computed for any set of geometrically similar structures as

$$\mu_R(D) = \int_0^{\infty} [1 - P_R(\sigma_N, D)]\,\mathrm{d}\sigma_N \tag{11.5}$$

$$\delta_R^2(D) = \int_0^1 \sigma_N^2 \mathrm{d}P_R(\sigma_N, D) - \mu_R^2(D) \tag{11.6}$$

As discussed in Chapter 7, a closed-form solution for $\mu_R(D)$ is impossible. An approximate expression for $\mu_R(D)$ can be constructed through asymptotic matching (i.e., Eq. 7.34):

$$\mu_R(D) = \left[\frac{N_a}{D} + \left(\frac{N_b}{D}\right)^{n_d\psi/m}\right]^{1/\psi} \tag{11.7}$$

Similarly, the size effect on the variance of σ_N can also be approximated by using a similar functional form:

$$\delta_R^2(D) = \left[\frac{C_a}{D} + \left(\frac{C_b}{D}\right)^{2qn_d/m}\right]^{1/q} \tag{11.8}$$

where n_d = number of the spatial dimensions of the structure to be scaled. Model parameters N_a, N_b, ψ, C_a, C_b, and q for Eqs. 11.7 and 11.8 can be related to the statistical properties of RVE strength, which include the Weibull modulus m, the Weibull scale parameter s_0, and the mean and standard deviation of the Gaussian core μ_G and δ_G (see Eqs. 5.84a and 5.84b). Asymptotic matching of the small- and large-size asymptotic behaviors of $\mu_R(D)$ and $\delta_R^2(D)$ is used for this purpose.

At the large-size limit (i.e., $D \to \infty$), the finite weakest-link model converges to the classical infinite weakest-link model, which leads to a two-parameter Weibull distribution for σ_N. With the well-known properties of the Weibull cdf (Weibull 1939), we can determine N_b and C_b from Weibull modulus m and scale parameter s_0:

$$N_b = D_0[s_0\Gamma(1 + 1/m)]^{m/n_d} \tag{11.9}$$

$$C_b = D_0[s_0\Gamma(1 + 1/m)]^{m/n_d}\left[\frac{\Gamma(1 + 2/m)}{\Gamma^2(1 + 1/m)} - 1\right]^{m/2n_d} \tag{11.10}$$

where $D_0 = l_0\left[\int_V \langle s(\xi)\rangle^m \mathrm{d}\xi\right]^{-1/n_d}$ and $\xi = \boldsymbol{x}/D$ = normalized coordinate.

After determining N_b and C_b, the parameters N_a, C_a, ψ, and q can then be obtained by matching four small-size asymptotes, i.e., $\mu_R(D_m)$, $\mathrm{d}\mu_R/\mathrm{d}D|_{D=D_m}$, $\delta_R^2(D_m)$, and

$\mathrm{d}\delta_R^2/\mathrm{d}D|_{D=D_m}$, with the predictions by the finite weakest-link model, where $D_m =$ the smallest structure size for which the finite weakest-link model makes physical sense. For example, for flexural failure, the structure should at least consist of about 4 RVEs across its depth (Le et al. 2012, 2013). At this small-size limit, the entire cdf of σ_N can be approximated by a Gaussian cdf with a mean of μ_G and standard deviation of δ_G for the purpose of calculating the mean and variance. Therefore, we can solve N_a, C_a, ψ, and q from the following set of equations:

$$\mu_R(D_m) = \int_0^\infty [1 - P_R(\sigma_N, D_m)]\,\mathrm{d}\sigma_N \tag{11.11}$$

$$\left.\frac{\mathrm{d}\mu_R}{\mathrm{d}D}\right|_{D=D_m} = \frac{n_d}{D_m}\int_0^\infty [1 - P_R(\sigma_N, D_m)]\ln[1 - P_R(\sigma_N, D_m)]\,\mathrm{d}\sigma_N \tag{11.12}$$

$$\delta_R^2(D_m) + \mu_R^2(D_m) = \int_0^\infty \sigma_N^2 p_R'(\sigma_N, D_m)\,\mathrm{d}\sigma_N \tag{11.13}$$

$$\left.\frac{\mathrm{d}\delta_R^2}{\mathrm{d}D}\right|_{D=D_m} + 2\mu_R(D_m)\left.\frac{\mathrm{d}\mu_R}{\mathrm{d}D}\right|_{D=D_m} = \frac{n_d}{D_m}\int_0^\infty \sigma_N^2\{1 + \ln[1 - P_R(\sigma_N, D_m)]\}p_R(\sigma_N, D_m)\,\mathrm{d}\sigma_N \tag{11.14}$$

where

$$P_R(\sigma_N, D_m) = 1 - \prod_{i=1}^{N_m}\{1 - \Phi[\varsigma(\xi_i)]\} \tag{11.15}$$

$$\varsigma(\xi_i) = [\sigma_N s(\xi_i) - \mu_G]/\delta_G \tag{11.16}$$

$$p_R(\sigma_N, D_m) = \mathrm{d}P_R(\sigma_N, D_m)/\mathrm{d}\sigma_N \tag{11.17}$$

$N_m =$ number of RVEs in the structure of size D_m and $\xi_i =$ normalized coordinate at the center of the ith RVE. It is clear that N_a, C_a, ψ, and q are all related to the Gaussian mean μ_G and standard deviation δ_G of the RVE strength.

Knowing the model parameters for the approximate scaling relations of the mean and variance of σ_N, we readily obtain a size-dependent Cornell index for quasibrittle structures, i.e.:

$$\beta = \frac{\left[(N_a/D) + (N_b/D)^{\psi n_d/m}\right]^{1/\psi} - \mu_L}{\sqrt{\left[(C_a/D) + (C_b/D)^{2qn_d/m}\right]^{1/q} + \delta_L^2}} \tag{11.18}$$

It is clear that, for quasibrittle structures, the Cornell index would not yield the exact failure probability (i.e., $\Phi(-\beta) \neq P_f$) since the probability distribution of σ_N, at least, is not Gaussian. Nevertheless, the Cornell index provides a qualitative description of the risk level (Haldar & Mahadevan 2000*b*), and as will be shown later in Section 11.4, in some cases $\Phi(-\beta)$ can give a reasonable approximation of the failure probability for quasibrittle structures.

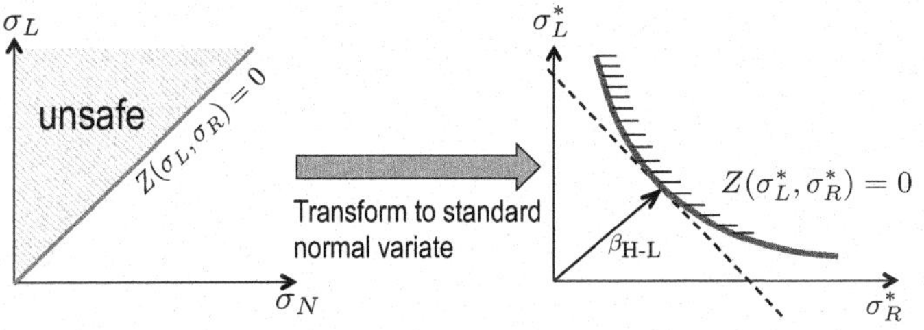

Figure 11.1. Transformation of failure surface described by non-Gaussian variables.

11.2 Size Effect on the Hasofer–Lind Reliability Index

A more accurate and robust method to compute the structural failure probability is to use the Hasofer–Lind index $\beta_{\text{H-L}}$, i.e.,

$$P_f = \Phi(-\beta_{\text{H-L}}) \tag{11.19}$$

Similar to the Cornell index, $\beta_{\text{H-L}}$ provides a first-order approximation of the failure probability. The Hasofer–Lind index was first developed for Gaussian variables as an AFOSM method. This method can more accurately handle the nonlinear failure state function (Hasofer & Lind 1974). It was later extended to non-Gaussian variables (Rackwitz & Fiessler 1976, 1978).

The essential idea of the Hasofer–Lind index is to transform non-Gaussian random variables to equivalent standard Gaussian variables. $\beta_{\text{H-L}}$ is then determined as the minimum distance between the origin and the limit state surface in the transformed variable space, as shown in Figure 11.1. Owing to the transformation of the non-Gaussian variables, the limit state function expressed by the equivalent Gaussian variables is usually nonlinear. Therefore, the search for $\beta_{\text{H-L}}$ forms a constrained optimization problem.

Based on the Lagrange multiplier method, various numerical algorithms have been developed to solve this optimization problem, e.g., Rackwitz (1976), Ayyub and Haldar (1984), Haldar and Mahadevan (2000*b*). During each iteration, the mean and standard deviation of the equivalent Gaussian variables need to be evaluated, based on the current values of the original random variables, usually referred to as the checking points. A detailed description of the optimization algorithms can be found in Haldar and Mahadevan (2000*a*) and Haldar and Mahadevan (2000*b*). These algorithms can typically converge in very few iterations (about 5–10) depending on the nonlinearity of the limit state function.

Therefore, computing $\beta_{\text{H-L}}$ is much more efficient than the direct numerical integration of Eq. 11.2, especially for large-size quasibrittle structures. Unlike the foregoing analysis of the Cornell index, where an approximate size effect equation is available, the size effect on $\beta_{\text{H-L}}$ can be evaluated only numerically. The question is how to incorporate the size effect on the cdf of structural strength into the computation of $\beta_{\text{H-L}}$.

Previous studies have suggested different ways to determine the mean μ^N and the standard deviation δ^N of the equivalent Gaussian variable (Rackwitz & Fiessler 1976). When the original distribution function is not significantly skewed, μ^N and δ^N can be determined by equating the cdfs and pdfs of the original random variable and the equivalent Gaussian variable at the checking point for each iteration, i.e.,

$$\mu^N = x^* - \Phi^{-1}[P_i(x^*)]\delta^N, \quad \delta^N = \phi\{\Phi^{-1}[P_i(x^*)]\}/p_i(x^*) \tag{11.20}$$

where x^* = value of original random variable for each iteration during the optimization process; $\phi(x) = e^{-x^2/2}/\sqrt{2\pi}$ = standard Gaussian pdf, $P_i(x)$ $(i = L, R)$ = cdfs of the applied nominal stress and structural strength, respectively; and $p_i(x)$ $(i = L, R)$ = pdfs of the applied nominal stress and structural strength, respectively.

As the skewness of the pdf of the original random variable increases, it has been suggested that μ^N and δ^N should be determined by two different conditions at each checking point (Rackwitz & Fiessler 1978): (1) the mean of the equivalent Gaussian variable is equal to the median of the original random variable, and (2) the cdf of the equivalent Gaussian variable is identical to the cdf of the original random variable, i.e.,

$$\mu^N = P_i^{-1}(0.5), \quad \delta^N = (x^* - \mu^N)/\Phi^{-1}[P_i(x^*)] \tag{11.21}$$

For the applied nominal stress, the probability distribution can usually be approximated by some empirical cdfs, such as Gaussian and lognormal distributions for normal loading and Gumbel distribution for extreme loading. Based on the knowledge of $P_L(\sigma_L)$, one can choose either Eq. 11.20 or Eq. 11.21 to transform the random applied nominal stress.

To transform the random structural strength to the equivalent Gaussian variable, we note that the skewness of $P_R(\sigma_N)$ increases with an increasing structure size. This indicates that the choice of the transformation method must depend on the structure size. Recall that in Section 7.1 we have discussed the qualitative behavior of the probability distribution of structural strength based on the concept of equivalent number of RVEs, N_{eq}, which is defined as

$$N_{\text{eq}} = \int_V \langle s(\boldsymbol{x})\rangle^m \, \mathrm{d}V(\boldsymbol{x})/V_0 \tag{11.22}$$

where $s(\boldsymbol{x})$ = dimensionless stress field such that $\sigma_N s(\boldsymbol{x})$ represents the maximum elastic principal stress at the material point with coordinate $\boldsymbol{x}$; $\langle x\rangle = \max(x, 0)$, $V_0 = l_0^{n_d}$ and l_0 = RVE size.

Depending on the value of N_{eq}, three asymptotic regions of strength distribution $P_R(\sigma_N)$ can be identified (Le et al. 2011): (1) When N_{eq} is small, i.e., $N_{\text{eq}} \in (1, 20)$, $P_R(\sigma_N)$ is almost Gaussian. (2) For the intermediate range of N_{eq} values, i.e., $N_{\text{eq}} \in (50, 500)$, the major part of $P_R(\sigma_N)$ can be modeled as a finite chain of Gaussian elements. (3) For structures with a very large value of N_{eq}, i.e., $N_{\text{eq}} \geq 2000$, $P_R(\sigma_N)$ approaches the Weibull distribution.

Instead of the direct use of the skewness of $P_R(\sigma_N)$, here we use N_{eq} as a criterion to select the transformation method for the Hasofer–Lind index based on the aforementioned asymptotic behavior of $P_R(\sigma_N)$ (Le 2015). It can be reasonably considered that, as N_{eq} is greater than 50, $P_R(\sigma_N)$ would start to have a significant skewness and, therefore, Eq. 11.21 should be used to transform σ_N to an equivalent Gaussian variable, whereas for $N_{\text{eq}} \leq 50$ Eq. 11.20 would be more appropriate. Such a choice of N_{eq} will be demonstrated later in a case study (Section 11.4.2). It should be emphasized that, as implied by the definition of N_{eq}, both the structure size and geometry (or stress field) would affect the choice of the transformation method.

To use Eq. 11.20, we just need to know the cdf and pdf of σ_N at the current checking point, which can be easily calculated by using the finite weakest-link model (Eq. 8.6). By contrast, it is not convenient to use Eq. 11.21 because direct computation of the median of σ_N is not straightforward. Therefore, we propose an approximate size effect on the median structural strength that has the same functional form as that for the mean strength (Eq. 11.7):

$$P_R^{-1}(0.5) = \left[\frac{N_1}{D} + \left(\frac{N_2}{D}\right)^{n_d p/m}\right]^{1/p} \tag{11.23}$$

where N_1, N_2, and p can be determined by using the same asymptotic conditions as we used for N_a, N_b, and ψ (Eqs. 11.9, 11.11, and 11.12). Since, at the small-size limit, $P_R(\sigma_N)$ is almost Gaussian, the mean value of σ_N is approximately equal to the median value of σ_N. At the large-size limit, the median value of σ_N can be easily calculated based on the Weibull distribution. Therefore, these three asymptotic conditions can be written as

$$N_2 = D_0 s_0^{m/n_d} (\ln 2)^{1/n_d} \tag{11.24}$$

$$\mu_R(D_m) = \left[\frac{N_1}{D_m} + \left(\frac{N_2}{D_m}\right)^{n_d p/m}\right]^{1/p} \tag{11.25}$$

$$\left.\frac{\mathrm{d}\mu_R}{\mathrm{d}D}\right|_{D=D_m} = -\frac{1}{p}\left[\frac{N_1}{D_m^2} + \frac{n_d p}{m D_m}\left(\frac{N_2}{D_m}\right)^{p n_d/m}\right]\left[\frac{N_1}{D_m} + \left(\frac{N_2}{D_m}\right)^{p n_d/m}\right]^{1/p-1} \tag{11.26}$$

From the foregoing analysis, it is clear that the size effect on $P_R(\sigma_N)$ of quasibrittle structures will directly lead to the size dependence of the equivalent Gaussian variable for σ_N, which in turn influences the failure limit state in the transformed coordinate system. This eventually introduces a size effect on the Hasofer–Lind index or, equivalently, on the structural failure probability.

11.3 Approximate Equation for Scaling of Safety Factors

The Cornell and Hasofer–Lind reliability indices provide an efficient means of evaluating the failure risk of quasibrittle structures, which is an essential step of reliability-based design of engineering structures. A more straightforward approach for reliability-based structural design is to use the concept of safety factors. The failure state is characterized by either the nominal values or the mean values of the structural strength and nominal applied stress. If we use the mean values, the failure domain can be written as

$$\mu_R < \bar{\xi} \mu_L \tag{11.27}$$

where $\bar{\zeta}$ is referred to as the central safety factor. If we use the nominal values, the failure domain becomes

$$\mu_{RN} < \zeta \mu_{LN} \tag{11.28}$$

where $\mu_{RN} =$ nominal value of $\sigma_N = \mu_R - k_R \delta_R$, $\mu_{LN} =$ nominal value of $\sigma_L = \mu_L + k_L \delta_L$; $k_R, k_L =$ constants, and ζ is the nominal safety factor. The safety factors $\zeta, \bar{\zeta}$ are determined to ensure that the overall failure probability of the structure would be equal to a prescribed value P_c, which is typically on the order of 10^{-6} [Nordic Committee for Building Structures (NKB) 1978; Melchers 1987; Duckett 2005]. Since the failure probability of quasibrittle structures depends strongly on the structure size, the safety factors corresponding to a prescribed failure probability should also depend on the structure size. In this section, we investigate this size dependence by considering that the applied nominal stress follows a Gaussian distribution, which is a common assumption for many external loads in most current design codes (American Concrete Institute Committee 318 2011; American Institute of Steel Construction 2011).

Based on the formulation of probability distribution of RVE strength discussed in Chapter 5 (Eqs. 5.84a and 5.84b), it is clear that the strength distribution can be characterized by four parameters, i.e., Weibull modulus m, Weibull scale parameter s_0, mean RVE strength μ, and the coefficient of variation (CoV) of RVE strength ω. Here we fix the values of m, ω as well as the ratio between μ and s_0 ($r = \mu/s_0$), and leave s_0 to be determined to achieve a prescribed structural failure probability P_c.

To compute the safety factors, it is convenient to consider an artificial material RVE, where the grafted strength cdf has the following parameters: Weibull modulus $= m$, Weibull scale parameter $s_0 = 1$, mean strength $= r$, and CoV $= \omega$. It is noted that, since the power-law tail of the grafted cdf of RVE strength is very short, the mean of the Gaussian core μ_G for this artificial material RVE is approximately equal to the mean RVE strength r, and the standard deviation of the Gaussian core δ_G is approximately equal to $r\omega$.

For any given structure, let μ_r, δ_r denote the mean and standard deviation of the strength cdf by assuming that the structure is made of the aforementioned artificial RVE. It is clear that the mean and standard deviation of the strength cdf of the actual structure are simply equal to $s_0\mu_r$ and $s_0\delta_r$, respectively. Therefore, the central and nominal safety factors of the structure can be computed as

$$\bar{\zeta}(D) = \frac{s_0(D)\mu_r(D)}{\mu_L} \tag{11.29}$$

$$\zeta(D) = \frac{s_0(D)[\mu_r(D) - k_R\delta_r(D)]}{\mu_L + k_L\delta_L} \tag{11.30}$$

where $\mu_r(D)$ and $\delta_r(D)$ can be approximated by Eqs. 11.7 and 11.8, where the parameters in Eqs. 11.7 and 11.8 can be related to m, r, and ω through Eqs 11.9 to 11.14. Now the crux of the problem is to determine $s_0(D)$ for the target failure risk P_c.

First consider the small-size limit. Based on the finite weakest-link model, the strength cdf can approximately be modeled as the Gaussian distribution. Since we limit our attention to the case where σ_L is Gaussian, it is expected that the Cornell index would give a reasonable estimation of the failure probability. Therefore, we have

$$\beta_c = \frac{s_0(D)\mu_r(D) - \mu_L}{\sqrt{s_0^2(D)\delta_r^2(D) + \delta_L^2}} \tag{11.31}$$

where $\beta_c = -\Phi^{-1}(P_c)$. Following Haldar and Mahadevan (2000*a*), we can write

$$\sqrt{s_0^2(D)\delta_r^2(D) + \delta_L^2} \approx \epsilon[s_0(D)\delta_r(D) + \delta_L] \tag{11.32}$$

where ϵ is around 0.75 for most cases. With such an approximation, we may express $s_0(D)$ as

$$s_0(D) = \frac{\epsilon\beta_c\delta_L + \mu_L}{\mu_r(D)}\left[1 - \frac{\epsilon\beta_c\delta_r(D)}{\mu_r(D)}\right]^{-1} \tag{11.33}$$

$$\text{or} \quad \mu_r(D)s_0(D) = (\epsilon\beta_c\delta_L + \mu_L)\left[1 - \frac{\epsilon\beta_c\delta_r(D)}{\mu_r(D)}\right]^{-1} \tag{11.34}$$

At the large-size limit, the probability distribution of structural strength must converge to the Weibull distribution (Eq. 7.3), i.e.,

$$P_R(\sigma_N) = 1 - \exp\left[-N_{\text{eq}}(\sigma_N/s_0)^m\right] \tag{11.35}$$

The Weibull distribution has a power-law tail, which extends to the point corresponding to the probability of 0.1 (with a 5% error in terms of the probability). Since we are interested only in a low structural failure probability ($P_c \sim 10^{-6}$), what matters for the overall structural failure probability is the power-law tail of $P_R(\sigma_N)$:

$$P_f = \int_0^\infty f_L(x)N_{\text{eq}}(x/s_0)^m \,\mathrm{d}x \tag{11.36}$$

Therefore, to ensure failure probability P_c, we have

$$s_0(D) = (D/D_0)^{n_d/m}\left[P_c^{-1}\int_0^\infty f_L(x)x^m \,\mathrm{d}x\right]^{1/m} \tag{11.37}$$

$$\text{or} \quad \mu_r(D)s_0(D) = \left[P_c^{-1}\int_0^\infty f_L(x)x^m \,\mathrm{d}x\right]^{1/m}\Gamma(1 + 1/m) \tag{11.38}$$

To achieve asymptotic matching of the aforementioned small- and large-size asymptotes, we propose the following approximate equation for $\mu_r(D)s_0(D)$:

$$\mu_r(D)s_0(D) = C\left[1 + \frac{\gamma\beta_c\epsilon\delta_r(D)}{\mu_r(D)}\right]^{1/\gamma} \tag{11.39}$$

where C, γ are constants. C and γ can be determined by matching the small-size and large-size asymptotic behavior of $\mu_r(D)s_0(D)$. With Eqs. 11.29, 11.30, and 11.39, we obtain the size-dependent central and nominal safety factors:

$$\bar{\zeta}(D) = \mu_L^{-1} C \left[1 + \frac{\gamma \beta_c \epsilon \delta_r(D)}{\mu_r(D)}\right]^{1/\gamma} \tag{11.40}$$

$$\zeta(D) = C \left[1 + \frac{\gamma \beta_c \epsilon \delta_r(D)}{\mu_r(D)}\right]^{1/\gamma} \frac{1 - k_R \delta_r(D)/\mu_r(D)}{\mu_L + k_L \delta_L} \tag{11.41}$$

It should be emphasized that, in the foregoing analysis, the safety factors are calculated with respect to the mean and nominal values of strength of the actual-size structure to be designed. If we want to extrapolate the small-scale laboratory test results to reliability-based design of large structures, the present size effect on safety factors (Eqs. 11.40 and 11.41) needs to be further combined with the size effect on the mean structural strength (Eq. 11.7). Consider that the design of a full-size structure (size $= D$) is based on a small-scale prototype of size D_0. The central safety factor with respect to the mean strength μ_{R0} of the small-scale prototype can be written as: $\bar{\zeta}_s = \mu_{R0}/\mu_L$. The value of $\bar{\zeta}_s$ that would guarantee the target failure probability P_c of the actual full-size structure can be computed as

$$\bar{\zeta}_s = \bar{\zeta}(D) \frac{\left[(N_a/D_0) + (N_b/D_0)^{n_d \psi/m}\right]^{1/\psi}}{\left[(N_a/D) + (N_b/D)^{n_d \psi/m}\right]^{1/\psi}} \tag{11.42}$$

It is also worthwhile to point out that, in engineering designs, the applied loads are sometimes modeled by lognormal distribution or Gumbel distribution. In such cases, the large-size asymptote remains unchanged since Eq. 11.37 is not limited to any form of probability distribution of external loads. However, the analysis of the small-size limit needs to be revised since the Cornell index is not directly applicable and therefore a closed-form solution does not exist. Instead, the Hasofer–Lind index could be used to calculate the value of $s_0(D)$ such that $\beta_{\text{H-L}} = \Phi^{-1}(P_c)$.

11.4 Analysis of Failure Statistics of the Malpasset Arch Dam

We now use the aforementioned formulation to analyze the failure statistics of the ill-fated Malpasset arch dam. It should be emphasized that the objective of the present analysis is not to redesign the dam, but to demonstrate the necessity of including the size effect in reliability-based structural analysis and design.

The Malpasset arch dam was built across the canyon of the Reyran Valley in France in 1954, and it failed in 1959 after the first full filling of the reservoir by 5 days of heavy rain. It was generally accepted that the failure was due to the development of vertical flexural cracks engendered by the lateral displacement of the abutment (Bartle 1985; Levy & Salvadori 1992). A recent study (Duffaut 2013) examined several geological issues that could possibly cause the large displacement of the abutment. Meanwhile,

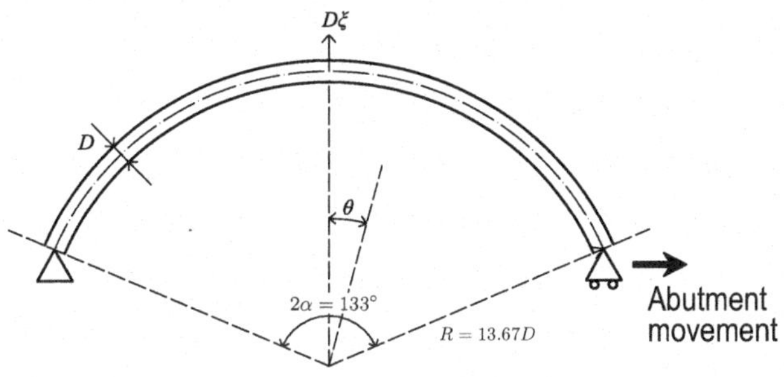

Figure 11.2. Simplified 2D analysis of failure of an arch dam.

there has also been a great interest in understanding how the dam itself responds to the abutment displacement (Bažant et al. 2007; Le et al. 2012).

In the present analysis, both random structural strength and abutment movement are considered to evaluate the size effects on the overall failure statistics of the dam as well as on the associated design safety factors.

11.4.1 Model Description

In the present simplification, the dam is modeled as a 2D arch (Bažant et al. 2007), shown as Figure 11.2. The dimension of the arch corresponds to the mid-height cross section of the actual dam, i.e., $R_a = 92.68$ m, $D_a = 6.78$ m, and $\alpha = 66.5°$. The arch is loaded by a displacement at its right support. To investigate the size effect on the failure statistics of the dam, we consider a series of geometrically similar arches of different sizes (Fig. 11.3), where D = depth of arch cross-section and R = arch radius = $13.67D$.

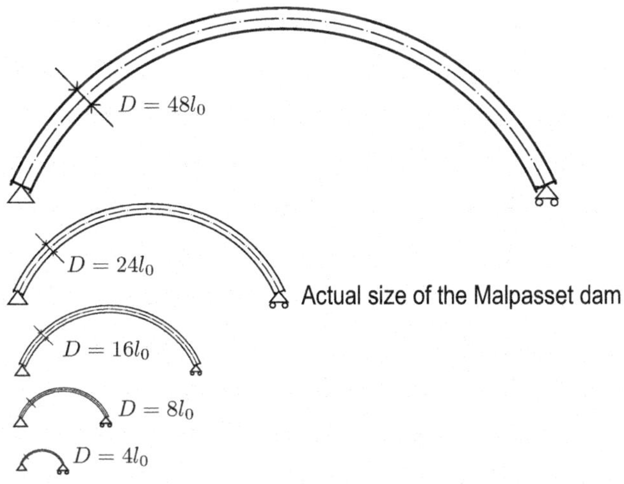

Figure 11.3. Geometrically similar arch dams of different sizes for the analysis. (The arch dam with $D = 96l_0$ is not shown in this figure.)

Based on the engineering beam theory, the elastic bending stress can be written as

$$\sigma(\theta,\xi) = \sigma_N \frac{2\xi(\cos\theta - \cos\alpha)}{1-\cos\alpha} \tag{11.43}$$

where $\xi = x/D$; $x =$ coordinate across arch thickness measured from the center line; $\alpha =$ central angle of half-arch (Fig. 11.2); and $\sigma_N =$ nominal structural strength of the arch dam, which is defined as the maximum elastic stress in the arch considered as a 2D ring. The probability distribution of σ_N can be calculated using the finite weakest-link model:

$$P_R(\sigma_N) = 1 - \exp\left\{\frac{1}{l_0^2}\int_0^{1/2}\int_{-\alpha}^{\alpha} \ln\{1 - P_1[\sigma_N s(\theta,\xi)]\} RD\,\mathrm{d}\theta\,\mathrm{d}\xi\right\} \tag{11.44}$$

where the dimensionless stress field $s(\theta,\xi)$ can be written as

$$s(\theta,\xi) = \frac{2\xi(\cos\theta - \cos\alpha)}{1-\cos\alpha} \tag{11.45}$$

Note that Eq. 11.44 is written by using the integral form of the original finite weakest-link equation (Eq. 7.1).

To calculate the nominal applied stress in the arch, which is equal to the maximum elastic stress induced by the abutment movement Δ, we can use Castigliano's theorem:

$$\sigma_L = \frac{E(1-\cos\alpha)D\Delta}{2R^2\int_{-\alpha}^{\alpha}(\cos\theta - \cos\alpha)^2 \mathrm{d}\theta} \tag{11.46}$$

It is clear that the failure state of the arch can simply be written as $\sigma_N - \sigma_L < 0$. Previous studies (Bažant et al. 2007; Le et al. 2012) suggested the following parameters for the RVE strength distribution for concrete: $m = 24$, $s_0 = 2.12$ MPa, $\mu_G = 2.91$ MPa, $\delta_G = 0.44$ MPa, $\omega = 0.15$, and as RVE size of 280 mm. However, the cdf of σ_L, which is determined by the statistics of abutment movement, is not available. For the present size effect analysis, we estimate the cdf of σ_L by assuming (1) σ_L follows a Gaussian distribution with a CoV of 40%, and (2) the failure probability of the dam with a laboratory test scale (i.e., $D = 4l_0$) is equal to 10^{-6}; i.e., we assume that the dam was designed based on a laboratory prototype and standard tests of flexural strength. Based on these assumptions, we obtain the mean value of σ_L to be 0.76 MPa.

11.4.2 Discussion of Cornell and Hasofer–Lind Indices

Based on the aforementioned probability distributions of σ_N and σ_L, we can use Eq. 11.2 to calculate the exact failure probability P_{FL} for geometrically similar dams of six sizes $D = 4l_0, 8l_0, 16l_0, 24l_0, 48l_0$, and $96l_0$, among which $D = 24l_0$ corresponds to the actual size D_a of the Malpasset dam. As shown in Fig. 11.4, the structural failure probability increases from 10^{-6} to 6×10^{-3} as dam size D increases from $4l_0$ to $96l_0$. We now compare these exact failure probabilities with the predictions by using the Cornell and Hasofer–Lind reliability indices.

To compute the Cornell reliability index β, we first determine the size effect on the mean and variance of σ_N using Eqs 11.7 to 11.14. Figure 11.5 compares the approximate

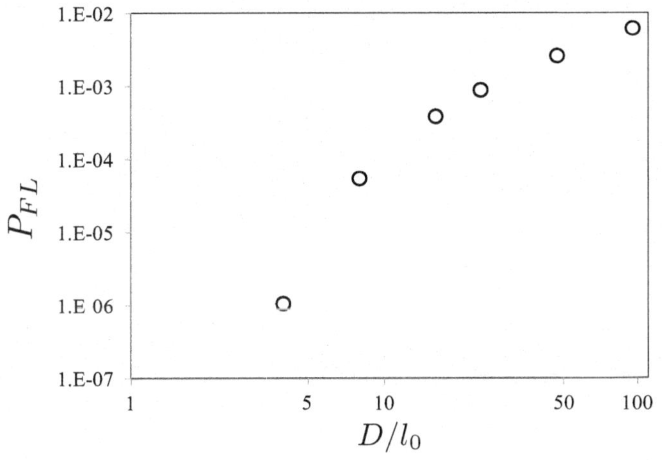

Figure 11.4. Calculated failure probabilities of a set of geometrically similar dams of different sizes.

equations (Eqs. 11.7 and 11.8) with the exact mean value and variance of σ_N calculated from the finite weakest-link model. It can be seen that they agree with each other very well for all six dam sizes. Based on Eqs. 11.7 and 11.8, we can readily calculate the Cornell index based on Eq. 11.18, from which we obtain the structural failure probabilities for different dam sizes.

Table 11.1 presents the exact failure probabilities as well as those calculated from the Cornell index (i.e., $P_{FL} = \Phi(-\beta)$). It can be seen that the Cornell index gives a reasonable estimation of the failure probability even though the probability distribution of σ_N is non-Gaussian. To explain this, we plot the value of the integrand of Eq. 11.2 normalized by the failure probability, i.e., $P_R(x)f_L(x)/P_{FL}$, for all six dam sizes, as shown in Fig. 11.6. When the structure size is small, most of $P_R(x)$ follows a Gaussian cdf and the

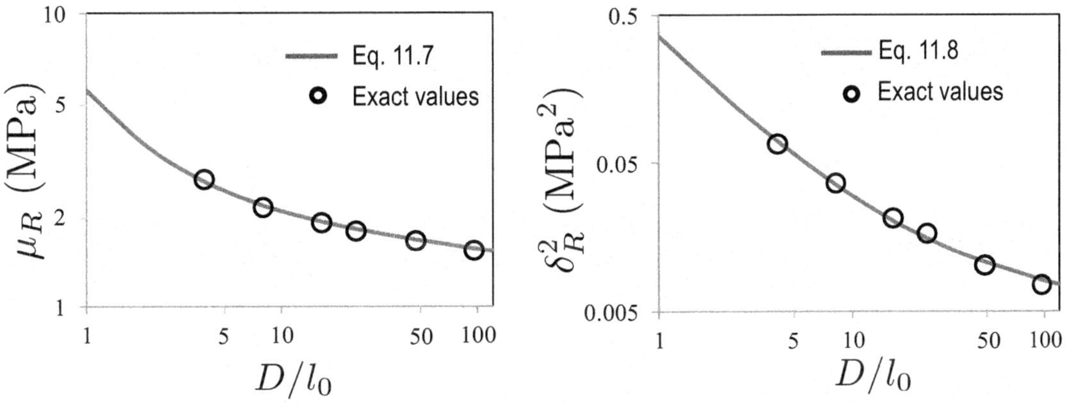

Figure 11.5. Approximations of size effect on the mean and variance of nominal strength of the geometrically similar dams of different sizes, similar to the Malpasset dam.

Table 11.1 Comparison of calculated structural failure probabilities P_f

Dam size	Exact P_{FL}	$\Phi(-\beta)$	$\Phi(-\beta_{\mathrm{H-L},1})$	$\Phi(-\beta_{\mathrm{H-L},2})$
$D = 4l_0$ ($N_{eq} = 2.26$)	1.05×10^{-6}	1.28×10^{-6}	1.20×10^{-6}	1.92×10^{-6}
$D = 8l_0$ ($N_{eq} = 9.05$)	5.37×10^{-5}	2.66×10^{-5}	4.94×10^{-5}	8.13×10^{-5}
$D = 16l_0$ ($N_{eq} = 36.2$)	3.84×10^{-4}	2.04×10^{-4}	3.29×10^{-4}	4.44×10^{-4}
$D = 24l_0$ ($N_{eq} = 144.8$)	8.72×10^{-4}	5.11×10^{-4}	7.02×10^{-4}	9.23×10^{-4}
$D = 48l_0$ ($N_{eq} = 579.2$)	2.57×10^{-3}	1.81×10^{-3}	2.24×10^{-3}	2.63×10^{-3}
$D = 96l_0$ ($N_{eq} = 2316.8$)	6.12×10^{-3}	4.89×10^{-3}	5.53×10^{-3}	5.88×10^{-3}

power-law tail makes a negligible contribution to the overall structural failure probability (for $D = 4l_0$, $\int_0^{\sigma_{gr}} f_L(x)P_R(x)\,dx = 4.80 \times 10^{-8} = 0.045P_{FL}$). Therefore, the Cornell index is expected to give a good estimation of the overall failure probability at the small-size limit. At the large-size limit, Fig. 11.6 indicates that the entire $P_R(x)$ contributes to the structural failure probability. Although $P_R(x)$ would follow a Weibull distribution, we note that, for the same mean and variance, Gaussian and Weibull distributions are close to each other except for the left and right tails (see Fig. 1.14). Therefore, the Cornell index can also predict the failure probability reasonably well. Based on Table 11.1, it can be seen that, for the intermediate size range, the failure probabilities predicted by the Cornell index are within the same order of magnitude of the exact values.

It should be emphasized that the foregoing analysis is based on the assumption that the applied nominal stress σ_L follows a Gaussian distribution. For a non-Gaussian distribution of σ_L, the Cornell index may not yield a good estimation of P_{FL}. Furthermore, the present calculation also indicates that the Cornell index cannot be directly used to calculate the safety factors that correspond to a very low failure probability for large-size structures (e.g., $P_{FL} \leq 10^{-6}$). This is because what matters for such a low failure probability is the left tail part of the Weibull distribution, which would

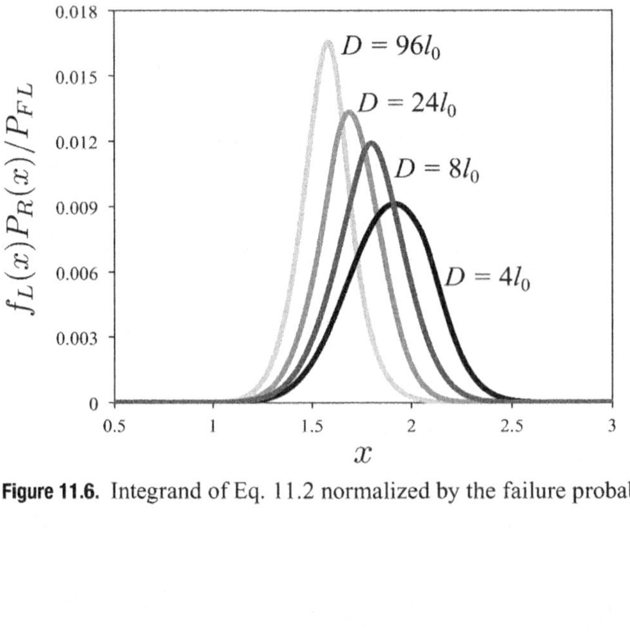

Figure 11.6. Integrand of Eq. 11.2 normalized by the failure probability of the dam.

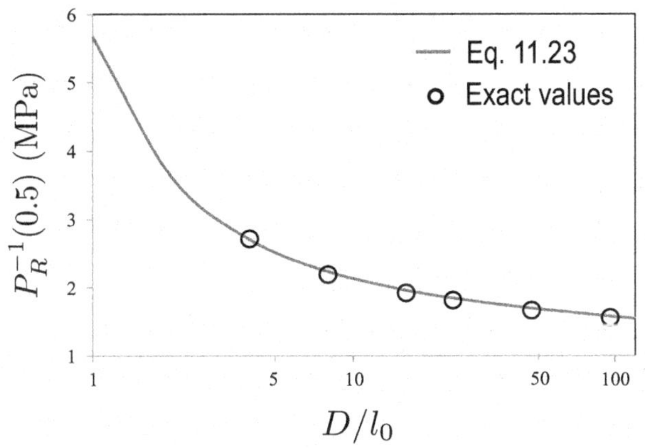

Figure 11.7. Approximation of size effect on the median strength of the dam.

significantly deviate from the Gaussian distribution that has the same mean and standard deviation.

To improve the accuracy of the estimation of P_{FL}, the Hasofer–Lind index is also used in the present analysis. As mentioned in Section 11.2, we consider two transformation methods for the non-Gaussian variables (i.e., Eqs. 11.20 and 11.21). We can use Eq. 11.20 to transform σ_L since it follows a Gaussian distribution. For structural strength σ_N, we note that the choice of the transformation method would depend on the structure size.

To study the appropriateness of the transformation method, we consider both methods. The corresponding values of the Hasofer–Lind index are denoted as $\beta_{\text{H-L},1}$ and $\beta_{\text{H-L},2}$, respectively. For computing $\beta_{\text{H-L},2}$, Eq. 11.23 is adopted to approximate the median strengths for all the dam sizes, where the parameters in Eq. 11.23 are calibrated using Eqs. 11.24 to 11.26. As shown in Figure 11.7, Eq. 11.23 agrees well with the exact median strengths calculated from the finite weakest-link model. Table 11.1 shows the failure probabilities computed by $\beta_{\text{H-L},1}$ and $\beta_{\text{H-L},2}$, i.e., $P_{FL} = \Phi(-\beta_{\text{H-L},i})$ $(i = 1, 2)$. By comparing these results with the exact values of P_{FL}, it can be seen that, when the structure size is small, $\beta_{\text{H-L},1}$ gives a better estimation of P_{FL} whereas, as the structure size increases, $\beta_{\text{H-L},2}$ outperforms $\beta_{\text{H-L},1}$.

In Table 11.1, it is seen that, for $D \leq 16l_0$ ($N_{\text{eq}} \leq 36$), Eq. 11.20 is more appropriate for the transformation and, for $D > 16l_0$ ($N_{\text{eq}} > 36$), Eq. 11.21 is better. This is consistent with the aforementioned discussion in Section 11.2 on the choice of transformation method as a function of structure size and geometry.

11.4.3 Discussion of Central and Nominal Safety Factors

The foregoing analysis has focused on the evaluation of the failure probability of the dam for a given RVE strength statistics. It is clear that the structural failure probability of the dam increases significantly with its size. In the actual design, we must ensure a prescribed failure probability P_c, regardless of the structure size. Following Section 11.3,

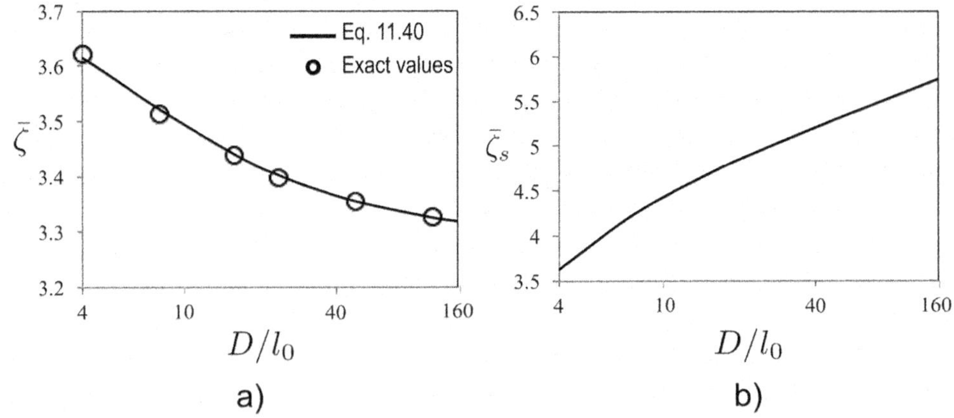

Figure 11.8. Size effect on the central safety factor computed on the basis of (a) mean structural strength $\bar{\sigma}_N(D)$, and (b) mean structural strength of the small-size dam $\bar{\sigma}_N(D = 4l_0)$.

we are interested in the central and nominal safety factors ($\bar{\zeta}$ and ζ) for these geometrically similar dams that correspond to a failure probability $P_c = 10^{-6}$.

Here we set the following parameters for computing $\bar{\zeta}$ and ζ: the Weibull modulus $m = 24$, the CoV of RVE strength $\omega = 0.15$, and the ratio between the mean RVE strength and the Weibull scale parameter $r = \mu/s_0 = 1.37$. The Weibull scale parameter s_0 of RVE strength distribution is determined, by using Eq. 11.2, so that the overall structural failure probability would be equal to 10^{-6}. We can then compute the mean structural strength and the corresponding values of $\bar{\zeta}$ and ζ for different dam sizes. The exact values of $\bar{\zeta}$ and ζ are shown as circles in Figures 11.8a and 11.9.

We now use the approximate size effect equations (Eqs. 11.40 and 11.41) to compute the central and nominal safety factors. To use Eq. 11.40, we first set $\epsilon = 0.75$ to compute the Weibull scale parameter s_0 using Eq. 11.33 for the small-size limit (i.e., $D = 4l_0$), and then update the value of ϵ. We find the final value of ϵ to be 0.72. Functions $\mu_r(D)$ and $\delta_r^2(D)$ describe the size effect on the mean and variance of the structural strength of the dam made of an artificial RVE with the following statistical parameters: $m = 24$, $s_0 = 1$, $\mu_G = 1.37$, and $\omega = 0.15$. The functions $\mu_r(D)$ and $\delta_r^2(D)$ can be obtained by using Eqs. 11.7 to 11.14. By matching the small- and large-size asymptotes of Eq. 11.39 with Eq. 11.34 and Eq. 11.38, we obtain $C = 2.835$ MPa and $\gamma = 2.4$.

Figure 11.8a shows that the central safety factors predicted by Eq. 11.40 agree well with the exact values for all the dam sizes. It is interesting to observe that the central safety factor decreases as the structural size increases. This is because, according to Eq. 11.40, the size dependence of the central safety factor is governed by the size effect on the CoV of σ_N, which has a decreasing trend in the present analysis. At the large-size limit, the structural strength follows a Weibull distribution and therefore the CoV becomes independent of the structure size. Therefore, the size dependence of the central safety factor must vanish as the structure size becomes sufficiently large.

Figure 11.8b presents the computed size effect on the central safety factor $\bar{\zeta}_s$ with respect to the mean strength of the small-size dam $D = 4l_0$, based on Eq. 11.42. Such

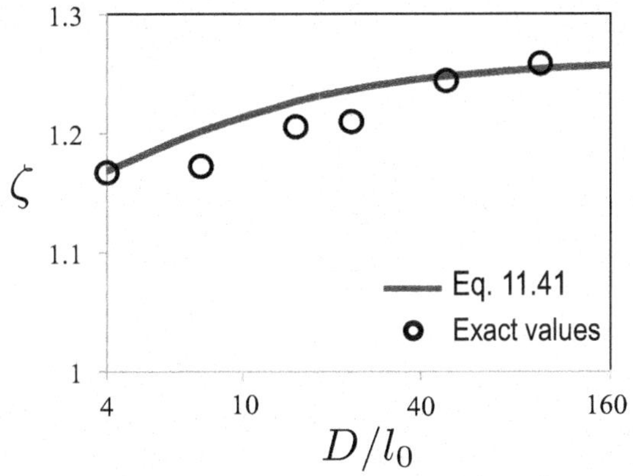

Figure 11.9. Size effect on the nominal safety factor.

size dependence is a combination of (1) the size effect on the central safety factor $\bar{\zeta}$ and (2) the size effect on the mean structural strength $\bar{\sigma}_N$. These two size effects have opposite trends, but the size effect on the mean structural strength is more pronounced compared to that on $\bar{\zeta}$. Therefore, $\bar{\zeta}_s$ increases with the structural size. At the large-size limit, the size effect on $\bar{\zeta}$ vanishes and the size effect on $\bar{\zeta}_s$ becomes governed purely by the inverse of the Weibull size effect on $\bar{\sigma}_N$, i.e., $\bar{\zeta}_s \propto D^{2/m}$.

To compute the nominal safety factor ζ, we choose $k_R = k_L = 3$. The prediction of Eq. 11.41 is compared with the exact values of ζ in Figure 11.9. It can be seen that Eq. 11.41 gives a reasonably good prediction. In contrast to the size effect on the central safety factor, ζ increases with the structure size.

The reason is that the decreasing in CoV of σ_N has two effects on the nominal safety factor: (1) it leads to a lower central safety factor, and (2) the ratio between the nominal and mean values of the structural strength becomes larger. As suggested by Eq. 11.41, these two effects counter each other and, for the k_R value chosen here, effect (2) prevails, which leads to an increasing ζ with the structure size. If we choose a smaller k_R value, effect (1) would prevail and ζ would then decrease as the structure size increases.

Overall, because of the competition between these two effects, the size effect on the nominal safety factor is less pronounced compared to that on the central safety factor.

12 Crack Length Effect on Scaling of Structural Strength and Type 1 to 2 Transition

As already explained in Chapters 1 to 3, there are two basic types of quasibrittle failure and size effect, types 1 and 2. They have different large-size asymptotes of the size effect and different statistical contributions of material randomness. Type 1 (Bažant & Li 1995*b*; Bažant 1997; Bažant & Novák 2000*a*, 2000*b*; Bažant 2005) occurs when the maximum load is attained already at macrocrack initiation, and type 2 (Bažant 1984*a*; Bažant & Pfeiffer 1987; Bažant & Kazemi 1990*a*, 1990*b*; Bažant 2005) occurs when, at maximum load, there is a large notch or a large stress-free (or fatigued) crack (note that in Bažant 2005 Type 1 is called Case 2, and Type 2 Case 1).

If the structure is much larger than the size of the representative volume element (RVE) or, equivalently, the size of the fracture process zone (FPZ), type 1 exhibits a large statistical effect of material randomness on the mean of structural strength, particularly on its coefficient of variation (CoV), while at small sizes the type 1 size effect is essentially deterministic. By contrast, the type 2 size effect on the mean structural strength is essentially deterministic at all sizes, owing to energy release and the associated stress redistribution, while the material randomness affects only the CoV of structural strength.

Progress in the modeling of concrete fracture and introduction of fracture concepts into the design codes and practice has been impeded by the unavailability of a comprehensive database for fracture. The literature features a vast amount of fracture data (e.g., Sabnis & Mirza 1979; Petersson 1981; Nallathambi 1986; Bažant & Pfeiffer 1987; Malvar & Warren 1988; Carpinteri, Chiaia, & Ferro 1995*a*; Rocco 1995; Tang, Bažant, Yang, & Zollinger 1996; Bažant & Planas 1998; Becq-Giraudon 2000; Bažant & Becq-Giraudon 2002; Karihaloo, Abdalla, & Xiao 2003, Hoover et al. 2013). But all these data except the last cover only rather limited ranges of specimen size, initial notch depth, and post-peak response, and have been performed on different concretes, on different batches of supposedly the same concrete, at different ages, at different environmental conditions, at different rates, with different test procedures, and on specimens of different types and dimensions. Combining all these data produces a database with enormous scatter and makes the modeling highly ambiguous because the effect of the aforementioned differences is understood much less than the fracture itself.

A universal size effect law that would include the type 1 and 2 size effects on the mean nominal strength as the limiting cases and describe the transition between them has proven to be a particular challenge, mainly because, in the type 1 to 2 transition, the problem is not linearizable, as the notch or crack is larger but not much larger than the

FPZ. Development of a universal size effect law was attempted in several previous studies (Bažant 1997; Bažant & Yu 2004, 2009), with progressive improvements. But it was hampered by insufficient experimental data, until the U.S. Department of Transportation funded a comprehensive series of 142 fracture tests at Northwestern University, reported and analyzed in a series of papers (Hoover et al. 2013; Hoover & Bažant 2013, 2014*a*, 2014*b*). Minimal statistical scatter (with CoV of errors of only 2.3%) was achieved by casting all the specimens from one and the same batch of modern ready-mixed concrete, curing them under the same conditions, and testing them at the same time. The specimens were geometrically similar three point bend beams of the same width. The beam depth varied within the range 1 : 12.5 and the notch depths from 0 to 30% of cross-sectional depth; 18 different combinations of notch depth and beam depth were used. A stiff loading frame, crack-mouth opening control, and fast hydraulics allowed measuring nearly complete post-peak softening on all the specimens.

In this chapter we review the results of a recent paper (Hoover & Bažant 2014*c*) in which these data and asymptotic matching were used to verify and calibrate a universal size effect law.

12.1 Type 1 Size Effect in Terms of Boundary Strain Gradient

Recall that the type 1 size effect is observed in structures of the so-called positive geometry, which fail under controlled load as soon as the macrocrack initiates from a smooth surface (Bažant & Planas 1998). As discussed in Chapter 3, the type 1 size effect curve for the small- and intermediate-size ranges can be derived by using a deterministic approach whereas the large-size limit of type 1 size effect can be described by the Weibull statistics. In some cases, where the zone of high stress is concentrated (even in the absence of a notch), the Weibull statistics, grounded in the statistical scatter of strength among many RVEs, becomes unimportant since the stress field has dictated the location of damage zone.

Depending on whether the potential macrocrack initiation zone occupies a large region, the type 1 size effect can be described as (Bažant 2005)

$$\sigma_N = f_{r\infty}\left[1 + \frac{rD_b}{\bar{D} + l_p}\right]^{1/r} \qquad \text{(deterministic, energetic)} \qquad (12.1)$$

$$\sigma_N = f_{r\infty}\left[\left(\frac{l_s}{l_s + D}\right)^{rn/m} + \frac{rD_b}{\bar{D} + l_p}\right]^{1/r} \qquad \text{(energetic-statistical)} \qquad (12.2)$$

Here $f_{r\infty}$, D_b, l_p, l_s, and r are empirical constants to be determined from tests; $f_{r\infty}$ = nominal strength for very large structures, assuming no statistical size effect (in the special case of very large beams, $f_{r\infty}$ represents the flexural strength, also called the modulus of rupture); and D_b = depth of the boundary layer of cracking (roughly equal to the FPZ size). In all previous works, $\bar{D} = D$ = the same characteristic structure size as used for the type 2 size effect (Eq. 3.37). The reason for placing an overbar on D is that in the next section $\bar{D}$ will be redefined in terms of the strain gradient.

In Eqs. 12.1 and 12.2, l_s denotes the statistical characteristic length, and l_p denotes the material characteristic length, which is related to the maximum aggregate size d_a. Introducing l_p is necessary for mathematical reasons, as a means to satisfy the asymptotic requirement of a finite plastic limit for $D \to 0$ while ensuring that the effect of l_p is negligible for $D \gg l_p$. However, if the structure is larger than $10l_p$, one can set $l_p \approx 0$, which corresponds to the original formulation of the type 1 law. Note that l_p differs from Irwin's characteristic length l_0. Also note that in the limit of $m \to \infty$, the statistical Eq. 12.2 reduces to the deterministic Eq. 12.1. In fact, Eq. 12.2 was constructed so as to satisfy all the asymptotic behaviors.

In Chapter 3, we showed that Eq. 12.1 can be derived based on the framework of equivalent LEFM (Eqs. 3.41–3.44). An asymptotic expansion of the dimensionless energy release rate function $g(\alpha)$ for initiating cracks and consideration of the limit $\alpha_0 \to 0$ indicated that

$$D_b = -c_f g''(0)/4g'(0), \quad f_{r\infty} = \sqrt{E'G_f/c_f g'(0)} \tag{12.3}$$

where typically $g''(0) < 0$. However, these expressions must be suspected of poorer accuracy since $g(0) = 0$ and approximations by higher-order derivatives always have much larger errors.

The comprehensive tests of Hoover et al. (2013) confirmed this suspicion. The tests included specimens with and without notches. When the type 2 size effect law (Eq. 3.37) was optimally fitted to the comprehensive test results (Hoover & Bažant 2013) for $\alpha_0 \geq 0.15$, the value of $f_{r\infty}$ predicted by Eq. 12.3 using function $g(\alpha)$ is 1.33 times smaller than necessary to fit the no-notch test results and the value of D_b is 2.1 times smaller. Thus, although the form of Eq. 12.1 obtained as the limit case of equivalent LEFM is correct, Eq. 12.3 gives parameter values that are too inaccurate (Hoover et al. 2013). It means that Eq. 12.3 cannot be used to determine G_f. In other words, the only way to determine G_f is to perform size effect tests on beams with a large size range and with deep notches.

To improve the formulation of type 1 size effect equation, we note that the type 1 size effect law was originally derived in Bažant and Li (1995*b*) by considering the stress redistribution in beam bending caused by the finite depth D_b of the boundary layer of cracking, which is roughly the same as the size of the FPZ of deep cracks; see Section 1.7.1 [for the tests in Hoover & Bažant (2013), $D_b \approx 7d_a$]. This stress redistribution is essentially local and depends on how much the normal strain ϵ_n at surface, in the direction parallel to surface, changes over the depth D_b of the boundary layer of cracking.

As suggested in Bažant (2005), this change may be characterized in terms of the strain gradient $\epsilon_{,n} = \partial\epsilon/\partial x_n$ where x_n is the coordinate normal to the surface, oriented outwards. The ratio $\epsilon/\epsilon_{,n}$ has the dimension of length and is proportional to the structure size D when geometrically similar structures are considered. This suggests redefining $\bar{D}$ in Eqs. 12.1 and 12.2 as follows:

$$\bar{D} = \frac{2\epsilon}{\psi \epsilon_{,n}} \tag{12.4}$$

The reason for the factor 2 is to make $\bar{D}$ coincide with the actual depth D of a slender beam subjected to bending moment M alone (to verify this fact, note that ϵ at the surface is equal to $\epsilon_{,n}D/2$, according to the plane cross section hypothesis). For beams under flexural loading, we may introduce a geometrical factor $\psi(S/D)$ (S/D = span-to-depth ratio), which is defined as 1 for very slender beams. Note also that the strain gradient should be obtained by elastic analysis with no crack, and not by fracture analysis with a crack, because it represents the modification of elastic analysis needed to approximate the cohesive fracture effect.

The beams tested in Hoover et al. (2013) are not slender. They have a span-to-depth ratio $S/D = 2.176$, and so $\psi(S/D) \neq 1$. According to the exact elastic solution [e.g., Timoshenko & Goodier (1951), p. 101], the stresses at mid-span cross section vary with transverse coordinate y as follows:

$$E\epsilon(y) = \frac{3P}{2c^3}\left(\frac{S}{4} - \frac{c}{\pi}\right)y + \frac{P}{2\pi c} + \frac{P}{\pi c}\left(\frac{y^3}{2c^3} - \frac{3y}{10c}\right) \tag{12.5}$$

$$\text{Hence} \quad \frac{\epsilon}{\epsilon_{,n}} = \frac{\epsilon(c)}{\epsilon_{,y}(c)} = \frac{\psi}{2}D \quad \text{with } \psi = 0.896 \tag{12.6}$$

where $c = D/2$ and S is the span. Calculation of the factor $\psi(S/D)$ for different S/D makes it possible to infer the size effect for beams with S/D other than 2.176.

12.2 Universal Size Effect Law

When the crack at failure is neither negligible nor large, the size effect trend is expected to be some sort of a transition between the types 1 and 2. The first attempt to describe this transition led to a formula that was not smooth (Bažant 1996; Bažant & Li 1996). A smooth formula was derived in Bažant and Yu (2009) but it could not be checked against proper experiments.

The comprehensive fracture tests in Hoover and Bažant (2014*c*) made this check possible and a better universal size effect law, presented in the test that follows, has been obtained. Although this law can be derived by a similar asymptotic matching procedure as in Bažant and Yu (2009), it is shorter to present it first and then demonstrate that is has the correct asymptotic properties. When the statistical type 1 component is negligible, the universal size effect law (USEL) has the form

$$\sigma_N = \left[\frac{E'G_f}{g_0 D + (1-\lambda)c_f g_0' + \lambda E' G_f / f_{r\infty}^2}\right]^{1/2} \left(1 + \frac{r\lambda D_b}{\bar{D} + l_p}\right)^{1/r} \tag{12.7}$$

and, more generally, when the Weibull statistical component is not negligible, it has the form

$$\sigma_N = \left[\frac{E'G_f}{g_0 D + (1-\lambda)c_f g_0' + \lambda E' G_f / f_{r\infty}^2}\right]^{1/2} \left[\left(\frac{\lambda l_s}{D + l_s}\right)^{rn/m} + \frac{r\lambda D_b}{\bar{D} + l_p}\right]^{1/r} \tag{12.8}$$

in which we introduce an empirical type 1 to 2 transition parameter defined as

$$\lambda = e^{-(\alpha_0^k/q)(\bar{D}/d_a)^p} \tag{12.9}$$

and $\bar{D}$ is given by Eq. 12.4; α_0 is the initial relative crack depth; $G_f, c_f, f_{r\infty}, D_b, r,$ k, p, q, l_p, and l_s are parameters to be calibrated by data fitting, while ψ is calculated

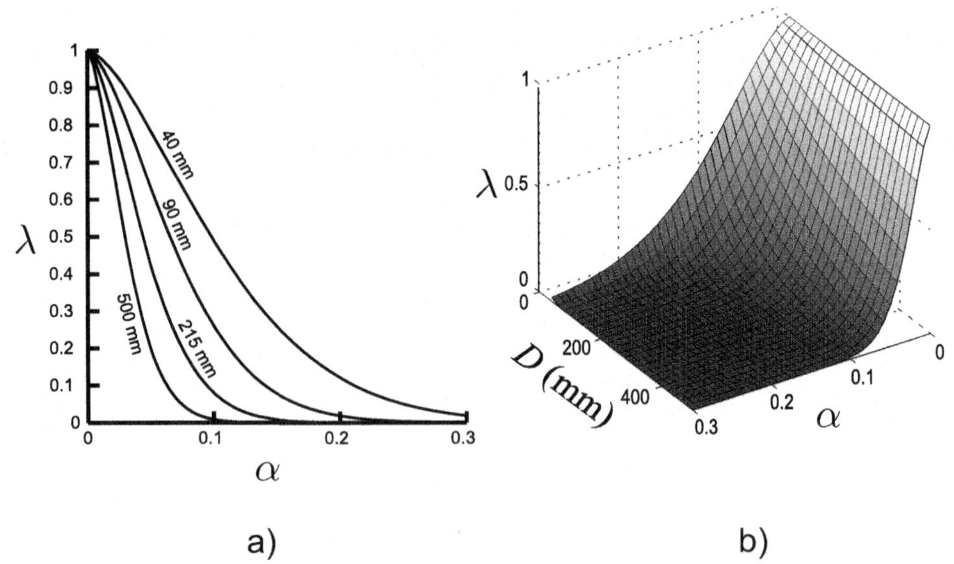

Figure 12.1. Trend of the transition function $\lambda(\alpha, D)$. (a) For each beam size used in Hoover and Bažant's test, and (b) for the entire surface.

from the strain profile. The only purpose of the transition parameter λ is to provide a smooth transition between the type 1 and 2 size effects. Thus it is justified as an asymptotic matching parameter, with asymptotic cases $\lambda = 1$ for no-notch specimens and $\lambda = 0$ for deep notch (or deep crack) specimens. The surface $\lambda(\alpha, D)$ is pictured in Figure 12.1.

Equations 12.7 and 12.8 are justified by checking that they match the following two basic asymptotic trends:

1. For no notch, upon setting $\alpha = 0$, $g_0 = 0$, $\lambda = 1$ and using $\psi = 0.896$, Eqs. 12.7 and 12.8 reduce, respectively, to the deterministic or statistical type 1 size effect laws in Eqs. 12.1 and 12.2.
2. For deep notches, on setting $\lambda = 0$, both Eqs. 12.7 and 12.8 reduce to the type 2 size effect law (Eq. 3.37).

Previously it was further shown that the type 1 and 2 size effect laws satisfy the large-size and small-size asymptotic properties of the cohesive crack model applied to type 1 and 2 failures (Bažant & Yu 2009). Furthermore, it was experimentally confirmed that, within the range of inevitable experimental scatter, the size effect law of type 2 gives about the same values of fracture energy G_f when applied to notched fracture specimens of different geometries [e.g., notched beam bending, edge-notched tension, eccentric compression, as checked in Bažant & Pfeiffer (1987)].

Equations 12.7 to 12.9 also give the correct prediction for an un-notched direct tension specimen. Indeed, in this case, $\epsilon_{,n} = 0$, $\bar{D} \to \infty$, $\lambda = 1$, and $g_0 \to 0$. Equation 12.8 thus reduces to $\sigma_N = f_{r\infty}[l_s/(D + l_s)]^{m/n}$, which correctly gives, for large D, the pure Weibull statistical size effect $\sigma_N = f_{r\infty} D^{-m/n}$.

12.3 Verification of the Universal Size Effect Law by Comprehensive Fracture Tests

To calibrate the deterministic USEL (Eq. 12.7), the mean of the data is computed, separately for each family of identical specimens, from the comprehensive fracture tests (Hoover et al. 2013). Then all these means are optimally fitted by Eq. 12.7. The trust-region-reflective optimization algorithm (Coleman & Li 1994, 1996) is used.

It would not have been meaningful to fit all the unknown parameters concurrently because every one of them would then be influenced by all the data. In particular, the fracture parameters G_f and c_f should not be influenced by the data for beams with no notches (type 1 data) or shallow notches, and $f_{r\infty}$, D_b, l_p, and r should not be influenced by the data for deep notches. Therefore, these parameters are determined first by separate fitting of specimens with deep notches ($\alpha = 0.30, 0.15$) and specimens with shallow or no notches ($\alpha = 0$), and only the remaining ones, i.e., k, p, and q, are identified by simultaneous optimal fitting of all the data. Only the nonstatistical USEL in Eq. 12.7 is considered because the parameters m and l_s for the Weibull statistical size effect cannot be identified from the present data. They are nevertheless well known from other studies (Bažant et al. 2007). Nonlinear fitting of the type 1 SEL (Eq. 12.1) to the notchless ($\alpha = 0$) beams yields (Hoover & Bažant 2013)

$$D_b = 73.2 \text{ mm}, \quad l_p = 126.6 \text{ mm}, \quad f_{r,\infty} = 5.27 \text{ MPa}, \quad r = 1/2. \tag{12.10}$$

The size range 1:12.5 is large enough to identify all the fracture parameters in Eq. 12.1. G_f and c_f are identified by weighted nonlinear regression of the type 2 size effect (Eq. 3.37) run jointly for the beams with $\alpha = 0.30$ and 0.15, with different g_0 and g_0' for each α. The means in each family is assigned the weights $w_i = \delta_i^{-2} / \sum_{i=1}^{N} \delta_i^{-2}$, representing the inverse variance of each beam family normalized with the sum of the inverse variances for all N families ($N = 8$). The optimization gives

$$G_f = 56.25 \text{ N/m}, \quad c_f = 29.79 \text{ mm} \tag{12.11}$$

The optimization of the remaining parameters, all belonging to the transition function $\lambda(\alpha, D)$ (Eq. 12.9), furnishes $k = 1.57$, $p = 0.755$, and $q = 0.108$.

The surface of the optimized USEL is displayed in Figure 12.2, where the contour lines are the lines of constant D or constant α. The CoV of fit of the entire surface (defined as the root-mean-square error divided by the mean of all data) is only 2.30%. For simplicity, the optimized parameters in the transition function may be rounded as

$$k = 3/2, \quad p = 3/4, \quad q = 1/9 \tag{12.12}$$

which makes the CoV only slightly higher, 2.45%. Although it seems that these parameter values can be kept the same even for other specimen geometries, recalibration may be desirable for good accuracy.

The USEL can be drawn for a fixed α, which gives a size effect plot of $\log(\sigma_N / f_t')$ versus $\log(D/D_0)$, or for a fixed relative depth D/D_0, which gives the plot of $\log(\sigma_N / f_t')$ versus α. In Figures 12.3 and 12.4, these plots are created and compared to the data from Hoover et al. (2013). The mean values for all different specimen sizes are shown in these

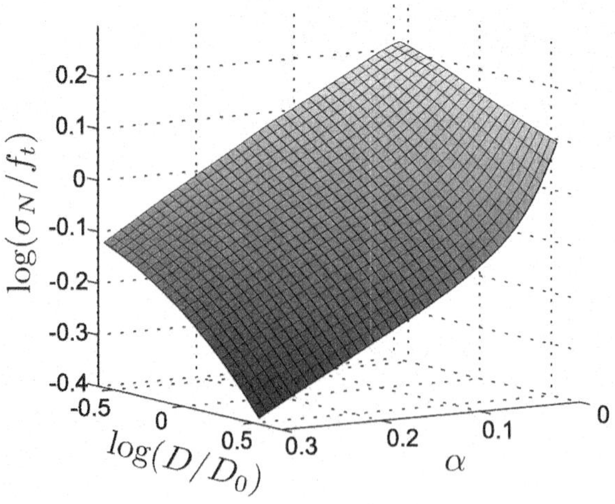

Figure 12.2. Plot of the entire USEL surface.

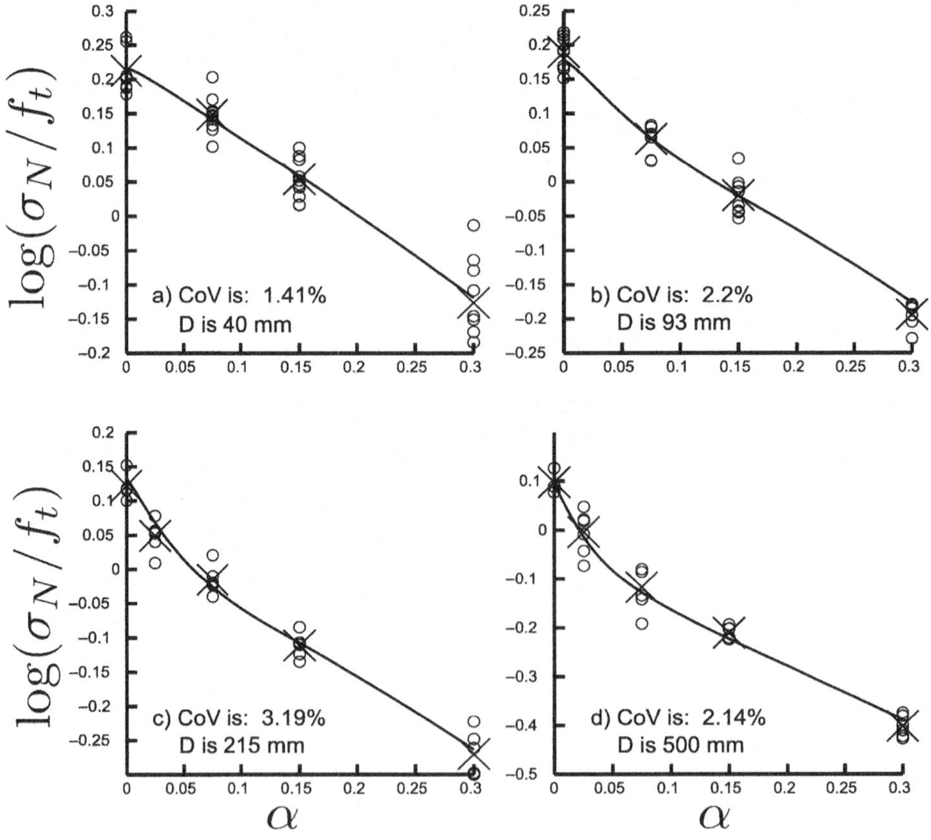

Figure 12.3. Effect of relative crack length on the normalized nominal strength of data from Hoover et al. (2013) optimally fitted by the USEL (Eq. 12.7).

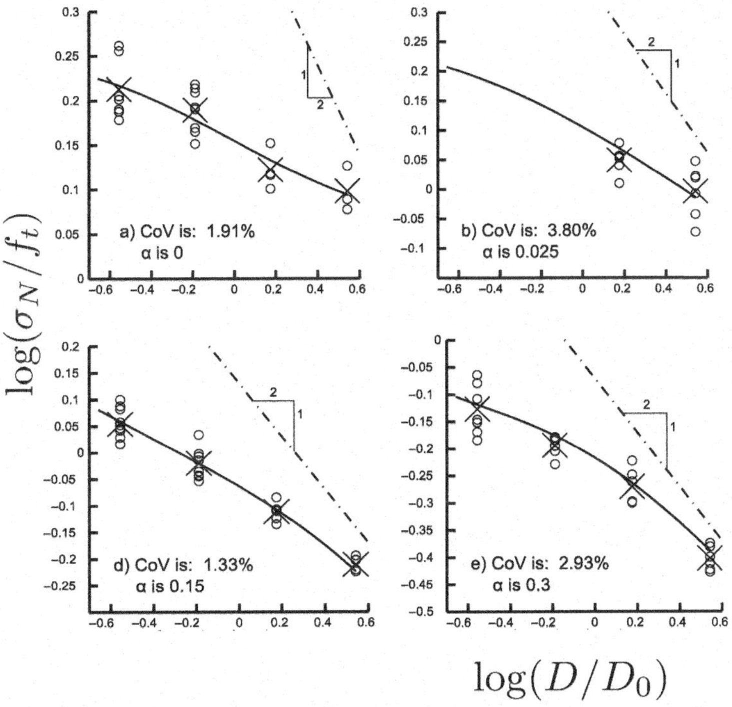

Figure 12.4. Effect of normalized structure size on the normalized nominal strength of the data from Hoover et al. (2013), optimally fitted by the USEL (Eq. 12.7).

figures as the large and thin x-points. The CoV for each curve is indicated in the figures. It is seen that the curves fit the data satisfactorily. The largest CoV, equal to 3.8%, comes from the size effect plot for $\alpha = 0.025$, which is in the transition. The remaining CoV values are less than 3.2%, and four of them less than 2%.

It must be admitted that the experimental verification of USEL is limited to beams of low slenderness. The USEL might have significant errors for fracture specimens of other geometries; e.g., for double-edge-notched tensile and edge-notched eccentric compression specimens [nevertheless, applicability of the type 2 size effect law to these specimens was verified in Bažant & Pfeiffer (1987)].

Because Eq. 12.4 describes a local effect, for which remote boundaries are irrelevant, this equation for $\bar{D}$ should, in theory, apply for any geometry. However, this remains to be proven experimentally. The transition function λ is not completely local, and so its applicability to other geometries is not assured. Extending this function to a very different material would surely necessitate recalibration.

13 Effect of Stress Singularities on Scaling of Structural Strength

In Chapter 12, we discussed the transition between type 1 and type 2 size effects in terms of the size of the preexisting crack. A universal size effect law (USEL) was proposed based on asymptotic matching and verified by a comprehensive set of size effect tests. Instead of preexisting crack length, we may also consider type 1 and type 2 size effects as two limiting cases in terms of the prevailing stress singularities. Type 1 size effect is applicable to structures of positive geometry, which fail at the macrocrack initiation from a smooth surface without stress singularities. By contrast, type 2 size effect is applicable to structures with a large preexisting crack, which is associated with a strong stress singularity.

This chapter discusses the size effect on strength of quasibrittle structures with weak stress singularities, which is expected to transition between the type 1 and type 2 size effects. Weak stress singularities can occur in many modern engineering structures, which consist of complex structural geometries or are made of dissimilar materials. Here we consider the type 1 size effect as a consequence of the finite weakest-link model as discussed in Chapter 7, whereas the type 2 size effect as energetic, dictated by the equivalent linear elastic fracture mechanics (LEFM). The transitional size effect for weak stress singularities is derived based on a generalized finite weakest-link model, in which the energetic scaling due to the prevailing stress singularities is combined with the finite weakest-link model.

In this chapter, we first derive this general energetic-statistical size effect equation for structures with a V-notch under mode I loading, and extend it to bimaterial hybrid structures with a weak interface. The analytical formulation is verified by a series of finite element simulations of fracture of both homogeneous and bimaterial structures.

13.1 Strength Scaling of Structures with a V-Notch under Mode I Loading

We consider structures of positive geometry containing a V-notch under mode I loading (Fig. 13.1), where the notch angle γ governs the order of the stress singularity. Here we also assume that the notch is sufficiently deep, i.e., $\alpha = a/D > 0.1$, where $a =$ notch depth and $D =$ depth of the structure. In general, the stress concentration at the V-notch tip is governed by two distinct stress singularities, which represent the symmetrical and antisymmetrical deformation modes (Williams 1952; Carpenter 1984; Sinclair, Okajima, & Griffin 1984; Liu, Suo, & Ma 1999). For mode I fracture, only one stress

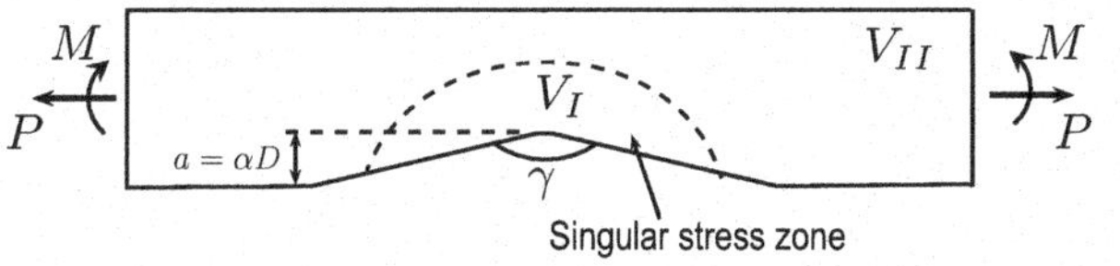

Figure 13.1. Structure with a V-notch under mode I fracture.

singularity λ prevails, which corresponds to displacements that are symmetric about the center-line of the notch.

13.1.1 Energetic Scaling of Strength of Structures with Strong Stress Singularities

First consider the case where the stress singularity induced by the V-notch is strong. In such a case, the fracture process zone (FPZ) must form at the notch tip, and a large volume of the structure, with many representative volume elements (RVEs), cannot be sampled. Therefore, the corresponding scaling mechanism is deterministic in nature.

The fracture of V-notch has been analyzed both theoretically and experimentally (Ritchie, Knott, & Rice 1973; Carpinteri 1987; Seweryn 1994; Dunn, Suwito, & Cunningham 1996; Dunn, Suwito, & Cunningham 1997; Gomez & Elices 2003). Three commonly used fracture criteria include (1) the peak load is reached when the stress intensity factor reaches a critical value (Carpinteri 1987; Dunn et al. 1996; Gomez & Elices 2003); (2) the peak load is attained when the stress at a certain distance c_f from the notch tip reaches the material tensile strength (Ritchie et al. 1973; Bažant & Yu 2006); and (3) the peak load is realized when the energy release rate of an equivalent cohesive crack that represents the FPZ reaches the fracture energy (Le, Bažant, & Yu 2010; Le 2011).

These criteria are based on the framework of LEFM, which can be used to derive the large-size asymptote of the size effect. For mode I loading, all three criteria yield essentially the same form of the power-law size effect. Among these criteria, the second criterion is relatively straightforward to use because the first criterion adopts a geometry-dependent critical stress intensity factor, which needs to be measured for every notch angle (Bažant & Yu 2006). The third criterion requires determination of the energy release at the tip of an equivalent crack through the solution of an ancillary boundary value problem (Leguillon 2002; Le et al. 2010; Le 2011).

Based on Williams' solution, the stress field near the notch tip under mode I loading can be expressed as

$$\sigma_{ij} = H r^{\lambda} f_{ij}(\theta, \gamma) \tag{13.1}$$

where r = radial distance from the notch tip, $f_{ij}(\theta, \gamma)$ = a dimensionless function describing the angular dependence of the stress, λ = order of stress singularity, and H = stress intensity factor. Dimensional analysis allows the stress intensity factor H to be written as

$$H = \sigma D^{-\lambda} h(\gamma) \tag{13.2}$$

where σ = nominal stress = cP/bD, P = applied load, c = a constant, b = width of the structure in the transverse direction, and $h(\gamma)$ = dimensionless stress intensity factor, which depends on the structure geometry. Considering the second failure criterion, we can calculate the nominal strength as

$$\sigma_N = f_t \psi(\gamma)(D/c_f)^{\lambda} \tag{13.3}$$

where $\psi(\gamma) = h^{-1}(\gamma) f_{\theta\theta}^{-1}(0, \gamma)$ and f_t = tensile strength.

Equation 13.3 represents the large-size asymptote of the energetic size effect law. At the small-size limit, we may consider that the FPZ occupies the entire crack ligament and consequently the ligament must behave like a crack filled with plastic glue. At this plastic limit, the size effect must vanish. An approximate equation that bridges these two asymptotes can be written as (Bažant 2005; Le 2011)

$$\sigma_N = \sigma_s \left[1 + \left(D/D_{0\gamma}\right)^{1/\beta_\gamma}\right]^{\lambda\beta_\gamma} \tag{13.4}$$

where σ_s = nominal strength at the small-size limit, β_γ = model parameter, and $D_{0\gamma}$ = transitional size. Equation 13.4 clearly indicates the effect of stress singularity on the scaling law for the case where the stress singularity is sufficiently strong. When $\lambda = -1/2$ and $\beta_\gamma = 1$, Eq. 13.4 converges to the classical type 2 size effect (i.e., Eq. 3.37) (Bažant 1984*b*, 2004*b*), which applies to structures with a large preexisting crack, as discussed in Chapter 3.

13.1.2 Generalized Finite Weakest-Link Model

Now consider that the structure has a wide-angle V-notch. In this case, the stress singularity at the notch tip could be very weak, and therefore there is no guarantee that the FPZ would form at the tip of the V-notch. This means that the failure location is random. Therefore, as discussed in Chapter 7, we can use the finite weakest-link model to describe the overall failure statistics of the structure.

On the other hand, note that there does exist a singular stress field at the V-notch tip even though the degree of stress concentration is weak. Furthermore, the fracture of the V-notch itself is associated with an energetic scaling law described by Eq. 13.4, which cannot be represented by the finite weakest-link model presented in Chapter 7. This prompts us to generalize the previous finite weakest-link model to include the energetic scaling of fracture of the V-notch.

In the generalized weakest-link model, we isolate the singular stress zone from the remaining part of the structure (Fig. 13.1), where the singular stress zone can be determined by comparing Eq. 13.1 to the numerically simulated elastic stress field. Since the singular stress zone is influenced by the presence of the V-notch, whose fracture exhibits an energetic scaling (i.e., Eq. 13.4), it is proposed to include this energetic scaling in the calculation of the failure probability of the singular stress zone

(Le, Pieuchot, & Ballarini 2014):

$$P_{f,V_I}(\sigma_N) = 1 - \prod_{i=1}^{N_1} \{1 - P_1[\mu(D)\sigma_N s(\boldsymbol{x}_i)]\} \tag{13.5}$$

$$\text{and} \quad \mu(D) = \left[1 + (D/D_{0\gamma})^{1/\beta_\gamma}\right]^{-\lambda\beta_\gamma} \tag{13.6}$$

where $P_1(x)$ = probability distribution function of the strength of one RVE of the material, as defined by Eqs. 5.84a and 5.84b; $s(\boldsymbol{x}_i)$ = dimensionless stress field such that $\sigma_N s(\boldsymbol{x}_i)$ is equal to the maximum elastic principal stress at a point of coordinate $\boldsymbol{x}$; $D_{0\gamma}$ and β_γ follow the same definition as those in Eq. 13.4; and N_1 = the number of RVEs in the singular stress zone. For the remaining part of the structure, the original finite weakest-link model can be used, i.e.

$$P_{f,V_{II}}(\sigma_N) = 1 - \prod_{i=1}^{N_2} \{1 - P_1[\sigma_N s(\boldsymbol{x}_i)]\} \tag{13.7}$$

where N_2 = the number of RVEs in the region outside the singular stress zone.

Therefore, the failure probability of the entire structure can be written as

$$P_f(\sigma_N) = 1 - \left[1 - P_{f,V_I}(\sigma_N)\right]\left[1 - P_{f,V_{II}}(\sigma_N)\right] \tag{13.8}$$

from which we can calculate the mean structural strength $\bar{\sigma}_N$. It is clear that a closed-form solution for $\bar{\sigma}_N$ is not possible. Similar to our previous analysis of the mean size effect implied by the finite weakest-link model, here we seek an approximate scaling equation for $\bar{\sigma}_N$ through asymptotic matching.

At the large-size limit, the failure probability of the structure is governed by the tail part of the strength cdf of one RVE. Based on the fact that $\ln(1 - x) \approx -x$ for $x \to 0$, we can rewrite the weakest-link model for 2D structures as

$$P_f(\sigma_N) = 1 - \exp\left\{-\int_{V_I} \frac{\mu^m(D)\sigma_N^m \langle s(\boldsymbol{x})\rangle^m}{s_0^m} \frac{\mathrm{d}V(\boldsymbol{x})}{l_0^2} - \int_{V_{II}} \frac{\sigma_N^m \langle s(\boldsymbol{x})\rangle^m}{s_0^m} \frac{\mathrm{d}V(\boldsymbol{x})}{l_0^2}\right\} \tag{13.9}$$

where l_0 = RVE size. Since $s(\boldsymbol{x})$ represents the dimensionless elastic stress field, we can use the linear transformation of the coordinate, i.e., $\xi = \boldsymbol{x}/D$, to rewrite Eq. 13.9 as

$$P_f(\sigma_N) = 1 - \exp\left\{-[\mu^m(D)\Psi_1 + \Psi_2]\frac{D^2}{l_0^2}\left(\frac{\sigma_N}{s_0}\right)^m\right\} \tag{13.10}$$

$$\text{where} \quad \Psi_1 = \int_{V_I} \langle s(\boldsymbol{x})\rangle^m \,\mathrm{d}V(\xi); \quad \Psi_2 = \int_{V_{II}} \langle s(\boldsymbol{x})\rangle^m \,\mathrm{d}V(\xi) \tag{13.11}$$

Based on Eq. 13.10, the mean strength can be calculated as

$$\bar{\sigma}_N = s_0[\mu^m(D)\Psi_1 + \Psi_2]^{-1/m}\Gamma\left(1 + \frac{1}{m}\right)\left(\frac{l_0}{D}\right)^{2/m} \tag{13.12}$$

where $\Gamma(x)$ = Eulerian gamma function. Equation 13.12 indicates that the large-size asymptote of the size effect curve differs from the classical Weibull size effect because the energetic scaling term is introduced for calculating the failure statistics of the singular stress zone.

It should be pointed out that here we use the integral form for the weakest-link model. In principle, Ψ_1 is infinite due to the singular stress field. However, it was shown that the singular stress should not be included in the weakest-link model (Bažant & Xi 1991). When evaluating Ψ_1, we must exclude the region where the radial distance from the notch tip is less than a small distance d_c.

At the small-size limit, the entire structure consists of a small number of RVEs. Therefore, it is expected that the RVEs in the singular stress zone would govern the failure of the entire structure. Chapter 7 presented the mean size effect equation for structures without stress singularities (Eq. 7.34) implied by the finite weakest-link model, i.e.,

$$\bar{\sigma}_N = \left[\frac{N_a}{D} + \left(\frac{N_b}{D}\right)^{rn_d/m}\right]^{1/r} \tag{13.13}$$

where n_d = dimension of scaling, and N_a, N_b, and r are constants. It is clear that, for structures without stress singularities, the small-size asymptote of the mean size effect can be expressed as $\bar{\sigma}_N \propto (D/D_b)^{-1/r}$, where D_b and r can be determined from the mean and standard deviation of the Gaussian part of the cumulative distribution function (cdf) of RVE strength (Cannone Falchetto et al. 2014). In the present model, the stress that governs the failure of the singular stress zone is scaled by the energetic scaling term. Therefore, the corresponding size effect at the small-size limit can be expressed as $\bar{\sigma}_N \propto \mu^{-1}(D)(D/D_b)^{-1/r}$.

Since the present model is developed within the framework of the weakest-link model, it is expected that the entire size effect curve can be approximated by a function similar to Eq. 13.13. Meanwhile, it is also clear that, as the stress singularity is sufficiently strong, all the scaling terms associated with the weakest-link model should vanish. This transition is expected to occur in a very narrow range of stress singularities, which is approximated by a Gaussian function. Therefore, the following scaling equation is proposed to bridge the limiting cases of strong and zero stress singularities (Le, Pieuchot, & Ballarini 2014):

$$\bar{\sigma}_N = \sigma_0 \left\{ C_1\left[\mu^m(D)\Psi_1 + \Psi_2\right]^{-r/m} \left(\frac{D + l_s}{l_0}\right)^{-2r/m} \exp[-(\lambda/\lambda_1)^2] + \frac{\mu^{-r}(D)D_b}{\exp[-(\lambda/\lambda_2)^2]D + l_p} \right\}^{1/r} \tag{13.14}$$

where σ_0 = reference stress and $C_1, r, \lambda_1, \lambda_2, l_s, l_p, D_b$ = constants. Note that, similar to Eq. 12.2, we purposely introduce constants l_s and l_p to regularize the functional behavior as $D \to 0$. Based on the large-size and small-size limits, we must have

$$\sigma_0 C_1^{1/r} = s_0^{1/r}\Gamma(1 + 1/m) \tag{13.15}$$

$$\sigma_s = \sigma_0 \left\{ D_b/l_p + C_1(\Psi_1 + \Psi_2)^{1/m}(l_s/l_0)^{-2r/m} \exp[-(\lambda/\lambda_1)^2] \right\}^{1/r} \tag{13.16}$$

The small-size strength limit σ_s ($D \to 0$) can usually be obtained by simple plastic analysis, where the ligament is imagined to be filled up with a plastic glue. It is clear that

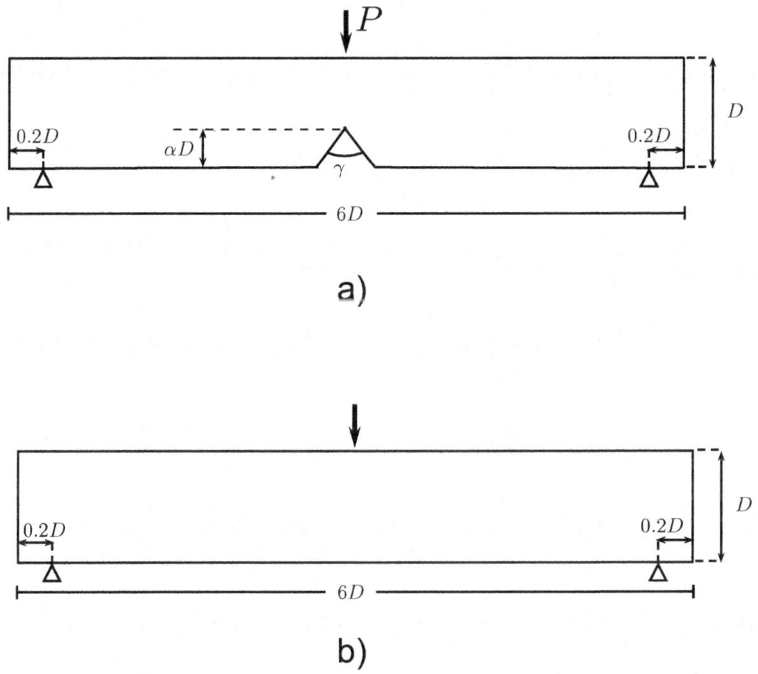

Figure 13.2. Numerical simulation of flexural failure of beams with a V-notch.

Eq. 13.14 converges to Eqs. 13.4 and 13.13 as the two limiting cases. For the transition between these two limits, the size effect consists of both energetic and statistical components.

13.2 Numerical Simulation of Mode I Fracture of Beams with a V-Notch

13.2.1 Model Description

To verify the proposed analytical model, we numerically investigate the size effect on the nominal strength of concrete beams with a V-notch under three point bending (Fig. 13.2a). The beam has a 6:1 span-to-depth ratio and the notch depth is 20% of the beam depth. In the simulation, we consider five different notch angles, i.e., $\gamma = 0^\circ, 90^\circ, 120^\circ, 135^\circ, 170^\circ$, and a series of geometrically similar specimens with the size range of 1:2:4:8:16:32:64:128 (the depths of the smallest and largest specimens are 37.5 mm and 4.8 m, respectively). Based on Williams' solution, these notch angles correspond to the following orders of mode I stress singularity: $\lambda = -0.5, -0.4555, -0.3843, -0.3264, -0.0916$. In addition to this set of specimens, we also include the size effect simulation of flat beams with a maximum size ratio 1:64 (Fig. 13.2b), which corresponds to the case of zero stress singularity.

In these simulations we use the default damage-plasticity model in ABAQUS (Lee & Fenves 1998), which is sufficient for the purpose of the present study. The material properties are chosen as follows: Young's modulus $E = 30$ GPa, Poisson ratio $\nu = 0.2$, tensile strength $f_t = 3$ MPa, compressive strength $f_c = 30$ MPa, and mode I fracture energy $G_f = 100$ N m^{-1}. Although we specify the compressive strength, the compressive region of the beam is expected to remain elastic. Therefore, the nonlinear part of the compressive behavior is not of particular interest for the present study. All the specimens undergo displacement-controlled loading.

In this study, the numerical simulation is performed within a deterministic framework. Previous studies have shown that the deterministic simulation with a strain-softening constitutive model can successfully capture the entire size effect for the case of strong stress singularity and the size effect for the small and intermediate structure sizes in the case of zero stress singularity (Bažant 2004*b*, 2005; Bažant et al. 2007). Therefore, we expect that, for the case of weak stress singularity, the deterministic numerical model is sufficient for simulating the size effect for the small and intermediate structure sizes. For the large-size asymptote, the deterministic simulation cannot yield the Weibull statistical scaling.

In finite element modeling, the notch tip is considered to have a very small width, i.e., 5 mm, which is a constant for all the geometries and sizes. For the deterministic simulation, the damage occurs near the mid-span of the beam. Therefore, to reduce computational efforts, we model the middle portion of the beam with a refined mesh (i.e., 5 mm) and the damage plasticity model whereas the remaining part of the beam is modeled by a coarse mesh with a purely elastic model. For each specimen, the assumed region, where the nonlinear material model is used, is further checked as part of the simulation. As the notch angle increases, this nonlinear region becomes larger.

13.2.2 Results and Discussion

Figure 13.3 presents the simulated curves of nominal stress versus relative deflection for specimens of different sizes and notch angles (the load-deflection curves of beams with $D = 4.8$ m are not shown here because they are qualitatively similar to those of beams with $D = 2.4$ m). The nominal stress is defined as $\sigma = P/bD$ and the relative displacement is defined as $\delta = \Delta/D$ (Δ = load-point displacement). It is observed that, as the structure size increases, the post-peak softening portion of the load-deflection curve becomes steeper, which implies a more brittle failure behavior. It should be noted that, for large specimens (i.e., $D = 1.2$, 2.4, and 4.8 m), the post-peak behavior is not captured, which indicates that a snap-back instability may have occurred.

The size-dependent failure behavior can also be explained by examining the fracture process at the notch tip. Figure 13.4 shows the normal stress profile along the notch ligament at the peak load for beams with a sharp notch (i.e., $\gamma = 0°$). Similar stress profiles are observed for beams with other notch angles. The FPZ can be determined as the region where strain-softening occurs. As seen, at the large-size limit, the FPZ becomes negligible compared to the structure size whereas at the small-size limit the FPZ occupies a large part of the structure and the stress in the FPZ exhibits a nearly plastic (or uniform) stress profile. This implies that the structure would behave in a

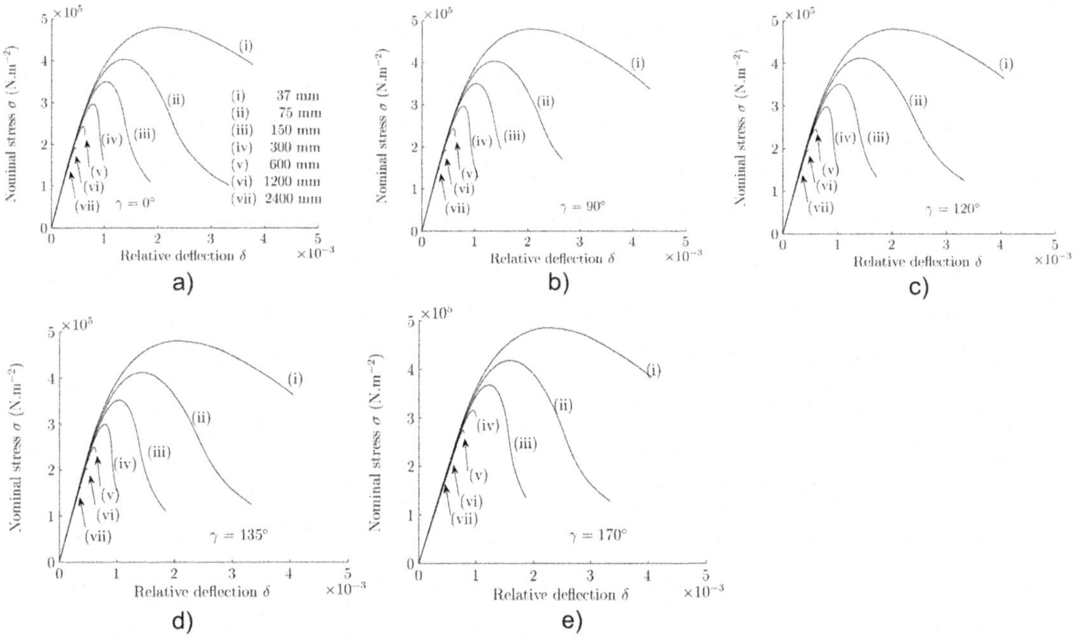

Figure 13.3. Simulated curves of nominal stress versus relative displacement.

quasi-plastic manner at the small-size limit and in a brittle manner at the large-size limit, which is consistent with the conclusion drawn from the simulated post-peak behavior of the load-deflection curves.

For the 2D specimens studied here, we define the nominal strength of the beam simply as $\sigma_N = P_{\max}/bD$, where $b = 1$. Figure 13.5 presents the simulated size effects on the nominal strength for different notch angles, and Figure 13.6 presents the simulated

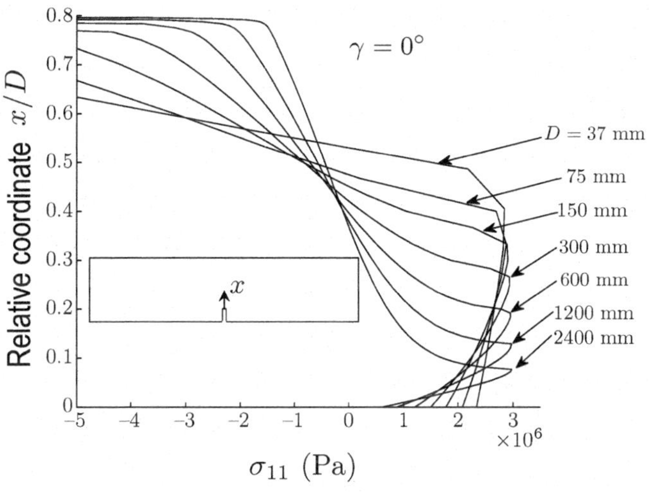

Figure 13.4. Simulated normal stress profile along the ligament.

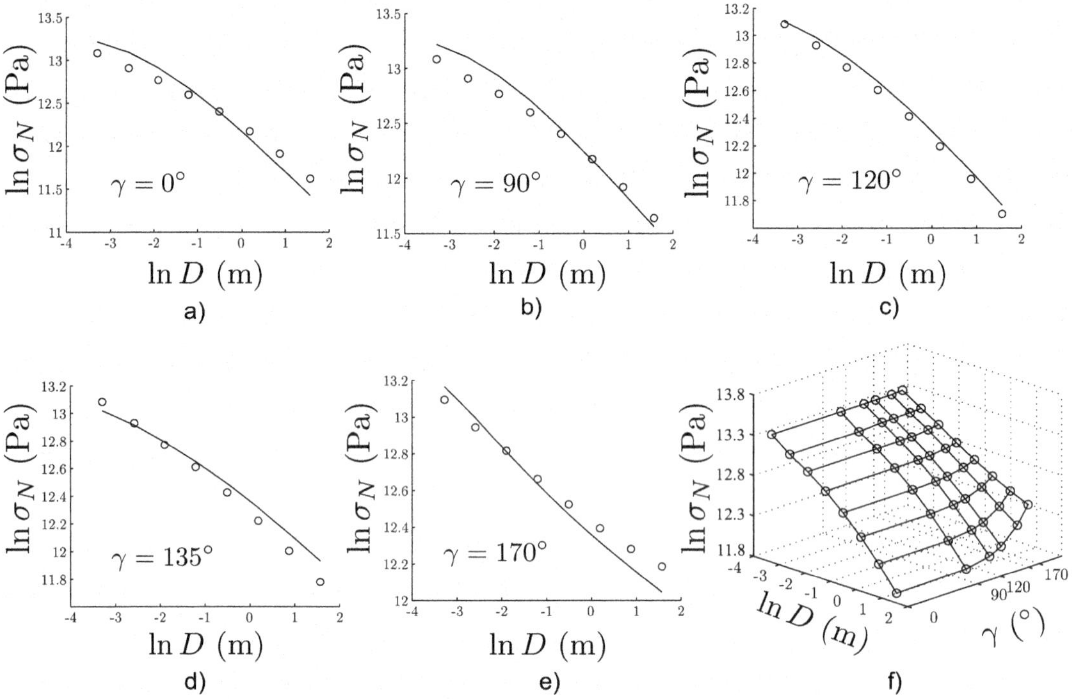

Figure 13.5. Simulated size effect curves for notched beams. (a–e) Size effect curves with optimum fits by Eq. 13.14. (f) Surface plot of effects of structure size and notch angle on nominal structural strength (σ_N is measured in Pa, and D is measured in m).

size effect curve for specimens without notches. At the small-size limit, the nominal strengths of all the specimens are almost the same while the large-size asymptote of the size effect varies with different notch angles (Fig. 13.5f). As the notch angle increases, which implies that the stress singularity weakens, there is a clear change in the curvature of the size effect curve.

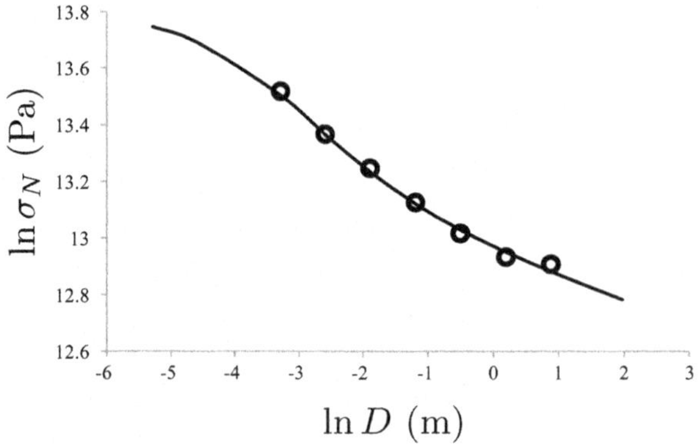

Figure 13.6. Simulated size effect curve for unnotched beams with optimum fit by Eq. 13.14.

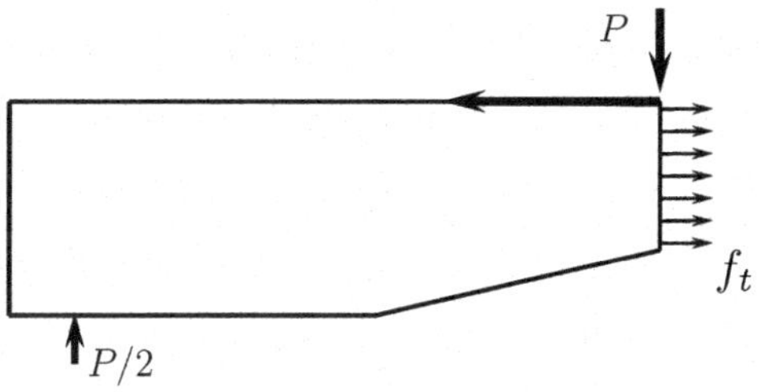

Figure 13.7. Plastic analysis of the small-size asymptote.

The simulated size effect curves are now compared with the proposed approximate size effect equation. As mentioned earlier, Eq. 13.14 contains the energetic and statistical scaling terms. For the energetic scaling term $\mu(D)$, a previous study by Bažant and Yu (2006) showed that parameter $D_{0\gamma}$ varies with the notch angle as

$$D_{0\gamma} = D_0 \psi(\gamma)/\psi(0) \tag{13.17}$$

where $D_0 = D_{0\gamma}$ at $\gamma = 0$. Elastic analysis directly yields $\psi = 2$, 1.836, 1.624, 1.485, and 1.089 for $\gamma = 0°, 90°, 120°, 135°$, and 170°, respectively. Parameter β_γ is introduced for a better fitting of size effect data. Therefore, in principle, β_γ may vary with the notch angle. For the case of zero notch angle, the classical type 2 size effect indicates that β_γ must be equal to 1. However, for other notch angles, β_γ may take other positive values. Here we leave β_γ as a calibration constant for every notch angle except for the zero notch angle.

For the statistical scaling term, the constants Ψ_1 and Ψ_2 for the Weibullian part can be easily determined by linear elastic analysis. Here we assume the RVE size l_0 to be 37.5 mm, which is about three times the size of the typical maximum aggregate d_a (Bažant & Pang 2007). Furthermore, when calculating Ψ_1, we exclude the notch tip region with a radius $d_c = D/400$ to avoid an infinite value of Ψ_1. It should be noted that we need only d_c for the size effect curves for the case of weak stress singularity. The choice of d_c is not particularly important because we further introduced two exponential functions to describe this transitional scaling mechanism, and because different values of d_c will be compensated by the exponential function $\exp[-(\lambda/\lambda_1)^2]$, which is empirically determined by the optimum fitting of the data.

Other parameters associated with the statistical scaling components can be determined on the basis of the size effect curve of the unnotched beams. The Weibull modulus m for concrete is known to be 24 (Bažant & Novák 2000*a*; Bažant & Pang 2007). Previous studies have shown that D_b is approximately equal to $4d_a$ (Bažant 2005), which is about 50 mm for our study. Parameters r, l_p, l_s, and s_0 can be determined by the optimum fitting of the size effect curve of the flat beam as shown in Figure 13.6.

We can then determine the reference stress σ_0 based on the nominal strength at the small-size limit. As mentioned earlier, the structural strength at the small-size limit can be calculated by plastic analysis, where we can assume that the FPZ behaves in the plastic manner as shown in Figure 13.7. It is noted that the calculated stress profile of the FPZ (Fig. 13.4) further verifies such a plastic model.

Therefore, regardless of the notch angle, the nominal strength at the small-size limit can be calculated as $\sigma_s = 16f_t/75 = 0.64$ MPa. Similarly, it can easily be shown that σ_s for the unnotched beams considered in the simulation is equal to 1 MPa. With Eq. 13.16, we can calculate σ_0, which is expected to vary with the notch angles, and Eq. 13.15 yields constant C_1. The last two parameters λ_1 and λ_2 can be determined by fitting the simulated size effect curve for the case of weak stress singularities.

We now apply the aforementioned procedure to calibrate Eq. 13.14 from the simulated size effect curves. By fitting the size effect curve for the case of unnotched beams, we obtain $l_p = 40$ mm, $l_s = 150$ mm, $r = 0.88$, and $s_0 = 0.448$ MPa. Based on the size effect curves for the series of V-notch specimens, we have $D_0 = 90$ mm, $\lambda_1 = 0.301$, $\lambda_2 = 0.208$, and $\beta_\gamma = 1, 1, 1.5, 2, 1.1$ for $\gamma = 0°, 90°, 120°, 135°, 170°$, respectively. Figures 13.5 and 13.6 show that the simulated size effect curves can be well fitted by Eq. 13.14.

It should be pointed out that the simulated size effect curve does not match Eq. 13.14 well at the large-size limit for the beams with a 170° V-notch and with no notch. This is due to the fact that the deterministic simulations cannot capture the associated large-size asymptote of the classical Weibull scaling relation. Furthermore, it is observed that such a difference occurs only for very large beam sizes (i.e., $D \approx 1.2$ m). This indicates that deterministic calculation is sufficient for most normal-size concrete beams.

13.3 Scaling of Fracture of Bimaterial Hybrid Structures

In the previous section, we discussed the general energetic-statistical size effect in fracture of structures with a V-notch under mode I loading. It is clear that the order of the stress singularity, which is governed by the V-notch geometry, dictates the energetic scaling of fracture of the V-notch as well as the randomness of the failure location. This naturally causes that, as the stress singularity weakens, the scaling mechanism transitions from energetic to primarily statistical (with some deterministic contribution of the finite FPZ, as in type I failures).

Another common reason that weakens the stress singularities is the combination of dissimilar materials, which is typically seen in many hybrid structures, such as single-lap and double-lap joints, butt joints, scarf joints, and so on. These structures have widely been used in many engineering designs such as large ship hulls, modern automobile components, and aircraft wings and fuselages (Barsoum 2003).

In these bimaterial hybrid structures, the strength of the corresponding stress singularities is generally weaker than the "$-1/2$" crack-like stress singularity. Over the past few decades, extensive efforts have been directed to analyzing the singular stress field at the tip of the bimaterial corner [e.g., Bogy (1971); Rice (1988); Hutchinson & Suo (1992); Desmorat & Leckie (1998); Qian & Akisanya (1998); Liu & Fleck (1999); Reedy Jr. (2000); Labossiere, Dunn, & Cunningham (2002)]. Most of these analyses were based on the framework of LEFM. For bimaterial hybrid structures, the adhesives as well as the base materials usually exhibit a strain-softening behavior, leading to the formation of a localized damage zone. The size of this damage zone is virtually independent of

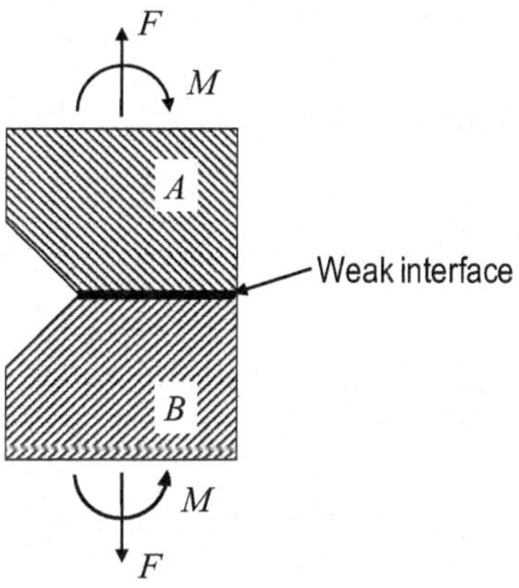

Figure 13.8. Geometry of a bimateral hybrid structure.

the structure size, which causes a size effect on the nominal structure strength. In the following, we extend the foregoing analysis to scaling of fracture of bimaterial hybrid structures with a weak bimaterial interface.

13.3.1 Energetic Scaling with Superposed Multiple Stress Singularities

Consider a general bimaterial hybrid structure shown as Figure 13.8. The stress singularities at the bimaterial corner can be calculated by either the Mellin transform for isotropic materials (Bogy 1971) or the complex potential method for anisotropic materials (Stroh 1958; Lekhnitskii 1963; Desmorat & Leckie 1998). In this study, we consider a power-law singular stress field, i.e., $\sigma \propto r^{\lambda}$, which is the asymptotic field for bimaterial corners. Depending on the corner geometry and material mismatch, the exponents of stress singularities $\lambda_k (k = 1, 2)$ could be either a pair of complex conjugates or two distinct real numbers.

In this section, we consider that the stress singularities are relatively strong so that the crack would always propagate from the bimaterial corner. Here we consider the bimaterial corner in which two distinct real stress fields with singularity exponents λ_1, λ_2 are superposed at the same singularity point. As will be shown later, the resulting scaling equation can also be used for structures with complex stress singularities.

In general, the near-corner stress field with two real stress singularities can be written as

$$\sigma_{ij}(r, \theta) = H_1 r^{\lambda_1} f_{ij}^1(\theta) + H_2 r^{\lambda_2} f_{ij}^2(\theta) \tag{13.18}$$

where $H_1, H_2 =$ stress intensity factors corresponding to stress singularities λ_1 and λ_2; $r, \theta =$ polar coordinates, and $f_{ij}^k(\theta)$ $(k = 1, 2) =$ dimensionless functions for the angular dependence of the near-corner stress field. Based on dimensional analysis, the stress intensity factors H_1, H_2 can generally be expressed as

$$H_k = \sigma D^{-\lambda_k} h_k \quad (k = 1, 2) \tag{13.19}$$

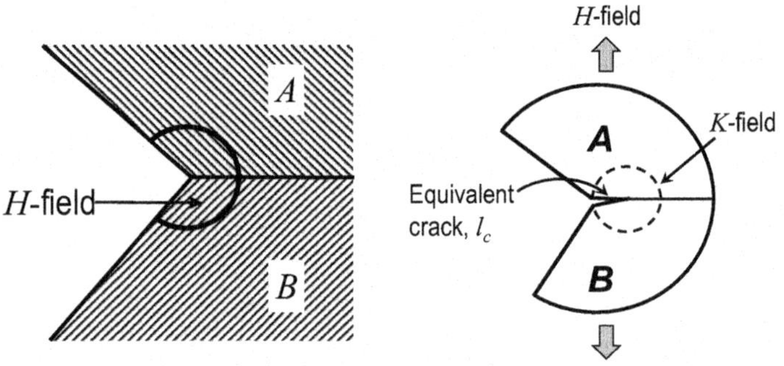

Figure 13.9. Analysis of equivalent linear elastic crack at the bimaterial corner.

where σ = nominal stress = P/bD, P = the applied load, and h_k ($k = 1, 2$) = dimensionless stress intensity factors determined by the geometry of the structure.

As the applied load increases, a FPZ, which contains many microcracks, starts to form at the bimaterial corner. At the large-size limit, this FPZ can be replaced by an equivalent discrete crack such that the remote fields of the actual structure coincide with the remote fields of this equivalent crack for the same loading. We may assume that the structure either has a weak bimaterial interface, or breaks in one material right next to the interface. Then the equivalent crack can be considered to be placed along the bimaterial interface. When the structure attains its peak load, the FPZ gets fully developed. Since the size of the fully developed FPZ is roughly a material constant, the length of the equivalent interfacial crack representing this FPZ must also be a material constant, l_c.

It is well known that the stress intensity factor of the bimaterial interfacial crack carries a dimension of $\mathrm{N/m^2} \times \mathrm{m}^{1/2-i\eta}$, where η = the imaginary part of the exponent of stress singularity of the interfacial crack, and $i = \sqrt{-1}$. Since we focus on the large-size limit, the equivalent interfacial crack must be fully enclosed by the singular stress field, which is governed by the stress intensity factors of the bimaterial corner. Therefore, the stress intensity factor K of the equivalent interfacial crack must be related to the stress intensity factors H_k of the bimaterial corner (Fig. 13.9).

Furthermore, at the large-size limit, the size of singular stress field is much larger compared to the length of the equivalent crack, which implies that the only relevant length scale involved is the length of the equivalent crack. Based on the dimensional analysis, K and H_k can be related as (Grenestedt & Hallstrom 1997; Liu & Fleck 1999; Dunn, Hui, Labossiere, & Lin 2001; Labossiere et al. 2002):

$$K = a_1 H_1 l_c^{\lambda_1+0.5-i\eta} + a_2 H_2 l_c^{\lambda_2+0.5-i\eta} \tag{13.20}$$

where a_1 and a_2 = dimensionless complex numbers, which can be calculated by solving an auxiliary small-scale crack problem (Akisanya & Fleck 1997; Liu & Fleck 1999; Labossiere et al. 2002; Le et al. 2010) (Fig. 13.9). The corresponding energy release rate at the tip of the equivalent crack can then be calculated as

$$\mathcal{G} = \frac{CK\bar{K}}{E} = \frac{1}{E}\left(A_1 H_1^2 l_c^{2\lambda_1+1} + A_2 H_2^2 l_c^{2\lambda_2+1} + A_3 H_1 H_2 l_c^{\lambda_1+\lambda_2+1}\right) \tag{13.21}$$

where E = elastic modulus of one of the dissimilar materials, C = a constant depending on the relative elastic properties of the two dissimilar materials, $A_1 = C|a_1|^2$, $A_2 = C|a_2|^2$, and $A_3 = 2C|a_1 a_2|\cos(\eta_{a1} - \eta_{a_2})$, and η_{a_1} and η_{a_2} are the phases of a_1 and a_2. Based on LEFM, the crack starts to propagate once the energy release rate reaches a critical value G_f. At the same time, the structure also reaches its peak load, i.e., the nominal stress σ attains its maximum value σ_N. By substituting H_1 and H_2 into Eq. 13.21, we have

$$\sigma_N = \frac{\sqrt{EG_f/l_c}}{\left[B_1(D/l_c)^{-2\lambda_1} + B_2(D/l_c)^{-2\lambda_2} + B_3(D/l_c)^{\ \lambda_1-\lambda_2}\right]^{1/2}} \tag{13.22}$$

where $B_1 = A_1 h_1^2$, $B_2 = A_2 h_2^2$, and $B_3 = A_3 h_1 h_2$.

Equation 13.22 represents the large-size asymptote of the size effect on the strength of bimaterial structures. By contrast, the small-size asymptote of the size effect is relatively easy to construct. At the small-size limit, the entire bimaterial ligament behaves as a crack filled by a perfectly plastic glue (Bažant 2005). In the case of plastic behavior, the size effect must vanish (Bažant & Planas 1998). Here we propose an approximate scaling equation, which bridges the small-size and large-size asymptotes:

$$\sigma_N = \sigma_s \left\{1 + \left[(D/D_1)^{-2\lambda_1} + (D/D_2)^{-2\lambda_2} + (D/D_3)^{-\lambda_1-\lambda_2}\right]^\gamma\right\}^{-1/2\gamma} \tag{13.23}$$

where γ = positive constant; D_1, D_2, and D_3 are constants determined by the fracture properties as well as the structure geometry; and σ_s represents the nominal strength at the small-size limit. By matching the large-size asymptote of the size effect equation (Eq. 13.22), we further obtain

$$D_1 = l_c\left(\sigma_s\sqrt{B_1 l_c/EG_f}\right)^{1/\lambda_1} \tag{13.24}$$

$$D_2 = l_c\left(\sigma_s\sqrt{B_2 l_c/EG_f}\right)^{1/\lambda_2} \tag{13.25}$$

$$D_3 = l_c\left(\sigma_s\sqrt{B_3 l_c/EG_f}\right)^{2/(\lambda_1+\lambda_2)} \tag{13.26}$$

There are three independent parameters in the proposed size effect equation (Eq. 13.23), namely σ_s, G_f, and l_c. They can be calibrated by the optimum fitting of the measured size effect curve. Note that, in this model, we use the equivalent crack to approximate the FPZ, which actually consists of the two dissimilar materials as well as the adhesive. Therefore, σ_s, G_f, and l_c depend not only on the property of the adhesive, but also on the fracture properties of both dissimilar materials.

If only one real stress singularity λ dominates the fracture process, then Eq. 13.23 reduces to

$$\sigma_N = \sigma_s\left[1 + (D/D_1)^{-2\gamma\lambda}\right]^{-1/2\gamma} \tag{13.27}$$

Note that Eq. 13.27 can also be applied to structures with complex stress singularities, i.e., $\kappa = \lambda \pm i\eta'$. The corresponding stress intensity factor of the bimaterial corner can be expressed as $H = \sigma D^{-\lambda}|h|\exp[i(\omega - \eta' \ln D)]$, where ω = phase of the dimensionless complex stress intensity factor h. Considering again the aforementioned auxiliary

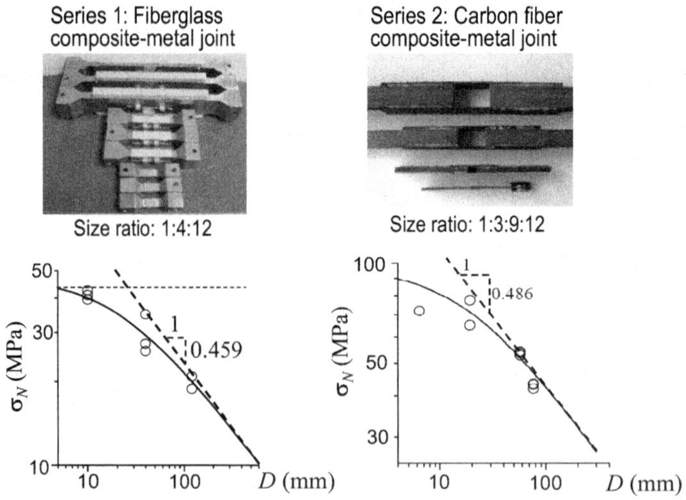

Figure 13.10. Experimental measured size effects in fracture of metal-composite hybrid joints.

small-scale cracking problem (Fig. 13.9), one can show that the energy release rate at the equivalent crack tip can be written as $\mathcal{G} \propto |H|^2 l_c^{2\lambda+1}$ (Le et al. 2010). With the same type of asymptotic matching, one obtains a size effect equation, which is identical to Eq. 13.27, except that λ now represents the real part of the exponents of stress singularities.

In a recent study (Yu et al. 2010), a systematic size effect test was performed on geometrically similar double-lap joints made of fiber composites and steel. Two series of specimens were prepared, where two types of fiber composites were considered: specimens series 1 used fiberglass G-10/FR4, and specimens series 2 used Newport NCT301 carbon laminates (Fig. 13.10). Based on the elastic properties of these two types of laminates, the dominant stress singularities for these two series of specimens were $-0.459 \pm 0.06i$ for specimens series 1 and -0.486 for specimens series 2. Figure 13.10 presents the measured size effect on σ_N with the optimum fitting of Eq. 13.27, for these geometrically similar joints. It is seen that Eq. 13.27 agrees very well with the test data.

13.3.2 Finite Weakest-Link Model for Failure of Bimaterial Interface

For bimaterial structures with weak stress singularities, there is no guarantee that fracture would initiate and propagate from the corner due to the randomness of interfacial strength. Note that here we do not consider the potential failure of the base materials because a weak interface is assumed. Similar to the formulation presented in Section 13.1.2, we describe the failure statistic of the bimaterial hybrid structures by combining the finite weakest-link model and the aforementioned energetic size effect in fracture of the bimaterial interface, i.e.,

$$P_f(\sigma_N) = 1 - \left[1 - P_{f,L_I}(\sigma_N)\right]\left[1 - P_{f,L_{II}}(\sigma_N)\right] \tag{13.28}$$

where P_{f,L_I} = failure probability of the part of the interface L_I that lies in the singular stress zone, and $P_{f,L_{II}}$ = failure probability of the interface outside the singular stress zone, denoted by L_{II}. The failure probability of the first region L_I can be calculated as

$$P_{f,L_I}(\sigma_N) = 1 - \prod_{i=1}^{N_1} \{1 - P_1[\chi(D)\sigma_N s(x_i)]\} \tag{13.29}$$

where

$$\chi(D) = \left\{1 + \left[(D/D_1)^{-2\lambda_1} + (D/D_2)^{-2\lambda_2} + (D/D_3)^{-\lambda_1-\lambda_2}\right]^{\gamma}\right\}^{1/2\gamma} \tag{13.30}$$

where the parameters in the scaling term $\chi(D)$ follow the same definition as those in Eq. 13.23, N_1 = number of interface RVEs in the singular stress zone, $s(x_i)$ = dimensionless stress field as defined for Eq. 13.5, and $P_1(x)$ = strength distribution of one RVE, as defined by Eqs. 5.84a and 5.84b.

For the remaining part of the interface, the failure probability can be calculated directly from the elastic stress:

$$P_{f,L_{II}}(\sigma_N) = 1 - \prod_{i=1}^{N_2} \{1 - P_1[\sigma_N s(x_i)]\} \tag{13.31}$$

where N_2 = number of interface RVEs in the second region.

It can be seen that the present formulation is essentially the same as the general finite weakest-link model derived for structures with a V-notch (Eqs. 13.5 to 13.8), except that here the energetic scaling term takes into account multiple stress singularities. Therefore, we may express the size effect on the mean structural strength $\bar{\sigma}_N$ by using an equation of a similar form as Eq. 13.14, i.e.,

$$\bar{\sigma}_N = \sigma_0 \left\{ C_1 \left[\chi^m(D)\Psi_1 + \Psi_2\right]^{-r/m} \left(\frac{D + l_s}{l_0}\right)^{-r/m} \exp[-(\Lambda/\Lambda_0)^2] + \frac{\chi^{-r}(D) D_b}{\exp[-(\Lambda/\Lambda_0)^2] D + l_p} \right\}^{1/r} \tag{13.32}$$

where σ_0 = reference stress; m = Weibull modulus of the strength distribution of the interface; l_0 = size of the RVE along the interface; Λ = magnitude of the dominant stress singularity for the case where there exist two distinct real stress singularities, or the magnitude of the real part of the complex stress singularities; $\Psi_1 = \int_{L_I} \langle s(x)\rangle^m \,\mathrm{d}x$, $\Psi_2 = \int_{L_{II}} \langle s(x)\rangle^m \,\mathrm{d}x$; and $C_1, r, \Lambda_0, l_s, l_p, D_b$ = constants. Similar to Eq. 13.14, evaluation of Ψ_1 must exclude the region with a distance from the notch tip less than a certain distance d_c, and the choice of d_c is not particularly important because different values of d_c can be compensated by adjusting the empirical constant Λ_0 in order to fit the size effect curve (Le & Xue 2013; Le, Pieuchot, & Ballarini 2014).

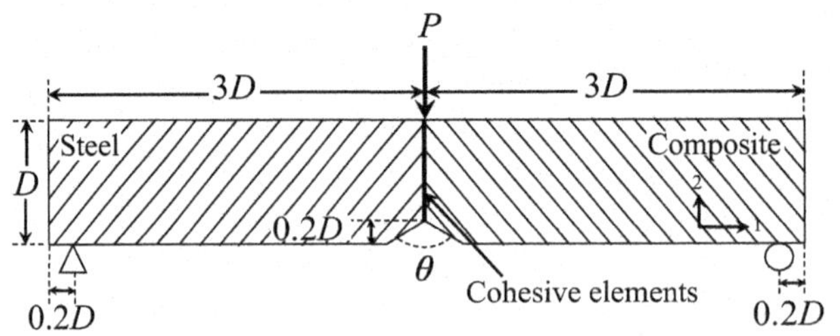

Figure 13.11. Numerical simulation of bimatertial hybrid beam under three point bending.

13.4 Numerical Analysis of Bimaterial Fracture

13.4.1 Description of Analysis

To verify Eq. 13.32, we perform a numerical study on a series of 2D metal-composite hybrid beams with a centered V-notch under three point bending. Let D denote the beam depth (Fig. 13.11). The steel is considered to be isotropic with Young's modulus $E = 200$ GPa and Poisson's ratio $\nu = 0.3$. The composite material is considered to be a unidirectional carbon/epoxy composite with the following properties: $E_1 = 147$ GPa, $E_2 = E_3 = 10.3$ GPa, $G_{12} = G_{13} = 7.0$ GPa, $G_{23} = 3.7$ GPa, $\nu_{12} = \nu_{13} = 0.27$, and $\nu_{23} = 0.54$ (G denotes the shear modulus).

Here we consider five different notch angles, i.e., $\theta = 0°, 120°, 135°, 150°$, and $170°$. Based on the complex potential method (Stroh 1958; Lekhnitskii 1963; Desmorat & Leckie 1998), the orders of stress singularities are found to be: $\lambda_{1,2} = -0.5 \pm 0.081i$ for $\theta = 0°$ ($i = \sqrt{-1}$), $\lambda_{1,2} = -0.396, -0.177$ for $\theta = 120°$, $\lambda_{1,2} = -0.363, -0.098$ for $\theta = 135°$, $\lambda_{1,2} = -0.305, -0.01$ for $\theta = 150°$, and $\lambda_{1,2} = -0.136, 0$ for $\theta = 170°$. It can be seen that, for the case of $\theta = 0°$, the beam has a pair of complex conjugate stress singularities, which is well known for a bimaterial crack. For other notch angles, the beam exhibits either two real stress singularities or just one real stress singularity. For each notch angle, a set of geometrically similar beams of a large size range, i.e., $D = 100, 200, 400, 800, 1600, 3200$ mm, is considered for the simulation.

Since we assume that the bimaterial interface is much weaker than the steel and the composite, it is reasonable to expect that fracture would always initiate and propagate along the bimaterial interface. This allows us to use the cohesive crack model for the bimaterial interface, and the steel and composite materials are treated as linear elastic. For the bimaterial interface, we consider that the adhesive layer is very thin and it does not vary with the beam size. Consequently, in the present simulation, we use a finite-thickness cohesive element (thickness $t = 1$ mm) provided by ABAQUS. Such a model has been used successfully to determine the fracture properties of bimaterial double-lap joints (Yu, Bažant, & Le 2013).

Before reaching its strength, the cohesive element is assumed to be linear elastic. The traction-separation law can be written as $T_i = K_{ij}\delta_j$ $(i, j = n, s)$, where T_i = traction, K_{ij} = stiffness, δ_j = separation, and n, s represent the normal direction and shear directions, respectively. The normal and shear tractions are considered to be uncoupled in the elastic regime, i.e., $K_{ij} = 0$ $(i \neq j)$. When the stress in the cohesive layer reaches

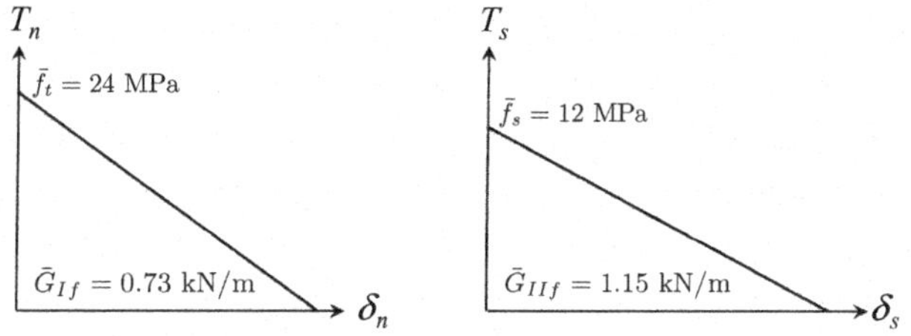

Figure 13.12. Cohesive traction-separation relations under pure tension and shear loading.

the strength criterion, the interfacial crack will initiate and propagate. In general, the cohesive interface is subjected to a combined normal and shear loading. Therefore, a mixed-mode fracture criterion for the cohesive layer is needed.

The mixed-mode damage initiation criterion is assumed to follow a simple quadratic form:

$$\frac{\langle T_n \rangle^2}{f_t^2} + \frac{T_s^2}{f_s^2} = 1 \tag{13.33}$$

Here f_t, f_s = tensile and shear strengths, respectively. The Maclaulay bracket, defined as $\langle x \rangle = \max(x, 0)$, is used here to ensure that the compressive stress would not contribute to the damage initiation. Once the damage criterion is met, the cohesive elements start to degrade. Here we consider an energetic-based damage evolution criterion, i.e.

$$\left(\frac{G_I}{G_{If}}\right)^2 + \left(\frac{G_{II}}{G_{IIf}}\right)^2 = 1 \tag{13.34}$$

where G_{If}, G_{IIf} = fracture energies corresponding to mode I and mode II loading, respectively; G_I, G_{II} are the mode I and II energy dissipations, i.e., $G_I = \int_0^{\delta_n} T_n(\delta')\,\mathrm{d}\delta'$; and $G_{II} = \int_0^{\delta_s} T_s(\delta')\mathrm{d}\delta'$. Furthermore, we consider a linear softening behavior in both normal and shear directions (Fig. 13.12).

To extend the aforementioned cohesive element formulation to stochastic simulation, we randomize four cohesive properties: f_t, f_s, G_{1f} and G_{IIf}. Following a recent study on stochastic simulation of quasibrittle fracture (Grassl & Bažant 2009), we assume that all these random properties are fully correlated. Therefore, the spatial randomness of these properties, denoted by B_i, can be described by scaling a single random field with the mean value of these properties:

$$B_i(x) = \zeta(x)\bar{B}_i \tag{13.35}$$

where x = spatial coordinate along the interface, B_i = any of these four random properties, $\zeta(x)$ = a random variable having a mean value of 1, and $\bar{B}_i$ = the mean value of the random property B_i. $\zeta(x)$ is considered to follow a general Gauss–Weibull grafted probability distribution as defined by Eqs. 5.84a and 5.84b. In the simulation, the following mean values are used for f_t, f_s, G_{If}, G_{IIf}: $\bar{f}_t = 24$ MPa, $\bar{f}_s = 12$ MPa, $\bar{G}_{If} = 0.73$ KN/m, and $\bar{G}_{IIf} = 1.15$ KN/m (Yu et al. 2013). For the cdf of ζ_i, we choose that the Weibull modulus $m = 12$, the Weibull scale parameter $s_0 = 0.54$, the mean value of $\zeta = 1$, and the coefficient of variation (CoV) = 0.136.

For stochastic simulation, another important consideration is the autocorrelation field for the random variables. Here we assume that the size of material RVE l_0, which is approximately equal to the autocorrelation length, is 10 mm for the interface between the composite and the steel. In the simulation, we choose 10 mm as the minimum cohesive element size so that the random properties of each cohesive element can be considered as uncorrelated random variables.

The random variable ζ is generated by using the Latin Hypercube Sampling (LHS) method. For each beam size and geometry, 20 realizations are computed to obtain the mean behavior. It should also be mentioned that, according to a recent study (Yu et al. 2013), which uses similar cohesive laws, a 10-mm element size is sufficient for capturing the overall failure behavior of the bimaterial interface between the composite and the steel.

13.4.2 Results and Discussion

Figure 13.13 presents the simulated curves of average nominal stress versus relative displacement for all specimens. The nominal stress is defined as $\sigma = P/bD$ (P = applied loading and b = beam width, which is chosen to be 1) and the relative displacement is calculated by normalizing the load-point displacement with respect to the beam depth D. It is clearly seen that as the beam size increases the load-deflection curve exhibits a steeper softening slope. This indicates that the beam becomes more brittle with an increasing size. For beam depth $D \geq 400$ mm, the load-deflection curve suddenly drops right after the peak load.

Here the mean nominal strength of hybrid beams is defined as $\bar{\sigma}_N = \bar{P}_{\max}/bD$. Figure 13.14 shows the simulated mean size effect curves for the hybrid beams with five different notch angles. It can be seen that the computed size effect is strongly affected by the magnitude of the stress singularities.

For beams with strong stress singularities (i.e., small notch angles), the mean size effect is very close to the classical energetic size effect of type 2, which is applicable to structures with a large crack formed prior to the peak load. For beams with weak stress singularities (i.e., large notch angles), the mean size effect resembles the classical type 1 size effect, which applies to structures that fail at crack initiation. This observation qualitatively agrees with the aforementioned energetic-statistical scaling model, which predicts a transition from the energetic size effect to the statistical size effect as the stress singularities weaken.

Now we use Eq. 13.32 to fit the simulated size effect curves. Based on the computation of Ψ_1 and Ψ_2, it is found that, for notch angles considered in this study, Ψ_1 is considerably larger than Ψ_2; therefore, we may approximate Eq. 13.32 as

$$\bar{\sigma}_N = \sigma_0 \chi^{-1}(D) \left\{ C_1 \Psi_1^{-r/m} \left(\frac{D + l_s}{l_0} \right)^{-r/m} \exp[-(\Lambda/\Lambda_0)^2] + \frac{D_b}{\exp[-(\Lambda/\Lambda_0)^2] D + l_p} \right\}^{1/r} \tag{13.36}$$

Note that, according to the aforementioned numerical simulation, here we consider the Weibull modulus $m = 12$ and the size of interface RVE $l_0 = 10$ mm.

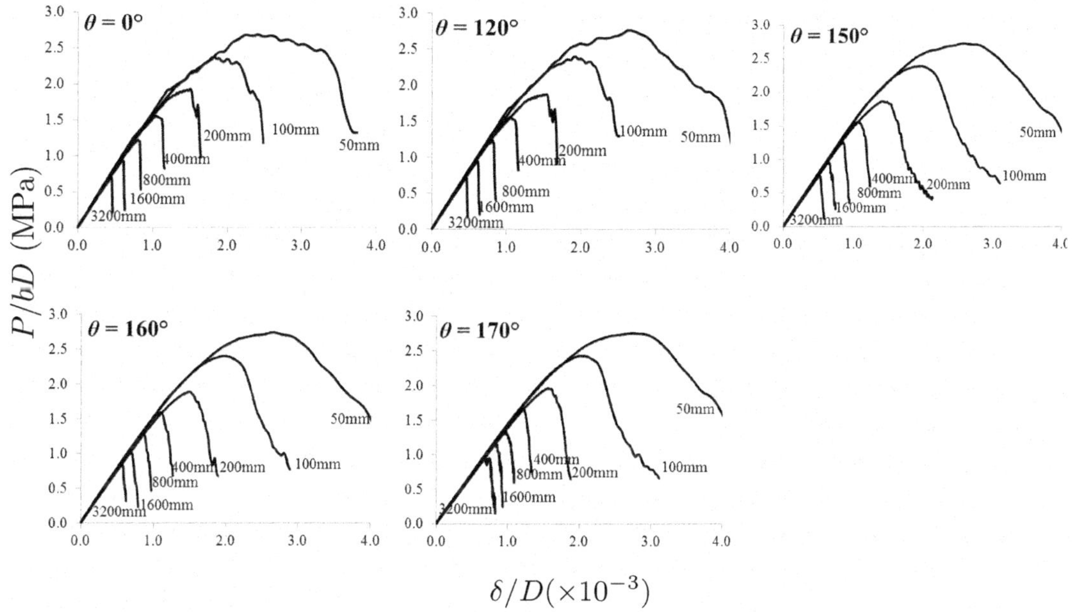

Figure 13.13. Simulated nominal stress–relative displacement curves of bimaterial hybrid beams.

For beams with a zero-angle notch, the energetic scaling term $\chi(D)$ simply reads $\chi(D) = [1 + (D/D_1)^{\gamma}]^{-1/2\gamma}$, according to Eq. 13.27, and Λ in Eq. 13.36 is equal to 1/2. For the other notch angles, the singular stress field is governed by either two distinct real stress singularities or one real stress singularity. For the cases of two distinct stress singularities, the magnitudes of these two singularities have a significant

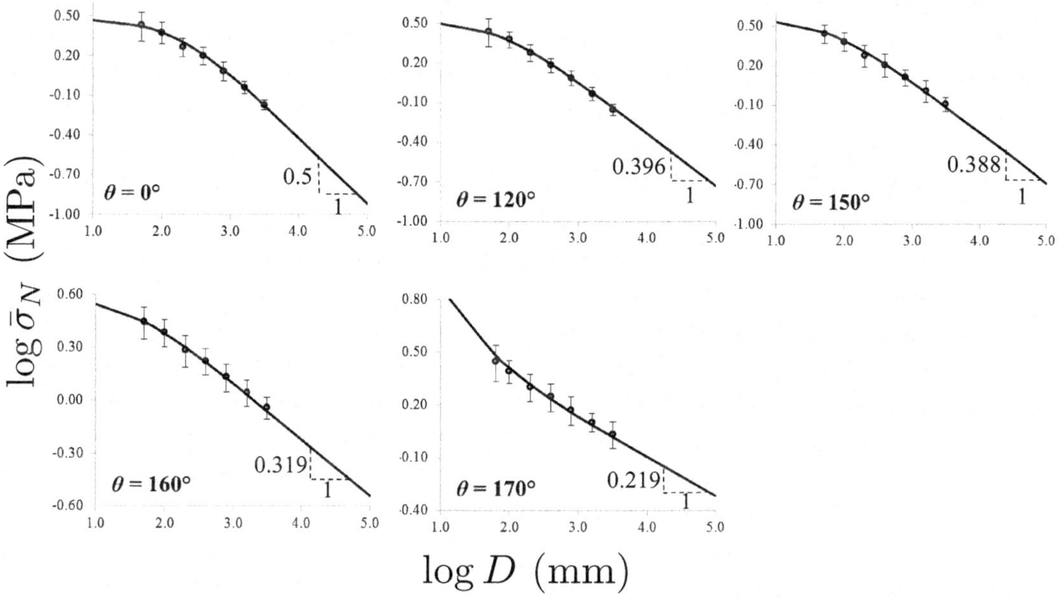

Figure 13.14. Simulated size effect curves with optimum fits by Eq. 13.36.

difference. Therefore, we may consider that the scaling of fracture of the material element at the notch tip is governed by the strongest stress singularity λ_1, and Λ in Eq. 13.36 is equal to $-\lambda_1$. The corresponding energetic scaling term $\chi(D)$ becomes $\chi(D) = [1 + (D/D_1)^{-2\gamma\lambda_1}]^{-1/2\gamma}$. Following Le et al. (2010) and Yu et al. (2010), we choose $\gamma = -1/(2\lambda_1)$. As mentioned, D_1 represents the transitional size for energetic scaling, which is determined by the structural geometry as well as the FPZ size.

For general mixed-mode fracture, the FPZ size is expected to depend on the mode mixity. However, little information is available on such dependence. Therefore, in this study, we consider D_1 as a fitting parameter for each notch angle. D_b, l_p, l_s, and r are related to the scaling behavior of structures with weak stress singularities, which correspond to a relatively narrow range of notch angles (e.g., $\theta \geq 150°$). So we may assume that these parameters are independent of the notch angles. C_1 and σ_0 are two fitting constants, which will vary with the notch angle.

Figure 13.14 shows that the simulated size effect curves can be well fitted by Eq. 13.36. Based on the fitting, we have $D_b = 25$ mm, $l_p = 10$ mm, $l_s = 20$ mm, $r = 1$, and $\Lambda_0 = 0.12$. It is interesting to see that the value of r and the fitted result $l_s = 2l_p$, $D_b \approx l_s$ are consistent with the findings of the previous study on the statistical size effect on concrete fracture (Bažant et al. 2007). The good agreement between the simulation results and Eq. 13.36 indicates that the general dependence of the size effect on the magnitude of stress singularities can be explained by the present energetic-statistical scaling model.

Based on Eqs. 13.32 and 13.36, it is clear that the large-size asymptote of the size effect can be described by a power-law function with an exponent p equal to the sum of the order of the dominant stress singularity and the classical Weibull scaling exponent (i.e., $p = \lambda_1 - 1/m$). This implies a combination of energetic and statistical scaling. Nevertheless, for the practical size range of beams with strong stress singularities, such a power-law exponent will not get revealed owing to the exponential transition function. Instead, the power-law exponent of the large-size asymptote will be equal to the order of the dominant stress singularity ($p = \lambda_1$), which is consistent with the classical energetic size effect. For beams with weak stress singularities, the combined energetic-statistical size effect will prevail, which leads to a compound power-law exponent at the large-size limit ($p = \lambda_1 - 1/m$).

The present simulation also indicates that, for beams with strong stress singularities (i.e., $\theta = 0°$, $120°$), the energetic power-law large-size asymptote (i.e., $\bar{\sigma}_N \propto D^{\lambda_1}$) is approached closely for beams with depth D larger than 1 m. This implies that, for most metal-composite hybrid beams, the failure is quasibrittle and nonlinear fracture mechanics is needed for the failure analysis.

14 Lifetime of High-*k* Gate Dielectrics and Analogy with Failure Statistics of Quasibrittle Structures

In chapters 4–9, we studied the statistics of strength and lifetime of quasibrittle structures. The underlying approach was anchored by the weakest-link theory, in which the structure is statistically modeled as a chain of representative volume elements (RVEs). Because the size of material inhomogenieties (or, equivalently, RVE size) is not negligible compared to the structure size, this chain is not infinitely long. This causes the probability distributions of structural strength and lifetime to deviate from the classical Weibull distribution. As a result of the finite-chain model, the probability distribution functions of structural strength and lifetime vary with the structure size and geometry.

The finite weakest-link model is a general mathematical model, which is not limited to the modeling of quasibrittle fracture. There exist other failure processes that follow the weakest-link model but require considering a finite number of links. One is the breakdown of high-k gate dielectrics, an important problem for the semiconductor industry. In this chapter, we use the same theoretical framework as developed in Chapters 4 to 7 to study the lifetime statistics of high-k gate dielectrics, based on a mathematical analogy between the breakdown of dielectrics and the fracture of quasibrittle structures (Le, Bažant, & Bazant 2009; Le 2012).

14.1 Deviation of Lifetime Histograms of High-k Dielectrics from the Weibull Distribution

High-k gate dielectrics, such as Al_2O_3, HfO_2, Si_3O_4, ZrO_2, and so on, have recently been adopted in the design of metal-oxide-semiconductor field effect transistor (MOSFET) as an attractive alternative to the conventional SiO_2 native oxide gate dielectrics, in order to reduce current leakage and increase the gate capacitance. These high-k dielectrics are known as "trap-rich" materials. The trapping of electrons in the gate oxide layer induces a trap-assisted tunneling process, which leads to the gate leakage current at a low voltage (Wilk, Wallace, & Anthony 2001; Kim & Lee 2004; Chatterjee et al. 2006). When the trap (or defect) density reaches a certain critical value, a weak localized breakdown path between the gate electrode and the substrate is formed, which is called the soft breakdown (SBD). The Joule heating in the local breakdown path then causes lateral propagation of the leakage spots and eventually leads to a significantly increased tunneling current passing through the layer. This is called the hard breakdown (HBD) (Chatterjee et al. 2006).

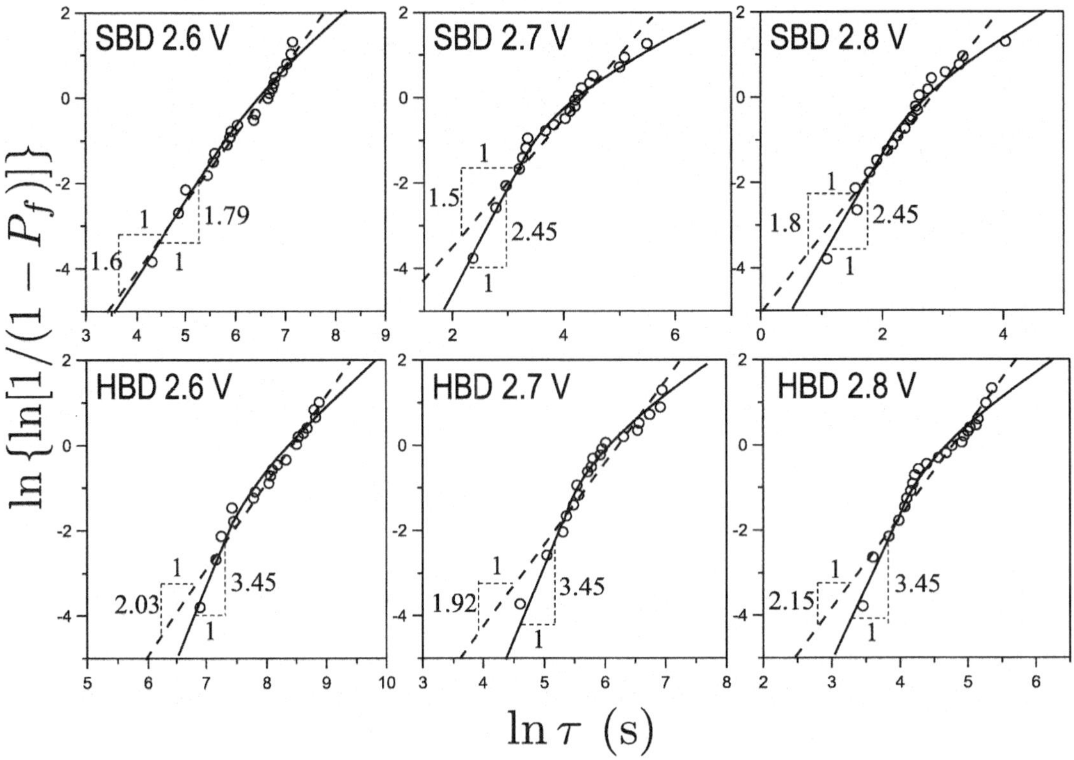

Figure 14.1. Measured lifetime histograms of high-k gate dielectrics under constant voltage stress with the optimum fits by the Weibull distribution and the finite weakest-link model.

The reliability of gate dielectrics is often assessed by the breakdown lifetime or the total charge-to-breakdown (i.e., the integral of the tunneling current over the lifetime) for a prescribed constant or AC gate voltage stress (Stathis 1999; Kim & Lee 2004). Similar to other engineering structures, the gate dielectrics must be designed against a certain tolerable breakdown probability during their entire service lifetime. Therefore, the understanding of statistics of breakdown lifetime becomes crucial.

Extensive efforts have been devoted to histogram testing of lifetime of high-k dielectrics under both constant and alternating voltage stresses. The data points in Figure 14.1, plotted in the Weibull scale, show, for both SBD and HBD under constant voltage, Kim and Lee's (2004) data for HfO_2 based gate dielectrics of area $A = 0.0016$ mm^2 and thickness about 4.8 to 5 nm, or the equivalent oxide thickness (EOT) $h = 1.4$ nm (EOT = thickness of SiO_2 gate oxide for which the gate capacitance would be the same as it is for the gate high-k gate dielectrics). It is clear that the measured histograms consistently deviate from the two-parameter Weibull distribution. Kim and Lee (2004) also observed the effect of gate area A, i.e., the size effect, on the median lifetime τ_{50} (time to 50% failure frequency), shown in Figure 14.2. According to the Weibull theory, the size effect should be a power law of exponent $-1/m$, i.e., the data plotted in double

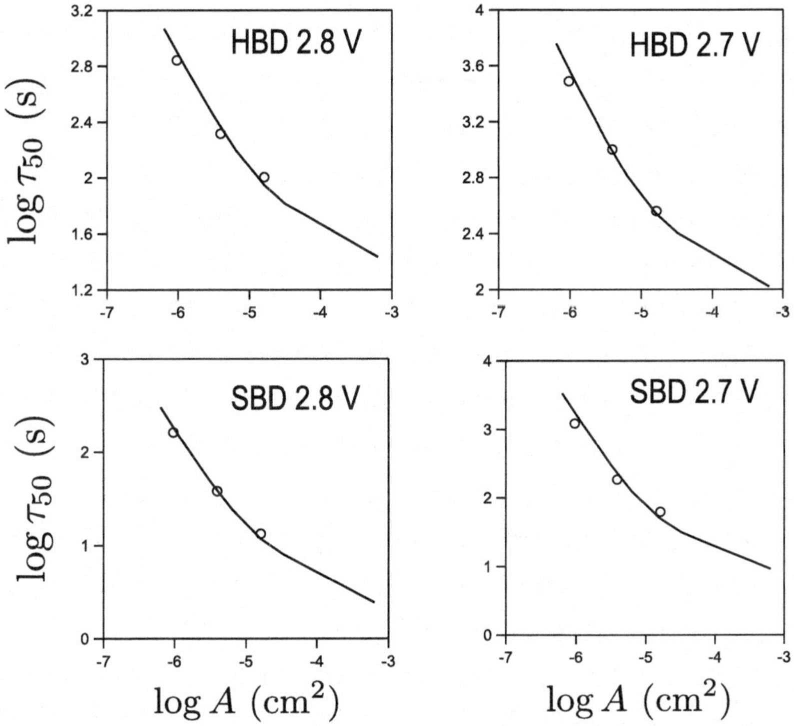

Figure 14.2. Effect of gate area on the median breakdown lifetime.

logarithmic scales should fit straight lines of slope $-1/m$. However, despite a relatively narrow range of A, we can discern systematic, though slight, deviations.

At the same time, Kim and Lee also performed histogram testing of lifetime under unipolar AC voltage stress on the same type of gate dielectrics, where the history of voltage stress is depicted in Figure 14.3. Figure 14.4 presents the measured lifetime histograms in the Weibull scale. Similar to the case of constant voltage stress, the lifetime histograms do not follow a straight line in the Weibull scale, despite the limited number of specimens.

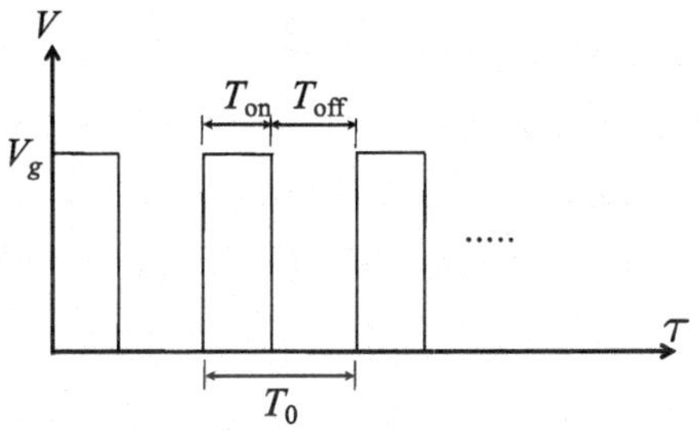

Figure 14.3. Time history of AC voltage stress.

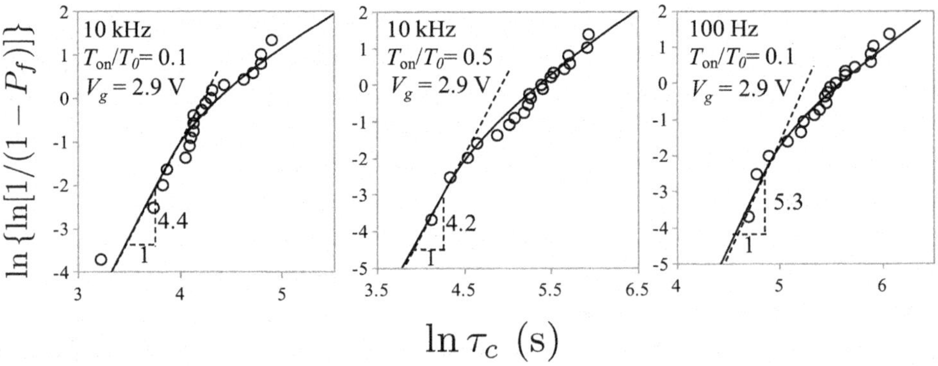

Figure 14.4. Measured lifetime histograms of high-*k* gate dielectrics under AC voltage stress with the optimum fits by the finite weakest-link model.

Interestingly, we observe that the lifetime histograms of gate dielectrics under both constant and AC voltage stresses (Figs. 14.1 and 14.4) are strikingly similar to those of quasibrittle structures (Figs. 8.10 and 8.12). As discussed in Chapters 4 to 8, such deviations can be successfully handled by the finite weakest-link model, which predicts both the cumulative distribution functions (cdfs) of strength and lifetime with their scaling aspects (Bažant, Le, & Bazant 2009).

Motivated by the aforementioned observation, here we exploit an analogy between the dielectric breakdown due to applied voltage and the failure of quasibrittle structures under applied stress to motivate a fiber-bundle–like statistical model for breakdown statistics. We first formulate this general analogy and apply it to the breakdown probability, and we extend the theory to lifetime statistics of gate dielectrics under constant and unipolar AC voltage stresses. The present theory is based on general notions of parallel and series couplings of capacitors, which become resistors on local breakdown (Beale & Duxbury 1988; Bowman & Stroud 1989). In principle, though, its key assumptions can also be connected to microscopic statistical models (Chakraborti & Benguigui 1997), such as the original Dielectric Breakdown Model (DBM) of Niemeyer, Pietronero, and Wiesmann (1984) as well as more recent percolation models developed for gate dielectrics (Degraeve et al. 1998; Stathis 1999; Miyakawa et al. 2007).

14.2 Breakdown Probability

14.2.1 Analogy with Strength of Quasibrittle Structures

It is now clear that the Weibull distribution of structural strength σ^* is based on the weakest-link model, in which the links (numbered $i = 1, 2, \dots N$), coupled in series, simulate the RVEs of material (only structure geometries for which the failure of one RVE causes the whole structure to fail are considered). So, $\sigma^* = \min_i(\sigma_i^*)$ where $\sigma_i^* =$ strengths of the individual RVEs. Likewise, the gate dielectric may be divided into a number of potential breakdown cells ($i = 1, 2, \dots N$) (Suné et al. 1990; Dumin, Maddux,

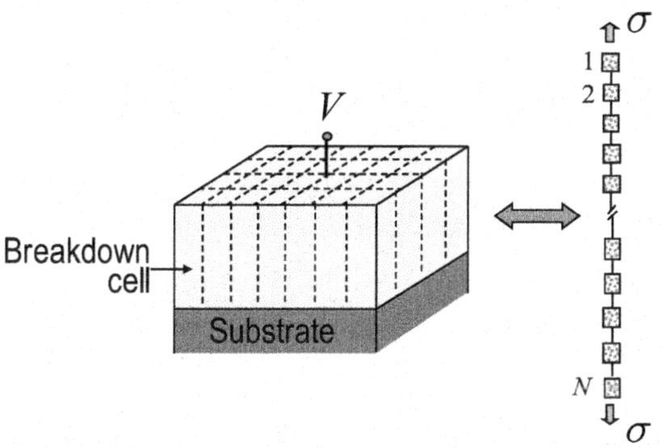

Figure 14.5. Weakest-link model for breakdown statistics of gate dielectrics.

Scott, & Subramoniam 1994) (Fig. 14.5), and applicability of the Weibull cdf implies that the gate fails when the weakest cell, with the smallest breakdown voltage V^*, allows the tunneling current to pass. This is similar to the weakest-link model since $V^* = \min_i(V_i^*)$ where $V_i^* =$ voltage required for breakdown of cell i.

These conditions of minima show that the series coupling in strength mechanics corresponds to the parallel coupling of potential electric breakdown cells. This analogy is further evidenced by comparing the relations $\epsilon = \sum_i(\epsilon_i)$ and $\sigma_i = \sigma$ (all i) for series coupling in strength mechanics, to the relations $I = \sum_i(I_i)$ and $V_i = V$ (all i) for parallel coupling of potential breakdown cells (in practice, of course, there are no perfect insulators; there is always some small current passing through the dielectric); $\sigma, \sigma_i, \epsilon, \epsilon_i =$ overall and local stresses and deformations in the chain and its links; and V, I, V_i, I_i are the overall voltage and current, and the voltage and current in the ith cell. To complete the analogy, note that, vice versa, the parallel coupling in structural mechanics corresponds to a series coupling of potential breakdown cells, which is evidenced by comparing the relations $\sigma = \sum_i(\sigma_i)$ and $\epsilon_i = \epsilon$ (all i) for parallel coupling in structural mechanics, to the relations $V = \sum_i(V_i)$ and $I_i = I$ (all i) for series coupling of resistors.

A chain survives if all its links survive, and likewise a gate dielectric suffers no breakdown if none of the cells suffers a breakdown. If the links are statistically independent (i.e., larger than the relevant autocorrelation lengths of the random breakdown voltage field in the material), the probability of mechanical failure and the probability of dielectric breakdown are given by similar equations:

$$P_f(\sigma, \tau) = 1 - [1 - P_1(\sigma, \tau)]^N \quad \text{(mechanical failure)} \tag{14.1}$$

$$P_f(V, \tau) = 1 - [1 - P_1(V, \tau)]^N \quad \text{(dielectric breakdown)} \tag{14.2}$$

where stress σ and voltage V are considered to be uniform for all the links and cells, and τ is a parameter representing the duration of applied load or voltage. For a fixed τ, the distributions $P_1(\sigma, \tau)$ and $P_1(V, \tau)$ represent the failure or breakdown probability as a function of σ or V in each individual RVE or in each individual cell. From the theory of statistics of extremes (Gumbel 1958; Castillo 1988), it is well known that the distributions P_f must tend to a Weibull form in the asymptotic limit of large N, but the weakest-link model with $N \to \infty$ is often an oversimplification. In the text that follows, we consider finite N as well as more general couplings of the independent cells, which lead to predictable deviations from the Weibull distribution.

In general, to determine the failure or breakdown probability P_f, we would need to know the entire functional form of $P_1(\sigma, \tau)$ or $P_1(V, \tau)$. In Chapter 5, we have shown that, in the case of mechanical strength, the strength distribution of one RVE, $P_1(\sigma, \tau)$ can be approximately described as a renormalized Gaussian cdf onto which a power-law tail is grafted (while preserving continuity of cdf and its slope) at grafting point $P_g \approx 10^{-4} - 10^{-3}$ (Eqs. 5.84a and 5.84b). Recall that this grafted cdf ensues from atomistic fracture mechanics, which dictates that the cdf of mechanical strength on the nanoscale must follow a power-law function with exponent 2 (see Chapter 4), and, for the purpose of statistics, the transition from the nanoscale to the RVE on the macroscale has been represented by a hierarchical series and parallel couplings (Fig. 5.10). As shown in Chapters 6 and 7, the lifetime statistics of quasibrittle structures under constant and cyclic loading can further be calculated through the power law for creep and fatigue crack growth. Now we attempt to use a similar approach to determine the probability distribution function, $P_1(V, \tau)$, of the breakdown voltage and lifetime for one cell.

14.2.2 Application to Dielectric Breakdown

The gate oxide layer may be imagined to consist of a large number of potential breakdown cells (Suné et al. 1990; Dumin et al. 1994; Stathis 1999) (Fig. 14.5). Upon voltage application, electrons or holes are injected into the oxide layer, and microscopic defects (electron or hole traps), whose size is set by typical distances for quantum tunneling, are generated in the layer. When the density of these defects first reaches a critical value in one of the cells, dielectric breakdown occurs. In other words, similar to an RVE, each cell is the smallest material volume whose breakdown triggers the breakdown of the whole gate dielectric. Various percolation models have been proposed to study gate dielectric breakdown numerically (Degraeve et al. 1998; Stathis 1999; Miyakawa et al. 2007). From optimum fitting of experimental observations, it has been inferred that the effective size of the defects, based on their minimum spacing, is about 2 to 3 nm, although the physical extent of the defect might be smaller (Stathis 1999).

Now imagine that each cell consists of $n = t_{ox}/l_0$ subcells, where t_{ox} = thickness of the oxide layer, and l_0 = effective thickness of a subcell (Fig. 14.6a). The breakdown of the cell occurs as soon as each subcell attains a failure criterion, e.g., a critical defect density relation to a percolating path of defects, which will be discussed later. Before the breakdown, each subcell functions as a capacitor (or, more precisely, as the parallel coupling of a capacitor and a resistor whose resistance is very large, in fact so large that the resistor could be neglected).

When the number of defects is sufficiently large, the subcell begins to function as a resistor (Bowman & Stroud 1989) [or a very large capacitance (Beale & Duxbury 1988)] that can conduct a significant current (Fig. 14.6b). Since the electric breakdown of one cell occurs when all these subcells break down, the cell can be modeled as a series coupling of capacitors. Each capacitor represents one subcell, and the capacitor

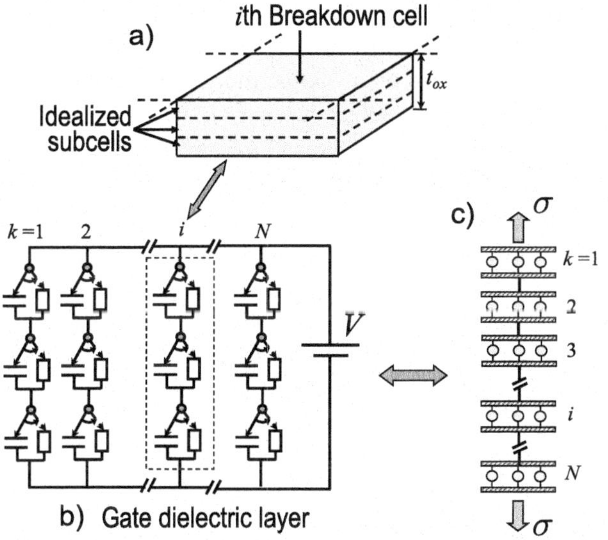

Figure 14.6. Analogy of dielectric breakdown and structural failure.

can be switched to a resistor if the number of defects in the subcell is large enough (Fig. 14.6b).

Note again an analogy between the series coupling of dielectric breakdown and the parallel coupling in the mechanical failure of structures. To get breakdown, all of the capacitors coupled in series must switch to resistors (Fig. 14.6b). Switching any capacitor to a resistor causes the electric field (voltage) to redistribute among the remaining capacitors. This voltage redistribution is analogous to the mechanics of failure in a parallel coupling of structural elements (called fiber-bundle) as we discussed in Chapter 5, where the failure of each element causes the applied load (or stress) to redistribute among the remaining elements (Fig. 14.6c).

Each potential breakdown cell can be modeled as a series coupling of capacitors. This is analogous to the fiber-bundle model for the structural failure, and so we can formulate the lifetime statistics of gate dielectrics by applying the same mathematical framework as used for strength and lifetime statistics of quasibrittle structures. It is clear that, to complete this formulation, we would need to understand the breakdown statistics of each subcell and the breakdown kinetics under both constant and alternating gate voltage stresses.

14.2.3 Microscopic Statistical Models

Our "mesoscopic" statistical theory of dielectric breakdown, based on the analogy with fiber-bundle models of mechanical strength, is very general, but it cannot predict the "microscopic" characteristics of individual subcells, which are the statistically independent elements of the bundle network. Here we use a microscopic physical model to understand the breakdown statistics of a single subcell, which can then be used to predict or extrapolate experimental data via the network couplings.

For gate dielectrics, some previous microscopic models have employed percolation theory to connect the electron-trapping defect density ρ to the probability of breakdown via a spanning defect cluster in a finite system (Degraeve et al. 1998; Stathis 1999; Miyakawa et al. 2007). For very thin gate dielectrics, recent mathematical results could also be used to describe the smoothed percolation transition from nonspanning to spanning clusters in a small, finite system (Bazant 2000; var der Hofstad & Redig 2006; Györgyi, Moloney, Ozogány, & Rácz 2008).

A difficulty with this approach, however, is that it requires additional input to describe the dynamics of breakdown ($\mathrm{d}\rho/\mathrm{d}t$) in response to the local applied voltage $V_a(t)$. Such time-dependence (discussed later) is obviously crucial for any theory of lifetime statistics, but the voltage dependence ($\mathrm{d}\rho/\mathrm{d}V_a$) would also be required to predict the statistics of the breakdown voltage during a voltage ramp, by analogy with the mechanical strength of a structure (defined as the failure stress after a loading ramp). Another general difficulty with applying the percolation models to dielectric breakdown is that they neglect the strong spatial-temporal correlations between discrete failure events, which lead to the strongly correlated growth of connected conducting clusters from an initial seed, in contrast to percolation clusters connecting independent breakdown sites.

To illustrate a possible microscopic input to our general mesoscopic theory, we employ the DBM (Niemeyer et al. 1984; Pietronero & Wiesmann 1988), which is the original and simplest dynamical model for the growth of defect conducting clusters. Consider an individual subcell in a gate dielectric, subject to an applied voltage V_a. In the DBM, a connected cluster of conducting defects (electron trapping sites) advances with a local growth probability measure (per cluster surface area, per time step) $p_g \propto |\nabla\phi|^\eta$, proportional to the normal electric field raised to power η.

For $\eta = 1$, the DBM reduces to the famous Diffusion Limited Aggregation (DLA) model (Witten & Sander 1981), which yields clusters with a fractal dimension $D = 1.71$ in two spatial dimensions [even with a broad class of external forcing (Bazant, Choi, & Davidovitch 2003)]. For $\eta < 1$, there is a reduced focusing of the growth events on protruding tips and, in the limit of the Eden Model ($\eta = 0$), the clusters have the same dimension as the device. In the more relevant limit of $\eta > 1$, DBM clusters look more like familiar lightning strikes with increasing η. For $\eta \geq 4$, a conjectured upper critical dimension, they are effectively linear objects aligned with the applied electric field of dimension $D = 1$ and negligible side branching (Hastings 2001).

For our purposes, we only need the scaling with voltage of the local breakdown probability, i.e., the probability that the conducting cluster spans the subcell. At each stage of growth, the electrostatic potential satisfies Laplace's equation $\nabla^2\phi = 0$ with the boundary conditions $\phi = 0$ on the cluster and $\phi = V_a$ at the other end of the subcell. Because this boundary-value problem is linear, $\phi \propto V_a$, the growth measure scales as $p_g \propto V_a^\eta$ regardless of the (evolving) cluster geometry. Since breakdown results from a sequence of discrete growth events, each of which has the same voltage dependence, we conclude that the probability of subcell breakdown in a time τ has the expected power-law tail,

$$P_l(V, \tau) \sim c_t V^p \quad \text{with } p = \eta \text{ for DBM} \tag{14.3}$$

It is interesting to compare this result to the statistical theory of mechanical failure on the nanoscale, which predicts a power law $p = 2$ from atomistic fracture mechanics (Chapter 4). So, in this sense, the nanoscale quasibrittle fracture is analogous to the dielectric breakdown with exponent $\eta = 2$. This analogy can also be understood in terms of the energy density $\Delta\varepsilon$ released by a breakdown event. In fracture mechanics for a linear elastic material (away from the crack), the exponent $p = 2$ comes from the scaling $\Delta\varepsilon \propto \sigma^2$ with the local stress σ. In dielectric breakdown for a linear material ($\mathbf{D} = \epsilon\mathbf{E}$), we have a similar scaling, $\Delta\varepsilon = DE/2 \propto E^2$ with the electric field strength E. It is important to note, however, that the local physics of time-dependent dielectric breakdown may reflect a variety of other nonlinear effects (Apte & Saraswat 1994; McPherson, Khamankar, & Shanware 2000), especially at the nanoscale in a gate dielectric layer, and this may alter the exponents in the model. As such, they are best left as fitting parameters, whose values are motivated, but not strictly constrained, by a particular microscopic model.

We can now extend the aforementioned model to the failure statistics of a single subcell under the unipolar AC voltage stress shown as Figure 14.3. It is clear that the probability of defect-cluster growth during the "on" period is described by Eq. 14.3. During the "off" period, the voltage is decreased by V_g, which could cause charge detrapping (i.e., the breakdown path gets healed). Here we assume that the probability of detrapping per time step also has a power-law tail: $p_h(V_g) \propto V_g^p$.

Furthermore, it is believed that there is a critical time T_{cr} for the enhancement in charge detrapping (Ghidini, Brazzelli, Clementi, & Pellizzer 1999; Kim & Lee 2004). If the "on" period is much longer than T_{cr}, then the detrapping process becomes ineffective. Therefore, the probability of detrapping p_h must depend on $T_{\mathrm{on}}/T_{\mathrm{cr}}$. For the sake of simplicity, we can use an exponential function to describe such dependence, i.e.,

$$p_h(V_g) = c_d \exp(-\alpha\lambda_{\mathrm{cr}})V_g^p \tag{14.4}$$

where α is a constant and $\lambda_{\mathrm{cr}} = T_{\mathrm{on}}/T_{\mathrm{cr}}$. By applying Eqs. 14.3 and 14.4 to the voltage stress shown in Figure 14.3, one could show that the net probability of the growth of defect cluster per voltage cycle has a power-law tail:

$$p_g(V_g) = c_l V_g^p \tag{14.5}$$

where $\lambda_{\mathrm{on}} = T_{\mathrm{on}}/T_0 =$ duty cycle, $c_l = \langle c_t\lambda_{\mathrm{on}} - c_d e^{-\alpha\lambda_{\mathrm{cr}}}(1 - \lambda_{\mathrm{on}})\rangle$, and $\langle x\rangle = \max(x, 0)$

Since the dielectric breakdown of one subcell consists of a sequence of discrete growth events and these events have the same voltage dependence, we conclude that the tail of the probability of subcell breakdown for a fixed number of voltage cycles N_0 can be written as

$$P_l = cV_g^p \tag{14.6}$$

where $c \propto c_l N_0$. Equation 14.6 shows that the cdf of breakdown voltage amplitude of one subcell has a power-law tail. The effect of characteristics of cyclic voltage history, i.e., λ_{on} and λ_{cr}, on the tail of the breakdown probability is implied by constant c_l. The constants c_t and c_d would depend on the properties of the oxide layer. Such dependence conceivably can be obtained through some sophisticated numerical model [e.g., Suné et al. (1990); Stathis (1999), and Miyakawa et al. (2007)].

14.2.4 Breakdown Voltage Distribution

The foregoing analysis has shown that the breakdown probability of a single subcell of gate dielectrics can be expressed as a power-law function of either the applied gate voltage for the case of a linear voltage ramp or the voltage amplitude for the case of a unipolar AC voltage stress. This result is analogous to the failure statistics of the nanoscale structure under either a linear ramped stress or a cyclic stress (see Chapter 4). Since the breakdown of each potential breakdown cell can be statistically modeled as a fiber-bundle model, we can directly use the statistical properties of the fiber-bundle model derived from Chapter 5 to determine the functional form of the breakdown voltage distribution under both ramped voltage stress and unipolar AC voltage stress.

Regardless of the specific microscopic model, we can reach some general conclusions from the fiber-bundle model for failure at the mesoscale. Since the power-law tail is indestructible in the fiber-bundle model as shown in Chapter 5, the distribution of breakdown voltage of each cell must have a power-law tail, $P_1(V) \sim c_0 V^m$. Moreover, the exponent of the power-law tail of the strength cdf of a bundle is additive, which implies

$$m = np \tag{14.7}$$

where n = number of the subcells in each potential breakdown cell. In the context of gate dielectrics, the additivity of the Weibull modulus m is meaningful when the thickness of the subcell is larger than the effective defect size.

Based on the asymptotic properties of the probability distributions of monotonic and fatigue strengths of the fiber-bundle model, the probability distribution of breakdown voltage of one cell under both linearly ramped voltage stress and unipolar voltage stress may be approximated as a power-law tail grafted from the left onto a Gaussian probability density function (pdf) at certain failure probability. The grafted probability distribution is analogous to the pdfs of monotonic strength and fatigue strength of one RVE (Eqs. 5.84a and 5.84b and 5.87a and 5.87b):

1. pdf of breakdown voltage under linearly ramped voltage stress:

$$\text{for } V < V_{\text{gr}}^L\text{:} \quad p_1(V) = (m_L/s_L)(V/s_L)^{m_L-1}e^{-(V/s_L)^{m_L}} = \phi_{\text{W}}^L(V) \tag{14.8a}$$

$$\text{for } V \geq V_{\text{gr}}^L\text{:} \quad p_1(V) = r_L e^{-(V-\mu_L)^2/2\delta_L^2}/(\delta_L\sqrt{2\pi}) = r_L\phi_{\text{G}}^L(V) \tag{14.8b}$$

2. pdf of breakdown voltage amplitude under unipolar AC voltage stress:

$$\text{for } V_g < V_{\text{gr}}^A\text{:} \; p_1(V_g) = (m_A/s_A)(V_g/s_A)^{m_A-1}e^{-(V_g/s_A)^{m_A}} = \phi_{\text{W}}^A(V_g) \tag{14.9a}$$

$$\text{for } V_g \geq V_{\text{gr}}^A\text{:} \; p_1(V_g) = r_A e^{-(V_g-\mu_A)^2/2\delta_A^2}/(\delta_A\sqrt{2\pi}) = r_A\phi_{\text{G}}^A(V_g) \tag{14.9b}$$

Here $p_1(V), p_1(V_g)$ = pdfs of breakdown voltage for the case of linearly ramped voltage stress, and breakdown voltage amplitude for the case of unipolar AC voltage stress, respectively; V, V_g = applied linearly ramped voltage and voltage amplitude, respectively; m_i (Weibull modulus) and $s_i (i = L, A)$ shape and scale parameters of the Weibull tail of $p_1(V)$ and $p_1(V_g)$; μ_i and δ_i, $(i = L, A)$ = means and standard deviations of the Gaussian cores if considered extended to $-\infty$; and r_i, $(i = L, A)$ = scaling parameters required to normalize the grafted cdfs of V and V_g such that $\int_{-\infty}^{\infty} p_1(V)\,\mathrm{d}V = 1$, $\int_{-\infty}^{\infty} p_1(V_G)\,\mathrm{d}V_G = 1$. Furthermore, continuity of the Gaussian and Weibull, pdfs at the grafting point requires that $\phi_W^L(V_{gr}^L) = r_L \phi_G^L(V_{gr}^L)$ and $\phi_W^A(V_{gr}^A) = r_A \phi_G^A(V_{gr}^A)$.

The grafting probabilities, i.e., $P_g^L = \int_0^{V_{gr}^L} \phi_W^L(V)\mathrm{d}V$ and $P_g^A = \int_0^{V_{gr}^A} \phi_W^A(V)\mathrm{d}V$ depend on the number of subcells in the cell, or the thickness of oxide layer. In practice, the thickness of layer is about 5 nm (Stathis 1999; Kim & Lee 2004), and it is believed that that each cell consists of about three or four layers, for an effective defect size of about 1.5 nm. This implies that each breakdown cell can be statistically modeled as a bundle of three or four elements. Although this is an assumption, it gives a reasonable range for the reach of power-law tail of a bundle, which is used to make an estimate of the grafting probability that would fit the histograms.

Based on the tail properties of fiber-bundle model as analyzed in Chapter 5, we thus conclude that the grafting probability is here about 10^{-7} to 10^{-10}. This probability is much lower than the grafting probability for the strength cdf of one RVE in the context of mechanical failure, which is about 10^{-4} to 10^{-2} as described in Chapter 5.

14.3 Breakdown Lifetime under Constant Voltage

14.3.1 Relation between Lifetime and Breakdown Voltage

What is of primary interest for gate dielectrics is the distribution of lifetime τ at a prescribed voltage application. As already noted, this requires a model of the time evolution of dielectric breakdown. But we will see that only minimal, generic, assumptions about the microscopic physics are required.

We start by considering percolation models for gate dielectrics (Degraeve et al. 1998; Stathis 1999; Miyakawa et al. 2007), which relate breakdown to the appearance of a spanning cluster of (uncorrelated) conducting defects. In this context, the simplest way to describe time evolution would be to postulate a differential equation for the mean defect density $\rho(t)$ in a potential breakdown cell. Motivated by the analogy with mechanical strength, it would be natural to postulate a power-law scaling of the defect production rate with the applied voltage V_a,

$$\frac{\mathrm{d}\rho}{\mathrm{d}t} = \alpha(\rho) V_a^{\kappa_L} \tag{14.10}$$

by analogy to the power law for crack growth rate (an empirical relation, which we have given a theoretical basis in atomistic fracture mechanics; see Chapter 6). In the text that follows, we will derive a similar equation from DBM. Its validity will be supported a posteriori by the success of our fitting of lifetime distributions for gate dielectrics.

First we show that a microscopic model is not required to relate the lifetime τ at constant voltage V_0 to the breakdown voltage V, defined by the onset of current in response to a linear voltage ramp, $V_a(t) = at$ (by analogy with the definition of mechanical strength, as noted earlier). To supplement Eq. 14.10 we make only one additional assumption, that the breakdown occurs when the defect density reaches a critical value ρ_c, starting from an initial value $\rho_0 < \rho_c$. Following Bažant et al. (2009), in both scenarios (measuring lifetime and breakdown voltage), we can then integrate the first-order, separable equation (Eq. 14.10) to obtain the identity

$$F(\rho_c) = \int_{\rho_0}^{\rho_c} \frac{\mathrm{d}\rho}{\alpha(\rho)} = V_0^{\kappa_L}\tau = \frac{V^{\kappa_L+1}}{a(\kappa_L+1)} \tag{14.11}$$

which implies the desired relationship

$$\tau = \beta \frac{V^{\kappa_L+1}}{V_0^{\kappa_L}} \tag{14.12}$$

where $\beta = [a(\kappa_L+1)]^{-1}$ is a constant.

14.3.2 Microscopic Physics

To illustrate a possible microscopic basis for Eq. 14.10, we turn again to the DBM (Niemeyer et al. 1984). To describe the spreading of a DBM cluster, we assign it a characteristic linear extent $\ell(t)$ (e.g., to its farthest tip), such that the spanning of a subcell of size l_0, and thus its breakdown, occurs with high probability as $\ell \to l_0$. As the conducting cluster advances, the typical electric field at its tip grows like $|\nabla\phi| \sim |V_a/(l_0-\ell)|$, and thus the DBM growth measure (for constant tip width) implies the scaling relation

$$\frac{\mathrm{d}\ell}{\mathrm{d}t} = c'\left(\frac{V_a}{l_0-\ell}\right)^{\eta} \tag{14.13}$$

which is analogous to the power law for the velocity of a subcritical crack. The defect density in the cell (fraction of sites in the conducting cluster) scales like $\rho \propto \ell^D$ with fractal dimension $D \geq 1$. We can write it as $\rho = \rho_c(\ell/l_0)^D$ using the spanning breakdown criterion. Substituting this into Eq. 14.13, we arrive at an expression similar to Eq. 14.10 with $\kappa_L = \eta$

$$\frac{\mathrm{d}\rho}{\mathrm{d}t} = \alpha(\rho)V_a^{\eta} \tag{14.14}$$

and

$$\alpha(\rho) = \frac{c''\rho^{(D^2-1)/D}}{l_0^{\eta+1}\left(\rho_c^{1/D} - \rho^{1/D}\right)^{\eta}} \tag{14.15}$$

where c'' = a constant. Note that the defect production rate diverges at a critical concentration ρ_c, but this is not a percolation transition. Instead, ρ_c is now the mean concentration of a spanning DBM cluster, although this microscopic detail does not affect the general relationship (Eq. 14.12).

In the mechanical failure, we have shown in Chapter 6 that the exponent of the power-law tail of strength cdf of nanoscale structure is equal to the exponent of the power law for the nanocrack growth rate (Bažant et al. 2009). Equation 14.14 presents a similar analogy where the exponent of power-law tail of breakdown voltage for each potential defect cluster is equal to the exponent of the power law for the rate of growth of the defect cluster.

14.3.3 Probability Distribution of Breakdown Lifetime

We now convert the pdf of the breakdown voltage V (Eqs. 14.8a and 14.8b) into a cdf for lifetime τ at voltage V_0, using the scaling relation (Eq. 14.12) between these two random variables. This directly leads to a lifetime distribution of one breakdown cell:

$$\text{for } \tau < \tau_{\text{gr}}:\ P_1(\tau) = 1 - \exp[-(\tau/s_\tau)^{m_L/(\kappa_L+1)}]; \tag{14.16a}$$

$$\text{for } \tau \geq \tau_{\text{gr}}:\ P_1(\tau) = P_{\text{gr}} + \frac{r_L}{\delta_L\sqrt{2\pi}} \int_{\gamma_L \tau_{\text{gr}}^{1/(\kappa_L+1)}}^{\gamma_L \tau^{1/(\kappa_L+1)}} e^{-(\tau'-\mu_L)^2/2\delta_L^2} \mathrm{d}\tau' \tag{14.16b}$$

where $\gamma_L = \beta^{-1/(\kappa_L+1)} V_0^{\kappa_L/(\kappa_L+1)}$, $\tau_{\text{gr}} = \beta V_0^{-\kappa_L} V_{\text{gr}}^{\kappa_L+1}$, $s_\tau = s_L^{\kappa_L+1} \beta V_0^{-\kappa_L}$, $P_{\text{gr}} = P_1(\tau_{\text{gr}})$.

According to this result, the Weibull moduli of the lifetime distribution and the breakdown voltage distribution are related by $m_l = m_L/(\kappa_L + 1)$. Using Eq. 14.7, we arrive at the following relation between the exponents;

$$m_l = \frac{pn}{\kappa_L + 1} \tag{14.17}$$

This equation indicates that, for the same dielectric material, the Weibull modulus of breakdown lifetime increases in proportion to the thickness of oxide layer (i.e., as n increases). Our simple microscopic model based on the DBM further implies

$$m_l = \frac{\eta n}{\eta + 1} \tag{14.18}$$

14.4 Breakdown Lifetime under Unipolar AC Voltage

The foregoing analysis is now extended to lifetime statistics of gate dielectrics subjected to the unipolar AC voltage stress. Again, we first need to describe the kinetics of spreading of DBM cluster. Extending the derivation of breakdown kinetics under constant voltage and considering the healing process of the breakdown path during the "off" period of the AC voltage stress (Eq. 14.5), we postulate that the net growth of the DBM cluster per voltage cycle can be written as

$$\frac{\mathrm{d}l}{\mathrm{d}N} = \langle c_g \lambda_{\text{on}} - c_h(1-\lambda_{\text{on}}) \exp(-\alpha \lambda_{\text{cr}}) \rangle \left(\frac{V_g}{l_0 - l} \right)^{\kappa_A} \tag{14.19}$$

where c_g, c_h = constants. Eq. 14.19 further leads to the growth rate of defect density:

$$\frac{\mathrm{d}\rho}{\mathrm{d}N} = \langle c_g \lambda_{\text{on}} - c_h(1-\lambda_{\text{on}}) \exp(-\alpha \lambda_{\text{cr}}) \rangle \Phi(\rho) V_g^{\kappa_A} \tag{14.20}$$

$$\text{where}\quad \Phi(\rho) = \frac{\rho^{(D^2-1)/D}}{l_0^{\kappa_A} + 1} (\rho_c^{1/D} - \rho^{1/D})^{\kappa_A} \tag{14.21}$$

The subcell suffers dielectric breakdown once the defect density grows from its initial value ρ_0 to a critical value ρ_c. For any two unipolar AC voltage stress histories $V_{gi}(N)$ and $V_{gj}(N)$ with the same λ_{on} and λ_{cr} but different voltage amplitudes, integrating Eq. 14.20 yields

$$\int_0^{N_i} V_{gi}^{\kappa_A}(N')\,\mathrm{d}N' = \int_0^{N_j} V_{gj}^{\kappa_A}(N')\,\mathrm{d}N' = \frac{\int_{\rho_0}^{\rho_c} \Phi^{-1}(\rho)\,\mathrm{d}\rho}{\langle c_g\lambda_{\text{on}} - c_h(1-\lambda_{\text{on}})\exp(-\alpha\lambda_{\text{cr}})\rangle} \tag{14.22}$$

where N_i and N_j are the breakdown cycles for these two voltage stress histories.

Now consider two tests performed on the same potential breakdown cell under unipolar AC voltage stress: (1) a breakdown voltage test, where we measure the breakdown voltage amplitude V_g for a prescribed number of voltage stress cycles N_0; and (2) a breakdown lifetime test, where we record the breakdown voltage stress cycles N_c for a prescribed voltage amplitude V_{g0}. In both tests, the "on" and "off" periods during each cycle are kept the same.

The power-law scaling for the growth rate of defect density (Eq. 14.20) is now applied to the potential breakdown cell. Consider that the current leakage occurs once the defect density in the cell reaches a critical value. Based on Eq. 14.22, we can then simply relate the breakdown voltage amplitude to the breakdown lifetime as: $V_g^{\kappa_A} N_0 = V_{g0}^{\kappa_A} N_c$, or

$$V_g = V_{g0}(N_c/N_0)^{1/\kappa_A} \tag{14.23}$$

By substituting Eq. 14.23 into Eqs. 14.9a and 14.9b and denoting the total breakdown time as $\tau_c = N_c T_0$, we obtain the lifetime distribution of one potential breakdown cell:

$$\text{for } \tau_c < \tau_{\text{gr}}: \quad P_1(\tau_c) = 1 - \exp[-(\tau_c/s_\tau)^{m_A/\kappa_A}]; \tag{14.24a}$$

$$\text{for } \tau_c \geq \tau_{\text{gr}}: \quad P_1(\tau_c) = P_{\text{gr},A} + \frac{r_A}{\delta_A\sqrt{2\pi}} \int_{\gamma_A \tau_{\text{gr}}^{1/\kappa_A}}^{\gamma_A \tau_c^{1/\kappa_A}} e^{-(\tau'-\mu_A)^2/2\delta_A^2}\,\mathrm{d}\tau' \tag{14.24b}$$

where $\gamma_A = \tau_0^{-1/\kappa_A} V_{g0}$, $\tau_{\text{gr}} = (V_{\text{gr}}^A/V_{g0})^{\kappa_A}\tau_0$, $s_\tau = s_A^{\kappa_A}\tau_0 V_{g0}^{-\kappa_A}$, $P_{\text{gr},A} = P_1(\tau_{\text{gr}})$, and $\tau_0 = N_0 T_0$. It is clear that the tail of lifetime distribution of the potential breakdown cell follows the Weibull distribution. The core of the cdf of breakdown lifetime follows a Gaussian distribution transformed by a power-law function. Note that the Weibull modulus of lifetime distribution m_τ is equal to m_A/κ_A. Since the Weibull modulus of breakdown voltage amplitude m_A is equal to np, the present model predicts that the Weibull modulus of breakdown lifetime under unipolar AC voltage stress would increase proportionally with the thickness of the gate oxide layer. This is consistent with the prediction of some existing analytical models as well as the numerical simulation by the percolation models (Suné 2001; Suné, Tous, & Wu 2009).

14.5 Experimental Validation

14.5.1 Breakdown under Constant Gate Voltage Stress

We start by testing Eq. 14.12, which implies a relation between the applied voltage V_0 and the mean lifetime $\bar{\tau}$ for a given gate dielectric, $V_0 \propto \bar{\tau}^{-1/\kappa_L}$. As shown in

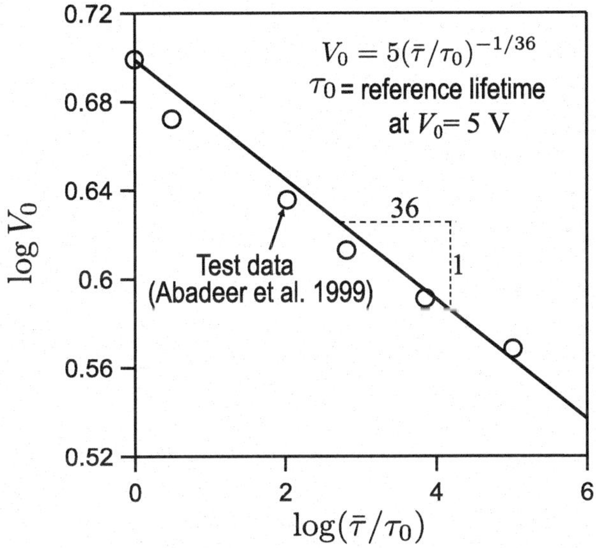

Figure 14.7. Voltage–mean lifetime relationship of gate dielectrics.

Figure 14.7, this power law gives a good fit of the experimental data (Abadeer et al. 1999), although the inferred exponent $\kappa_L = 36$ is much larger than the values of η used in simulations of fractal DBM clusters, as well as the value $\eta = 2$ discussed previously. We could also try to fit the data in Figure 14.7 by an exponential function (Abadeer et al. 1999; Kim & Lee 2004), but this would not be reconciled with the clear experimental evidence of Weibull statistics, and thus the presumed power-law tails of failure distributions, as noted earlier.

It is interesting that a similar paradox arises in structural failure, where the exponent κ_L inferred from lifetime statistics is also large, and much larger than would be expected from microscopic theories of fracture. In that case, a possible resolution is based on the hypothesis of hierarchical couplings of failure from the nanoscale to the mesoscale (Bažant et al. 2009). Admittedly, for thin gate dielectrics, which are already on the nanoscale (and thus can consist of at most several atomic-scale breakdown subcells), this argument is harder to justify. Perhaps the large fitted value $\eta = \kappa_L$ reflects an approximation to a microscopic exponential dependence, which could, e.g., arise from irreversible activated hopping over an energy barrier, biased by the electric field (McPherson et al. 2000).

In any case, regardless of the microscopic dynamical model, our mesoscopic statistical theory makes predictions in good agreement with experiments. For example, our prediction (Eq. 14.17) agrees very well with Kim and Lee's recent experimental observations on the dependence of Weibull modulus on the layer thickness (Kim & Lee 2004). Their data show that the Weibull modulus doubles (from 2 to 4) when the gate dielectric thickness is almost doubled (from 4.8 nm to 9.7 nm, or EOT of 1.4 nm to 2.5 nm), as shown in Figure 14.8.

The mesoscopic theory also provides a good fit of experimental lifetime distributions. The curves in Figure 14.1 show that the preceding distribution (Eqs. 14.16a and 14.16b),

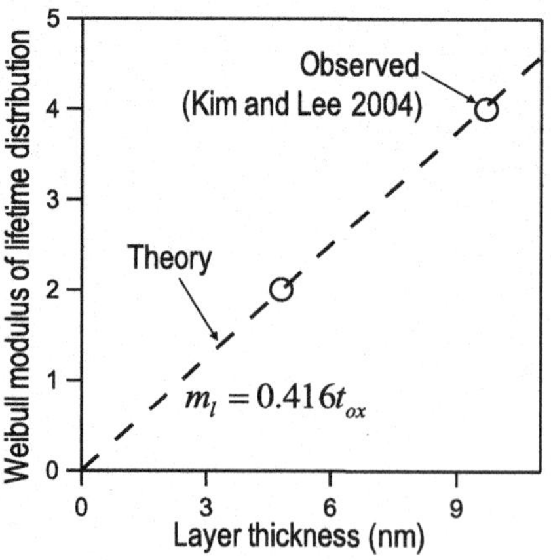

Figure 14.8. Effect of layer thickness on the Weibull modulus of lifetime distribution.

with Eq. 14.2 based on the analogy with mechanical strength, provides good fits of the lifetime histograms observed by Kim and Lee (2004) on high-*k* gate dielectrics with an oxide layer of thickness about 4.8 to 5 nm. Note that the curves have a kink that is centered at the grafting point separating two segments, of which the lower one is a straight line (representing the Weibull distribution) and the upper part deviates from the straight line systematically to the right. It follows from the location of the kink that the number of cells in this gate dielectric is on the order of 10^7 (i.e., the area A_0 of each cell is about 10^{-10} mm^2). This is consistent with the order of magnitude obtained from various percolation models (Suné et al. 1990; Dumin et al. 1994).

Based on the optimum fits by the present theory, the Weibull modulus m_l is found to be 2.5 to 3.5, which is approximately equal to the number of elements in the bundle (≈ 3) representing each potential breakdown cell in Figure 14.6a (it is also equal to the number of subcells n in each breakdown cell). Interestingly, this is predicted by our foregoing simple microscopic DBM calculations, which require $p = \kappa_L = \eta$, and thus for $\eta \gg 1$ (inferred from the preceding fitting): $p/(\kappa_L + 1) = \eta/(\eta + 1) \approx 1$. Thus, Eq. 14.17 becomes $m_l \approx n$.

The same authors also measured the effect of gate area A (cm^2) on the median lifetime τ_{50} (defined as the time to 50% failure frequency, in second). Based on the aforementioned calibration of the model, one can extrapolate the lifetime cdfs of the gate dielectrics for different gate areas by Eq. 14.2. Figure 14.2 shows that the predicted τ_{50} agrees very well with the τ_{50} observed in the experiments. In the log–log plot of τ_{50} and A, the experimentally observed points cannot be optimally fitted by a straight line (Kim & Lee 2004). This indicates an insufficiency of the two-parameter Weibull distribution, in which τ_{50} must be a power-law function of A [similar to the power law for the mechanical size effect on the lifetime of structures (Bažant et al. 2009)]. In view of Figure 14.2, the nonlinearity in this plot, i.e., the deviation from the power law, is caused by the effect of gate area of the dielectric on the type of lifetime cdf.

14.5.2 Breakdown under Unipolar AC Voltage Stress

Figure 14.4 presents the optimum fits of the measured lifetime histograms of HfO_2-based gate dielectrics under unipolar AC voltage stress (Kim & Lee 2004). Despite the limited number of specimens, it can be seen that the measured lifetime histograms do not follow a straight line in the Weibull scale, which implies that the the lifetime cdf does not strictly follow the two-parameter Weibull distribution. Similar to the lifetime histograms under constant voltage stress, the histogram kink separates into two segments. The present theory can fit these two segments simultaneously. Clearly, the deviation from the two-parameter Weibull distribution is due to the non-negligible gate area of the potential breakdown cell.

It should be pointed out that the range of experimentally measured lifetime data might not be sufficient to fully validate the present model for breakdown lifetime under the unipolar AC voltage stresses. Nevertheless, the optimum fitting demonstrates how the present model could capture the observed deviation of the lifetime cdf from the classical two-parameter Weibull theory. The complete check would be through the histogram testing of lifetime of gate dielectrics with different gate areas. Regrettably, no such data are available in the literature.

14.6 Size Effect on Mean Breakdown Lifetime

The effect of gate area on the mean breakdown lifetime for both constant voltage stress and unipolar AC voltage stress can be numerically calculated as $\bar{\tau}_c = \int_0^\infty [1 - P_1(V, \tau)]^{n_c} \, d\tau$. Exploiting the similarity between the lifetime distributions of gate dielectrics and of quasibrittle structures, we can express the mean size effect by the

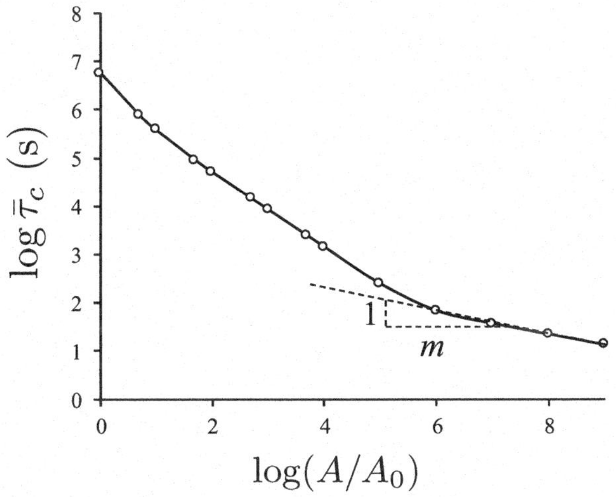

Figure 14.9. Calculated mean size effect curve of breakdown lifetime and its optimum fit by Eq. 14.25.

approximate formula of the same form as Eqs. 7.37 and 7.38:

$$\bar{\tau}_c = [(C_a/A) + (C_b/A)^{r/\mu}]^{1/r} \tag{14.25}$$

where μ = Weibull modulus of lifetime distribution of gate dielectrics under either constant voltage stress or unipolar AC voltage stress, parameters C_a, C_b, r can be determined by three asymptotic matching conditions: $[\bar{\tau}_c]_{A\to A_0}$, $[d\bar{\tau}_c/dA]_{A\to A_0}$, and $[\bar{\tau}_c A^{1/\mu}]_{A\to\infty}$. Figure 14.9 presents the mean size effect curve of the breakdown lifetime with its fitting by Eq. 14.25.

As discussed in Chapter 9, one important implication of the finite weakest-link model of quasibrittle fracture is that the strength and lifetime statistics of quasibrittle structures can be derived from the mean size effect analysis. With the mathematical analogy between quasibrittle fracture and dielectric breakdown, we can use the same mathematical framework as presented in Chapter 9 to determine the lifetime statistics of gate dielectrics from the mean size effect curve. The present theory is intended to provide the semiconductor industry with a more efficient and robust method for assessing the reliability of gate dielectrics.

Appendix A Power-Law Scaling of Boundary Value Problems

Scaled geometrically similar structures of different sizes can always be obtained from the reference structure (after proper positioning) by polar affinity transformation (Fig. A.1):

$$\overline{x}_i = \lambda x_i \tag{A.1}$$

where x_i are the Cartesian coordinates for the reference structure of characteristic dimension (size) D; $\overline{x}_i$ are the coordinates for a geometrically similar scaled structure; and $\lambda = \overline{D}/D$, where $\overline{D}$ is the characteristic dimension of the scaled structure. For the sake of brevity, we will denote $\partial/\partial x_i = \partial_i$, $\partial/\partial \overline{x}_i = \overline{\partial}_i$. From the chain rule of differentiation, $\partial_i = \lambda \overline{\partial}_i$, $\overline{\partial}_i = \lambda^{-1}\partial_i$.

For the reference structure of size D and the similar scaled structure of size $\overline{D}$, the field equations and the boundary conditions are

For D: For $\overline{D}$:

$$\partial_j \sigma_{ij} + f_i = 0 \qquad \overline{\partial}_j \overline{\sigma}_{ij} + \overline{f}_i = 0 \tag{A.2}$$

$$\varepsilon_{ij} = (\partial_j u_i + \partial_i u_j)/2 \qquad \overline{\varepsilon}_{ij} = (\overline{\partial}_j \overline{u}_i + \overline{\partial}_i \overline{u}_j)/2 \tag{A.3}$$

$$\sigma_{ij} n_j = p_i \quad \text{on } \Gamma_1 \qquad \overline{\sigma}_{ij} \overline{n}_j = \overline{p}_i \quad \text{on } \overline{\Gamma}_1 \tag{A.4}$$

$$u_i = U_i \quad \text{on } \Gamma_2 \qquad \overline{u}_i = \overline{U}_i \quad \text{on } \overline{\Gamma}_2 \tag{A.5}$$

in which σ_{ij} and ε_{ij} are the stresses and strains in Cartesian coordinates x_i (the strains being assumed to be small); u_i = displacements of material points; Γ_1 and Γ_2 are the portions of the boundary with prescribed surface tractions p_i and with prescribed displacements U_i; f_i = prescribed volume forces; and $n_j = \overline{n}_j$ = direction cosines of unit outward normals on the stress boundary.

From Eq. 1.5 we already know that the scaling law must be a power function. Let us now assume that the displacements are related by the scaling law

$$\overline{u}_i = \lambda^{s+1} u_i \tag{A.6}$$

where s is an unknown exponent. Substituting this into the differential equations and boundary conditions Eqs. A.2 to A.5, the following transformation rules then ensue:

$$\overline{\varepsilon}_{ij} = \varepsilon_{ij}\lambda^s, \quad \overline{\sigma}_{ij} = \sigma_{ij}\lambda^s, \quad \overline{\sigma}_N = \sigma_N \lambda^s \tag{A.7}$$

$$\overline{p}_i = p_i \lambda^s, \quad \overline{f}_i = f_i \lambda^{s-1}, \quad \overline{u}_i = u_i \lambda^{s+1} \tag{A.8}$$

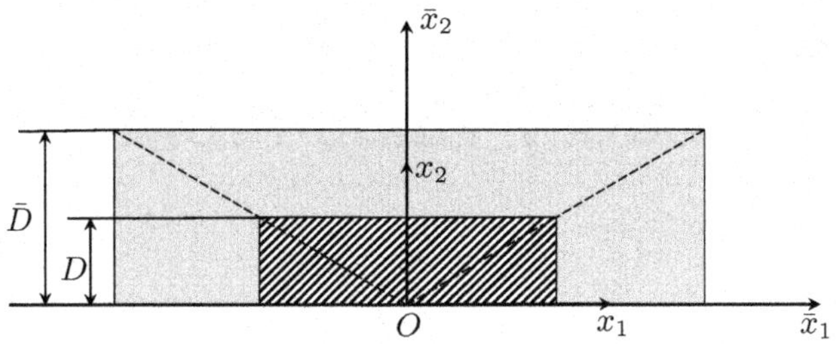

Figure A.1. Affinity transformation of structure from size D to size $\bar{D}$ (O is the center of affinity).

These rules indicate how a solution for one size can be transformed to a solution for another size. However, the value of s is indeterminate. To determine it, we cannot ignore the constitutive law and the failure condition. Next we consider in this regard two important special cases.

(a) Plastic or Elastic-Plastic Behavior

The constitutive relation and the condition of peak load (either the yield condition or the condition of allowable stress) have the general form:

$$\sigma_{ij} = \mathcal{F}_{ij}(\varepsilon_{km}), \qquad \phi(\sigma_{ij}, \varepsilon_{ij}) = \sigma_0 \tag{A.9}$$

in which $\mathcal{F}_{ij}$ are tensor-valued functions or functionals of a tensorial argument, ϕ is a nonlinear scalar function of tensorial argument, and σ_0 is the material yield limit or allowable stress limit. After transformation of scale, Eq. A.9 takes the form $\overline{\sigma}_{ij} = \mathcal{F}_{ij}(\overline{\varepsilon}_{km})$, $\phi(\overline{\sigma}_{ij}, \overline{\varepsilon}_{ij}) = \sigma_0$. Since at least function ϕ (and possibly also function $\mathcal{F}$) is nonlinear, these two equations and Eq. A.9 can hold true simultaneously if and only if $\overline{\sigma}_{ij} = \sigma_{ij}$ and $\overline{\varepsilon}_{km} = \varepsilon_{km}$, which means that $s = 0$. The transformation rules from Eqs. A.7 and A.8 then become:

$$\overline{u}_i = \lambda u_i, \qquad \overline{\varepsilon}_{ij} = \varepsilon_{ij}, \qquad \overline{\sigma}_{ij} = \sigma_{ij} \tag{A.10}$$

$$\overline{p}_i = p_i, \qquad \overline{f}_i = f_i/\lambda, \qquad \overline{u}_i = u_i\lambda \tag{A.11}$$

Also

$$\overline{\sigma}_N = \sigma_N \tag{A.12}$$

that is, the nominal stress at failure does not depend on the structure size. We say in this case that there is no size effect. This is characteristic for all failure analyses according to elasticity with allowable stress limit, plasticity, and classical continuum damage mechanics (as well as viscoelasticity and viscoplasticity, because time has no effect on this analysis).

(b) Linear Elastic Fracture Mechanics

In this case, the constitutive relation and the condition of peak load can be written as

$$\sigma_{ij} = E_{ijkl}\varepsilon_{kl}, \qquad J = G_f \tag{A.13}$$

in which E_{ijkl} is the fourth-order tensor of elastic constants, G_f is the fracture energy (a material property), and J is the J-integral (Rice 1968*b*), expressed as

$$J = \oint \left(\frac{1}{2} \sigma_{ij} \varepsilon_{ij} \, \mathrm{d}y - \sigma_{ij} n_j \partial_1 u_i \, \mathrm{d}s \right) \tag{A.14}$$

e.g., Rice (1968*b*), Knott (1973), and Kanninen and Popelar (1985). Using the transformation rules in Eqs. A.6 to A.8, we find that the J-integral transforms as

$$\begin{aligned} \overline{J} &= \oint \left[\frac{1}{2} (\lambda^s \sigma_{ij})(\lambda^s \varepsilon_{ij}) \lambda \, \mathrm{d}y - \lambda^s \sigma_{ij} n_j \lambda^{-1} \partial_1 (\lambda^{s+1} u_i) \lambda \, \mathrm{d}s \right] \\ &= \lambda^{2s+1} \oint \left(\frac{1}{2} \sigma_{ij} \varepsilon_{ij} \, \mathrm{d}y - \sigma_{ij} n_j \partial_1 u_i \, ds \right) = J \lambda^{2s+1} \end{aligned} \tag{A.15}$$

Since both $\overline{J}$ and J must satisfy the same equality, that is, $\overline{J} = G_f$ and $J = G_f$ in all cases, it is obviously necessary and sufficient that $2s + 1 = 0$, that is,

$$s = -1/2 \tag{A.16}$$

It is noted that, for more general cases where s is complex (e.g., interfacial crack), Eq. A.16 can be rewritten as $\Re(s) = -1/2$, where $\Re(x)$ denotes the real part of x.

Thus, according to Eqs. A.7 and A.8, the transformation laws for linear elastic fracture mechanics (LEFM) are

$$\overline{u}_i = u_i \sqrt{\lambda}, \quad \overline{\varepsilon}_{ij} = \varepsilon_{ij} / \sqrt{\lambda}, \quad \overline{\sigma}_{ij} = \sigma_{ij} / \sqrt{\lambda} \tag{A.17}$$

$$\overline{p}_i = p_i / \sqrt{\lambda}, \quad \overline{f}_i = f_i \lambda^{-3/2}, \quad \overline{U}_i = U_i \sqrt{\lambda} \tag{A.18}$$

$$\overline{\sigma}_N = \frac{\sigma_N}{\sqrt{\lambda}} \tag{A.19}$$

where $\lambda = D/\overline{D}$. So the nominal stress at failure depends on the structure size D, $\sigma_N \sim 1/\sqrt{D}$, or

$$\log \sigma_N = -\frac{1}{2} \log D + \text{constant} \tag{A.20}$$

In the plot of $\log \sigma_N$ versus $\log D$, the LEFM failures are represented by a straight line of slope $-\frac{1}{2}$, while all the stress- or strain-based failure criteria correspond to a horizontal line. The foregoing argument can be generalized to nonlinear elastic behavior, to which the J-integral is also applicable.

Appendix B Proof of Transitional Size Effects of Types 1 and 2 by Dimensional Analysis and Asymptotic Matching up to Second Order

B.1 Second-Order Type 2 Size Effect

This Appendix (based on Bažant 2004*b* and also Appendix A.1 in Bažant 2005) shows that the general approximate size effect law (up to second-order asymptotic terms) can be derived by combining dimensional analysis with asymptotic matching. From Vashy–Buckingham's Π theorem of dimensional analysis (Vashy 1892; Buckingham 1914, 1915; Barenblatt 1996, 2003), two special size effects can readily be proven: (1) If the failure depends on f_t (dimension N/m^2) but not G_f, then there is no size effect, i.e., $\sigma_N =$ constant (this is so for all elastoplastic failures). (2) If the failure depends on G_f (dimension J/m^2) but not on f_t, then there is a size effect of the type $\sigma_N \propto D^{-1/2}$ [this is so for linear elastic fracture mechanics (LEFM), if the cracks or notches are geometrically similar]. Nothing more can be deduced from dimensional analysis alone.

One can, however, deduce more information by considering the physical meaning of the material characteristic length governing the length of the fracture process zone (FPZ), $l_0 = EG_f/f_t^2$ (Irwin 1958). The FPZ length and width for $D \to \infty$ are ηl_0 and $\eta' l_0$, where η and η' are constants, usually of the order of 1, depending on microstructure characteristics (the parameter $l_e = G_f/E$ is also a length but is irrelevant for failure; it controls the ratio of the deformation to the stress intensity factor K_I in LEFM).

Depending on the ratio D/l_0, two asymptotic cases may be distinguished:

1. When $D/l_0 \to 0$, the body is much smaller than a fully developed FPZ. So, G_f cannot matter. Therefore, the case of no size effect, $\sigma_N =$ constant, is the small-size asymptote.
2. When $D/l_0 \to \infty$, the FPZ becomes a point in dimensionless coordinates $\xi_i = x_i/D$ and the stress field approaches a singularity. So, f_t cannot matter. Therefore, the LEFM scaling, $\sigma_N \propto D^{-1/2}$, is the large-size asymptote of quasibrittle failure (which is represented by a straight line of slope $-1/2$ in the plot of $\log \sigma_N$ vs. $\log D$).

For the intermediate sizes, the size effect curve may be expected to be a gradual transition between these two asymptotes. Following Bažant (2004*b*), it will now be shown that the approximate form of this transition can be deduced upon noting how the asymptotes are approached, i.e., by exploiting the higher-order asymptotic terms in Eqs. 3.30 and 3.32.

Transforming the formulation of the boundary value problem with cohesive crack to dimensionless coordinates, one concludes that [if the structure shape, a_0/D and the normalized cohesive curve $\sigma/f_t = \varphi(w/w_f)$ are fixed] σ_N depends on only three parameters, f_t, D, and K_c, where $K_c = (EG_f)^{1/2}$ = mode I fracture toughness for plane strain and E = Young's modulus. Thus, there are four governing parameters, σ_N, D, f_t, and K_c. Since they involve two independent physical dimensions (length and force), the Π theorem (Barenblatt 1996, 2003) implies that there can be only $4 - 2 = 2$ independent dimensionless parameters, Π_1 and Π_2.

So, the equation governing failure may be written as $F(\Pi_1, \Pi_2) = 0$ where function $F(\Pi_1, \Pi_2)$ is assumed to be sufficiently smooth. Although many diverse choices of Π_i $(i = 1, 2)$ are possible, the *key idea* here is to make a choice for which, *in each asymptotic case, all* Π_i *vanish except one* (Bažant 2004*b*). If consideration is limited to dimensionless monomials, this can be most generally achieved by choosing:

$$\Pi_1 = \left(\frac{\sigma_N}{f_t}\right)^p \left(\frac{D}{l_0}\right)^u, \quad \Pi_2 = \left(\frac{\sigma_N}{f_t}\right)^q \left(\frac{D}{l_0}\right)^v \tag{B.1}$$

where exponents p, q, u, v are four unknown real constants.

If we let $\Pi_1 = 0$ correspond to $D \to 0$, then $F(0, \Pi_2) = 0$, which implies that Π_2 = constant or ${\sigma_N}^q D^v$ = constant for $D \to 0$, and so one has the case of no size effect; hence $v = 0$.

If we let $\Pi_2 = 0$ correspond to $D \to \infty$, then $F(\Pi_1, 0) = 0$, which implies that Π_1 = constant or ${\sigma_N}^p D^u$ = constant, or $\sigma_N \propto D^{-u/p}$ for $D \to \infty$, and so one must have the LEFM scaling; hence $u/p = 1/2$ or $u = p/2$.

To find p and q, we truncate the Taylor series expansion of F after the linear terms:

$$F(\Pi_1, \Pi_2) \approx F_0 + F_1\Pi_1 + F_2\Pi_2 = 0 \tag{B.2}$$

$$\text{or} \quad F_1(\sigma_N\sqrt{D}/f_t\sqrt{l_0})^p + F_2(\sigma_N/f_t)^q = -F_0 \tag{B.3}$$

where $F_1 = [\partial F/\partial \Pi_1]_0$ and $F_2 = [\partial F/\partial \Pi_2]_0$ (evaluated at $\Pi_1 = \Pi_2 = 0$) and $F_0 = F(0, 0)$ $(F_0, F_1, F_2 \neq 0)$.

In general, the last equation cannot be solved for σ_N explicitly, but it can for D:

$$D = l_0 {f_t}^2(-F_0/F_1)^{2/p} {\sigma_N}^{-2}\left[1 + (F_2/F_0 {f_t}^q){\sigma_N}^q\right]^{2/p} \tag{B.4}$$

This inverse relation may be compared to the inverse expansion of the large-size asymptotic expansion of the cohesive crack model, Eq. 3.32, which has the form (Bažant 2004*b*)

$$D = B_2 {\sigma_N}^{-2}(1 - C_2 {\sigma_N}^2 + \cdots) \tag{B.5}$$

for $\sigma_N \to 0$ (B_2, C_2 = positive constants). Evidently, matching of the first two terms of this expansion requires $p = q = 2$. Then Eq. B.3 can be solved for σ_N. This yields, and verifies, the size effect law in Eq. 1.12 (Fig. 1.9b), in which $Bf_t = f_t(-F_0F_2)^{-1/2}$ and

$D_0 = l_0 F_2/F_1$. For $D \to 0$, Eq. 1.12 has the approximation

$$\sigma_N \approx B f_t (1 - D/2D_0) \tag{B.6}$$

which verifies that the form of the second term of small-size expansion (Eq. 3.30) can be matched, too. So, Eq. 1.12 is the simplest formula for Type 2 size effect that can match all the four asymptotic terms, two for $D \to 0$ and two for $D \to \infty$.

The type 3 size effect formula ensues similarly. The result is

$$\sigma_N = \sigma_0 [(1 + D/\chi D_0)^{-1} + D/D_0]^{-1/2} \tag{B.7}$$

$\sigma_0, D_0, \chi =$ constants. To ensure monotonicity, $\chi \geq 1$ (Bažant 2001, 2005).

B.2 Second-Order Asymptotic Matching Derivation for Type 1 Size Effect Transition

For general dimensional analysis of the deterministic type 1 size effect, we need a slightly different procedure than for type 2. Consider the **deterministic** type 1 first (i.e., the Weibull modulus is infinite). Because, according to Eqs. 3.30 and 3.31, both asymptotes in the plot of $\log \sigma_N$ versus $\log D$ are horizontal, the size effect curve must have an inflexion point. This suggests the existence of what was called by Barenblatt the *intermediate asymptote* (Barenblatt 2003), which consists of some unknown power law. Because it separates the large-size and small-size asymptotics, we now try to match the large-size asymptotic terms alone, and choose the large-size asymptote to correspond to $\Pi_1 = 0$, i.e., to $F(0, \Pi_2) = 0$. This means that $\Pi_2 =$ constant for $D \to \infty$, and so $v = 0$ in Eq. B.1. The linear approximation (Eq. B.2) cannot be solved for σ_N in general, but it can for D:

$$l_0/D = (-F_0/F_1)^{-1/u} [(f_t/\sigma_N)^p + (F_2/F_0)(f_t/\sigma_N)^{p-q}]^{-1/u} \tag{B.8}$$

The large-size expansion may be generalized as

$$\sigma_N = (b_1 + r c_1 D^{-1} + \cdots)^{1/r} \tag{B.9}$$

without violating Eq. 3.31 ($r =$ arbitrary constant $\neq 0$). The inverse expansion is $1/D = (-b_1 + {\sigma_N}^r + \cdots)/r c_1$, and matching of Eq. B.8 obviously requires $u = -1$ and $p = q = -r$ in Eq. B.1. So $\Pi_1 = (f_t/\sigma_N)^r l_0/D$ and $\Pi_2 = (f_t/\sigma_N)^r$. Now Eq. B.2 can be solved for σ_N. This yields, and thus also verifies, Eq. 3.46, in which $D_b = F_1 l_0 / r F_2$ and $f_{r_\infty} = f_t(-F_2/F_0)^{1/r}$.

For the **statistical** case (i.e., a finite Weibull modulus), we again choose the large size asymptote to correspond to $\Pi_1 = 0$, $\Pi_2 =$ constant. At the large-size limit, one has $\sigma_N = C_1 D^{-n/m} \propto f_t (D/l_0)^{-v/q}$; so, $v/q = n/m$. Not to lose the deterministic limit, we keep $p = q = -r$ and $u = -1$. Equation B.2 can now be solved for σ_N, which leads to the *mean* type 1 energetic-statistical size effect (Fig. 3.15):

$$\sigma_N = f_{r_\infty} \left[(\lambda l_0/D)^{rn/m} + r\kappa(\lambda l_0/D) \right]^{1/r} \tag{B.10}$$

where $f_{r_\infty} = \lambda^{-n/m}(-F_2/F_0)^{1/r}$, $\kappa = \lambda^{rn/m-1}F_1/F_2r$, and r is parameter of the order of 1, sensitive to structure geometry.

Equation B.10, however, violates the small-size asymptotics in Eq. 3.30, and must therefore be modified further, but without affecting the first two large-size asymptotic terms. To bridge the small-size and intermediate asymptotes, we could engage in similar arguments as we did for bridging the large-size and intermediate asymptotes. Suffice to say, the complete law for the mean size effect of type 1, matching the small-size asymptotics in Eq. 3.30, is obtained by replacing $\lambda l_0/D$ in Eq. B.10 with θ:

$$\sigma_N = f_r^0 \left(\theta^{rn/m} + rs\kappa\theta\right)^{1/rs}, \quad \theta = (1 + D/s\eta l_0)^{-s}\lambda/\eta \tag{B.11}$$

Here η = positive constant of the order of 1 and s may be taken as 1. Note that if we let $\eta \ll D/l_0 \ll \lambda$, or $\lambda/\eta \to \infty$, Eq. B.11 converges to the intermediate asymptote, which is given by the power law $\sigma_N = s_0(\kappa r \eta_0 l_0)^{1/r}D^{-1/r}$ and coincides with the small-size asymptote of Eq. B.10, as well as with the large-size asymptote of $f_r^0\theta^{n/m}$.

The physical reason for the existence of an intermediate asymptote is that normally the size of averaging domain in nonlocal Weibull theory ($\approx$ FPZ width) is much smaller than the FPZ length (Bažant 2004*a*). Equation B.11 is supported by finite element simulations with nonlocal Weibull theory, as well as the available test data (Fig. 3.16).

However, test data for concrete and composites show that the D values for which the difference between Eq. B.10 and Eq. B.11 is significant are less than the material inhomogeneities. This means that Eq. B.10 should mostly suffice in practice.

Appendix C Proof of Small-Size Asymptotics of Cohesive Crack Model up to Second Order

While the large-size asymptotic properties of the quasibrittle size effect can be derived by means of the equivalent LEFM (Sec. 3.9), the small-size asymptotic properties may be determined from the cohesive crack model. The numerical solutions with the cohesive crack model, crack band model, and nonlocal model show that, for a vanishing structure size D, the nominal strength σ_N of a quasibrittle structure approaches a finite value, σ_N^0. Accordingly, the size effect plot of $\log \sigma_N$ versus $\log D$ must approach on the left a horizontal asymptote. What we want to determine is how precisely this horizontal asymptote is approached. In other words, we want to obtain the second term of the small-size asymptotic expansion of the size effect. Following the exposition in Bažant (2001), we will present it now.

To this end, consider the static boundary value problem of linear elasticity. Its definition in Cartesian coordinates x_i ($i = 1, 2, 3$) is as follows:

$$\sigma_{ij} = E_{ijkl}\tfrac{1}{2}(u_{k,l} + u_{l,k}), \qquad \sigma_{ij,j} + f_i = 0 \qquad (\text{in } \mathcal{V}) \tag{C.1}$$

$$n_j\sigma_{ij} = p_i \quad (\text{on } \Gamma_s), \qquad u_i = u_{i0} \quad (\text{on } \Gamma_d) \tag{C.2}$$

Here σ_{ij} = stress tensor components, $\frac{1}{2}(u_{i,j} + u_{j,i}) = \epsilon_{ij}$ = strain tensor components, E_{ijkl} = elastic moduli, f_i = body forces, p_i = surface tractions, prescribed on surface domain Γ_s, n_i = unit normal of the surface, and Γ_d is the surface domain where the displacements are prescribed. We consider geometrically similar structures of various sizes D and introduce the dimensionless coordinates and variables, labeled by an overbar:

$$\bar{x}_i = x_i/D, \qquad \bar{u}_i = u_i/D, \qquad \bar{\sigma}_{ij} = \sigma_{ij}/f_t \tag{C.3}$$

$$\bar{p}_i = p_i/\sigma_N, \quad \bar{f}_i = f_iD/\sigma_N, \quad \bar{E}_{ijkl} = E_{ijkl}/f_t, \quad \bar{u}_{i0} = u_{i0}/D \tag{C.4}$$

The load magnitude is assumed to be characterized by the nominal strength σ_N as a single parameter, and so $\bar{p}_i$ is a size-independent (or dimensionless) distribution of the surface tractions on Γ_s, and $\bar{f}_i$ is a size-independent distribution of body forces in volume $\bar{\mathcal{V}}$. The surface normals n_i at homologous points are independent of size D (and thus need no overbar) and f_t denotes the material tensile strength.

With the notation $\partial_i = \partial/\partial\bar{x}_i$ = partial derivatives with respect to the dimensionless coordinates, and $\partial/\partial x_i = (1/D)\partial_i$, we can transform the foregoing equations to the following dimensionless form:

$$\bar{\sigma}_{ij} = \bar{E}_{ijkl}\tfrac{1}{2}(\partial_k\bar{u}_l + \partial_l\bar{u}_k), \qquad \partial_j\bar{\sigma}_{ij} + \bar{f}_i\,\sigma_N/f_t = 0 \quad (\text{in } \bar{\mathcal{V}}) \tag{C.5}$$

$$n_j\bar{\sigma}_{ij} = \bar{p}_i\,\sigma_N/f_t \quad (\text{on } \bar{\Gamma}_s), \qquad \bar{u}_i = \bar{u}_{i0} \quad (\text{on } \bar{\Gamma}_d) \tag{C.6}$$

where $\bar{\mathcal{V}}$ is the domain of structure volume in the dimensionless coordinates, and $\bar{\Gamma}_s$ and $\bar{\Gamma}_d$ are the surface domains in dimensionless coordinates corresponding to Γ_s and Γ_d.

We can always position coordinates x_i such that the crack would lie in the plane (x_1, x_3) and that the tip of the cohesive crack (and not the notch tip) would be at $x_1 = 0$. For a small enough D, the crack-bridging stress $\sigma > 0$ along the whole crack length L, and if D is small enough and if the compression strength is unlimited, the cohesive crack at maximum load will occupy the entire cross section or, in the case of a notch, the entire ligament; then the dimensionless crack length $\bar{L} = L/D =$ constant (if the compression strength is limited and the cross section is for instance subjected to bending, L/D will not necessarily be size independent but we may assume it to be such, as an approximation for small D).

Equations C.5 and C.6 must be supplemented by two conditions for the cohesive crack:

1. The dimensionless total stress intensity factor $\bar{K}_t = K_t/\sigma_N\sqrt{D}$ produced jointly by the applied load and the stresses $\bar{\sigma} = \bar{\sigma}_{22}$ acting on the crack faces must vanish to ensure the finiteness of the crack-tip stresses, i.e.,

$$\bar{K}_t = 0 \tag{C.7}$$

2. The cohesive (crack-bridging) stresses σ must satisfy the softening law of the cohesive crack, i.e., the curve relating σ to the opening displacement $w = 2u_2$ on the crack plane. At the small-size limit, what matters is the initial part of the cohesive law (i.e., w is small). We will consider the cohesive law with an initial descent as

$$\sigma = f_t[1 - (w/w_0)^p] \qquad (\text{for } -L \le x_1 < 0, x_2 = 0) \tag{C.8}$$

where p, $w_0 =$ positive constants. Using the dimensionless variables corresponding to Eq. C.3, we may write the dimensionless form of the assumed softening law as follows:

$$\bar{\sigma} = 1 - (\bar{D}\bar{w})^p \qquad (\text{for } -\bar{L} \le \bar{x}_1 < 0, \bar{x}_2 = 0) \tag{C.9}$$

with $\bar{\sigma} = \sigma/f_t$, $\bar{w} = w/D$, $\bar{D} = D/w_0$.

Consider now the dependence of the solution on the structure size, D. We assume that the dimensionless displacements, stresses, and total stress intensity factor approach their limit for $\bar{D} \to 0$ as power functions of $\bar{D}$ with exponent p, and try to verify the correctness of this hypothesis. So, for small enough $\bar{D}$, we set

$$\sigma_N = \sigma_N^0 + \sigma_N'\bar{D}^p, \qquad \bar{\sigma}_{ij} = \bar{\sigma}_{ij}^0 + \bar{\sigma}_{ij}'\bar{D}^p, \qquad \bar{\sigma} = \bar{\sigma}^0 + \bar{\sigma}'\bar{D}^p \tag{C.10}$$

$$\bar{u}_i = \bar{u}_i^0 + \bar{u}_i'\bar{D}^p, \qquad \bar{w} = \bar{w}^0 + \bar{w}'\bar{D}^p, \qquad \bar{K}_t = \bar{K}_t^0 + \bar{K}_t'\bar{D}^p \tag{C.11}$$

in which $\sigma_N^0, \sigma_N', \sigma^0, \sigma', \sigma_{ij}^0, \ldots, K_t'$ are size independent. These expressions may now be substituted into Eqs. C.5 to C.9. The resulting equations must be satisfied for not one but all different small sizes $\bar{D}$. For $\bar{D} \to 0$, the dominant terms in these equations are those of the lowest powers of D, which are those with $\bar{D}^0$ and $\bar{D}^p$. By collecting the terms without $\bar{D}$ and those with $\bar{D}^p$, two independent sets of equations ensue. It so happens that each of these two sets defines a physically meaningful boundary value problem of elasticity for a body with a crack. This proves the validity of our hypothesis made in Eqs. C.10 and C.11.

Elasticity Problem I: By isolating the terms that do not contain $\bar{D}$ (i.e., contain $\bar{D}^0$), we get

$$\bar{K}_t^0 = 0, \qquad \bar{\sigma}^0 = 1 \qquad (\text{for } -\bar{L} \le \bar{x}_1 < 0, \bar{x}_2 = 0) \tag{C.12}$$

$$\bar{\sigma}_{ij}^0 = \bar{E}_{ijkl}\, \tfrac{1}{2}\left(\partial_l \bar{u}_k^0 + \partial_k \bar{u}_l^0\right), \qquad \partial_j \bar{\sigma}_{ij}^0 + \bar{f}_i\, \sigma_N^0/f_t = 0, \quad (\text{in } \bar{\mathcal{V}}) \tag{C.13}$$

$$n_j \bar{\sigma}_{ij}^0 = \bar{p}_i\, \sigma_N^0/f_t \quad (\text{on } \bar{\Gamma}_s), \qquad \bar{u}_i^0 = 0 \quad (\text{on } \bar{\Gamma}_d) \tag{C.14}$$

Elasticity Problem II: By isolating the terms that do contain $\bar{D}^p$, we get

$$\bar{K}_t' = 0, \qquad \bar{\sigma}' = -(\bar{w}^0)^p \qquad (\text{for } -\bar{L} \le \bar{x}_1 < 0, \bar{x}_2 = 0) \tag{C.15}$$

$$\bar{\sigma}_{ij}' = \bar{E}_{ijkl}\, \tfrac{1}{2}\left(\partial_l \bar{u}_k' + \partial_k \bar{u}_l'\right), \qquad \partial_j \bar{\sigma}_{ij}' + \bar{f}_i\, \sigma_N'/f_t = 0, \quad (\text{in } \bar{\mathcal{V}}) \tag{C.16}$$

$$n_j \bar{\sigma}_{ij}' = \bar{p}_i\, \sigma_N'/f_t \quad (\text{on } \bar{\Gamma}_s), \qquad \bar{u}_i' = 0 \quad (\text{on } \bar{\Gamma}_d) \tag{C.17}$$

Note that parameter $\bar{w}'$ does not appear in this problem.

The role of stresses and displacements is played by $\bar{\sigma}_{ij}^0$ and $\bar{u}_i^0$ in problem I, and by $\bar{\sigma}_{ij}'$ and $\bar{u}_i'$ in problem II. In problem I, the crack faces are subjected to fixed uniform tractions whose value is 1. In problem II, in which σ' plays the role of the cohesive stress, the crack faces are subjected to tractions $-(\bar{w}^0)^p$ that vary along the crack faces but can be determined in advance from the $\bar{w}^0$ values obtained in solving problem I.

The fact that isolation of the terms with the zeroth power and the pth power of D happens to yield two separate boundary value problems of elasticity is essential for our objective. The rest of the argument is easy and may be stated as follows.

The magnitude of the loads (surface tractions and body forces) is proportional to σ_N^0 in problem I, and to σ_N' in problem II. These are elasticity problems, which are known to have a unique solution. If σ_N^0 were zero, i.e., if the applied load in problem I vanished, the crack face tractions equal to 1 would cause K_t^0 to be nonzero, in violation of Eq. C.12. Likewise, if σ_N' were zero, i.e., if the applied load in problem II vanished, the nonuniform crack face tractions $-(\bar{w}^0)^p$ in problem II would cause K_t' to be nonzero, in violation of Eq. C.15. If the loads for problems I and II were infinite, then K_t^0 or K_t' would be infinite as well, which would again violate Eq. C.12 or Eq. C.15. Therefore, the only possibility left is that both σ_N^0 and σ_N' are finite. So we have proven the following.

If the cohesive law of the cohesive crack model has a finite tensile strength and begins its initial postpeak descent as w^p, then the size effect law for nominal strength approaches, for $D \to 0$, a finite value and does so as D^p.

The case considered in all the work (Eq. 3.30) so far is $p = 1$. Thus, according to Eq. C.10, the size effect law must begin near zero size D as a linear function of D, which shows up as exponential in the logarithmic plot, because of the approximation $\ln \sigma_N - \ln \sigma_N^0 = \ln(1 + \sigma_N' \bar{D}/\sigma_N^0) \approx (\sigma_N'/\sigma_N^0)\, e^{\ln \bar{D}}]$.

The case $p > 1$ would mean that the softening law would begin its descent from a horizontal initial tangent, which is what might be assumed for ductile fracture of plastic

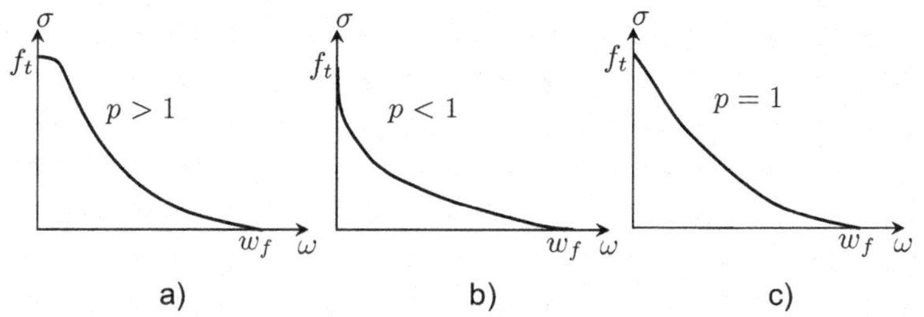

Figure C.1. Cohesive laws with different initial descending behaviors. (a) $p > 1$, (b) $p < 1$, and (c) $p = 1$.

materials [a very short horizontal segment might as well occur for quasibrittle fracture (Fig. C.1a), although no data exist to answer this question]. The case $p < 1$ would mean that the cohesive law would begin its descent with a vertical tangent (Fig. C.1b), which would be an unrealistic super-brittle behavior. Many quasibrittle materials such as concrete correspond to the case of $p = 1$ (Fig. C.1c), which agrees with both type 1 and type 2 size effect equations (Eqs. 3.37 and 12.1). Indeed, for small D we have

$$\sigma_N \propto \left(1 + \frac{rD_b}{l_p + D}\right)^{1/r} \approx A[1 - BD] \qquad \text{Type 1} \qquad (C.18)$$

$$\sigma_N \propto (1 + D/D_0)^{-1/2} \approx 1 - D/2D_0 \qquad \text{Type 2} \qquad (C.19)$$

where $D_0, D_b, l_p, r =$ constants, $A = (1 + rD_b/l_p)^{1/r}$, and $B = D_b l_p^{-2}(1 + D_b/l_p)^{-1}$.

It must be admitted that our imposition of the small-size asymptotic properties of the cohesive crack model on the size effect law might be regarded as fiction since, for cross-sectional thicknesses less than several aggregate sizes, the material is not a continuum. For this reason, it may well be considered admissible to have an infinite σ_N for $D \to 0$. This happens to be the property of the Hall–Petch formula (Petch 1954), which is widely used for the dependence of yield strength σ_Y on the crystal size d_c in metals and reads

$$\sigma_Y = \sigma_0 + k d_c^{-1/2} \qquad (C.20)$$

where $k, \sigma_0 =$ constants and $\sigma_0 =$ yield strength for large d_c.

Yet imposition of two-sided asymptotic properties is advantageous from the viewpoint of *asymptotic matching*, i.e., approximations that have two-sided exact asymptotic support [popularly, "interpolation between opposite infinities," Bender & Orszag (1978)]. In the spirit of asymptotic matching, an approximation for the middle range will be better if it satisfies both the small-size and large-size asymptotic properties of the theory applicable in that range (i.e., of the cohesive crack model).

Without a realistic probabilistic analysis, the mechanics of structural failure prevention is a fiction.

References

Abadeer, W. W., Bagramian, A., Conkie, D. M., Griffin, C. W., Langlois, E., Lloyd, B. F., Mallette, R. P., Massucco, J. E., Mckenna, J. M., Mittl, S. W., & Noel, P. H. (1999). Key measurements of ultrathin gate dielectric reliability and in-line monitoring. *IBM J. Res. Dev.* **43**(3), 407–416.

Abraham, F. F., Broughton, J. Q., Bernstein, N., & Kaxiras, E. (1998). Spanning the continuum to quantum length scales in a dynamical simulation of brittle fracture. *Europhys. Lett.* **44**(6), 783–787.

Aifantis, E. C. (1984). On the microstructural origin of certain inelastic models. *J. Eng. Mater. Technol. ASME* **106**, 326–330.

Akisanya, A. R., & Fleck, N. A. (1997). Interfacial cracking from the free-edge of a long bi-material strip. *Int. J. Solids Struct.* **34**(13), 1645–1665.

American Association of State Highway and Transportation Officials (2010). *Standard method of test for performance graded asphalt binder, AASHTO M320-10-UL*, Washington, D.C.

American Concrete Institute Committee 318 (2011). *ACI318-11 Building Code Requirement for Structural Concrete and Commentary*, Farmington Hills, MI.

American Institute of Steel Construction (2011). *Steel Construction Manual.*

Andersons, J., Hojo, M., & Ochiai, S. (2004). Empirical model for stress ratio effect on fatigue delamination growth rate in composite laminates. *Int. J. Fatigue* **26**, 597–604.

Ang, A. H. S., & Tang, W. H. (1984). *Probability Concepts in Engineering Planning and Design.* Vol II: *Decision, Risk and Reliability*. John Wiley & Sons, New York.

Apte, P. P., & Saraswat, K. C. (1994). Correlation of trap generation to charge-to-breakdown (q_{bd}): A physical-damage model of dielectric breakdown. *IEEE Trans. Electron Dev.* **41**, 1595–1602.

Atkinson, B. K. (1984). Subcritical crack growth in geological materials. *J. Geophys. Res.* **89**(B6), 4077–4114.

Award, M. E., & Hilsdorf, H. K. (1972). Strength and deformation characteristics of plain concrete subjected to high repeated and sustained loads. In *Proceedings of Abeles Symposium: Fatigue of Concrete*, Hollywood, FL and Atlantic City, NJ, pp. 1–13.

Ayyub, B. M., & Haldar, A. (1984). Practical structural reliability techniques. *J. Struct. Eng.-ASCE* **110**(8), 1707–1724.

Aziz, M. J., Sabin, P. C., & Lu, G. Q. (1991). The activation strain tensor: Nonhydrostatic stress effects on crystal growth kinetics. *Phys. Rev. B* **41**, 9812–9816.

Bao, G., & Suo, Z. (1992). Remarks on crack-bridging concepts. *Appl. Mech. Rev.* **45**, 355–366.

Barenblatt, G. I. (1959). The formation of equilibrium cracks during brittle fracture, general ideas and hypothesis, axially symmetric cracks. *Prikl. Mat. Mech.* **23**(3), 434–444.

Barenblatt, G. I. (1962). The mathematical theory of equilibrium cracks in brittle fracture. *Adv. Appl. Mech.* **7**, 55–129.

Barenblatt, G. I. (1979). *Similarity, Self-Similarity and Intermediate Asymptotics*. Consultants Bureau, New York.

Barenblatt, G. I. (1987). *Dimensional Analysis*. Gordon and Breach, New York.

Barenblatt, G. I. (1996). *Scaling, Self-Similarity, and Intermediate Asymptotics*. Cambridge University Press, Cambridge.

Barenblatt, G. I. (2003). *Scaling*. Cambridge University Press, Cambridge.

Barenblatt, G. I. (2014). *Flow, Deformation and Fracture*. Cambridge University Press, Cambridge, U.K.

Barenblatt, G. I., & Botvina, L. R. (1981). Incomplete self-similarity of fatigue in the linear range of crack growth. *Fatigue Fract. Eng. Mater. Struct.* **3**, 193–212.

Barsoum, R. S. (2003). The best of both worlds: Hybrid ship hulls use composites & steel. *AMPTIAC Q.* **7**(3), 55–61.

Bartle, A. (1985). Four major dam failures re-examined. *Int. Water Power Dam Constr.* **37**(11), 33–36, 41–46.

Basquin, O. H. (1910). The exponential law of endurance tests. *Proc. Am. Soc. Test. Mater. ASTE* **10**, 625–630.

Batdorf, S. B. (1982). Tensile strength of unidirectionally reinforced composites. *J. Reinf. Plast. Compos.* **1**, 153–164.

Bažant, Z. P. (1976). Instability, ductility, and size effect in strain-softening concrete. *J. Eng. Mech. Div. ASCE* **102, EM2**, 331–344.

Bažant, Z. P. (1982). Crack band model for fracture of geomaterials. In Z. Eisenstein (ed.), Proceedings of the 4th International Conference on Numerical Methods in Geomechanics, Vol. 3. Edmonton, Alberta, pp. 1137–1152.

Bažant, Z. P. (1984*a*). Imbricate continuum and progressive fracturing of concrete and geomaterials. *Meccanica* **19**, 86–93.

Bažant, Z. P. (1984*b*). Size effect in blunt fracture: Concrete, rock, metal. *J. Eng. Mech. ASCE* **110**(4), 518–535.

Bažant, Z. P. (1992*a*). Large-scale thermal bending fracture of sea ice plates. *J. Geophys. Res. (Oceans)* **97**(C11), 17739–17751.

Bažant, Z. P., ed. (1992*b*). *Fracture Mechanics of Concrete Structures: Proceedings of the First International Conference*, Elsevier, London.

Bažant, Z. P. (1993). Scaling laws in mechanics of fracture. *J. Eng. Mech. ASCE* **119**(9), 1828–1844.

Bažant, Z. P. (1996). Size effect aspects of measurement of fracture characteristics of quasibrittle material. *Adv. Cem. Based Mater.* **4**(3/4), 128–137.

Bažant, Z. P. (1997). Scaling of quasibrittle fracture: Asymptotic analysis. *Int. J. Fract.* **83**(1), 19–40.

Bažant, Z. P. (2001). Size effects in quasibrittle fracture: Apercu of recent results. In R. de Borst, J. Mazars, G. Pijaudier-Cabot, & J. G. M. van Mier (eds.), *Fracture Mechanics of Concrete Structures* (*Proceedings of FraMCos-4 International Conference*). A. A. Balkema, Lisse, pp. 651–658.

Bažant, Z. P. (2002*a*). Scaling of sea ice fracture – Part I: Vertical penetration. *J. Appl. Mech. ASME* **69**, 11–18.

Bažant, Z. P. (2002*b*). Scaling of sea ice fracture – Part II: Horizontal load from moving ice. *J. Appl. Mech. ASME* **69**, 19–24.

Bažant, Z. P. (2004*a*). Probability distribution of energetic-statistical size effect in quasibrittle fracture. *Prob. Eng. Mech.* **19**(4), 307–319.

Bažant, Z. P. (2004*b*). Scaling theory of quaisbrittle structural failure. *Proc. Nat. Acad. Sci. USA* **101**(37), 13400–13407.

Bažant, Z. P. (2005). *Scaling of Structural Strength*. 2nd edition. Elsevier, London.

Bažant, Z. P., & Becq-Giraudon, E. (2002). Statistical prediction of fracture parameters of concrete and implications for choice of testing standard. *Cem. Concr. Res.* **32**(4), 529–556.

Bažant, Z. P., Belytschko, T., & Chang, T.-P. (1984). Continuum model for strain softening. *J. Eng. Mech. ASCE* **110**(12), 1666–1692.

Bažant, Z. P., & Cedolin, L. (1979). Blunt crack band propagation in finite element analysis. *J. Eng. Mech. Div. ASCE* **105**, 297–315.

Bažant, Z. P., & Cedolin, L. (1980). Fracture mechanics of reinforced concrete. *J. Eng. Mech. Div. ASCE* **106**, 1257–1306.

Bažant, Z. P., & Cedolin, L. (1991). *Stability of Structures: Elastic, Inelastic, Fracture and Damage Theories*. Oxford University Press, New York.

Bažant, Z. P., & Chen, E.-P. (1997). Scaling of structural failure. *Appl. Mech. Rev. ASME* **50**(10), 593–627.

Bažant, Z. P., Daniel, I. M., & Li, Z. (1996). Size effect and fracture characteristics of composite laminates. *J. Eng. Mater. Technol, ASME* **118**(3), 317–324.

Bažant, Z. P., & Frangopol, D. M. (2002). Size effect hidden in excessive load factor. *J. Struct. Eng. ASCE* **128**(1), 80–86.

Bažant, Z. P., Gettu, R., & Kazemi, M. T. (1991). Identification of nonlinear fracture properties from size effect tests and structural analysis based on geometry-dependent R-curves. *Int. J. Rock Mech. Miner. Sci.* **28**(1), 43–51.

Bažant, Z. P., & Hubler, M. H. (2014). Theory of cyclic creep of concrete based on paris law for fatigue growth of subcritical microcracks. *J. Mech. Phys. Solids* **63**, 187–200.

Bažant, Z. P., & Jirásek, M. (2002). Nonlocal integral formulations of plasticity and damage: Survey of progress. *J. Eng. Mech. ASCE* **128**(11), 1119–1149.

Bažant, Z. P., & Kazemi, M. T. (1990*a*). Determination of fracture energy, process zone length and brittleness number from size effect, with application to rock and concrete. *Int. J. Fract.* **44**, 111–131.

Bažant, Z. P., & Kazemi, M. T. (1990*b*). Size effect in fracture of ceramics and its use to determine fracture energy and effective process zone length. *J. Amer. Ceramic Soc.* **73**(7), 1841–1853.

Bažant, Z. P., & Kazemi, M. T. (1991). Size effect on diagonal shear failure of beams without stirrups. *ACI Struct. J.* **88**, 268–276.

Bažant, Z. P., & Kim, J.-J. (1998*a*). Size effect in penetration of sea ice plate with part-through cracks. I. Theory. *J. Eng. Mech. ASCE* **124**(12), 1310–1315.

Bažant, Z. P., & Kim, J.-J. (1998*b*). Size effect in penetration of sea ice plate with part-through cracks. II. Results. *J. Eng. Mech. ASCE* **124**(12), 1316–1324.

Bažant, Z. P., Kim, J.-J. II., Daniel, I. M., Becq-Giraudon, E., & Zi, G. (1999). Size effect on compression strength of fiber composites failing by kink band propagation. *Int. J. Fract.* **95**, 103–141.

Bažant, Z. P., Kim, K. T., & Yu, Q. (2013). Non-uniqueness of cohesive-crack stress-separation law of human and bovine bones and remedy by size effect tests. *Int. J. Fract.* **181**, 67–81.

Bažant, Z. P., & Le, J.-L. (2009). Nano-mechanics based modeling of lifetime distribution of quasibrittle structures. *J. Eng. Fail. Anal.* **16**, 2521–2529.

Bažant, Z. P., Le, J.-L., & Bazant, M. Z. (2009). Scaling of strength and lifetime distributions of quasibrittle structures based on atomistic fracture mechanics. *Proc. Natl. Acad. Sci. USA* **106**, 11484–11489.

Bažant, Z. P., Le, J.-L., & Hoover, C. G. (2010). Nonlocal boundary layer model: Overcoming boundary condition problems in strength statistics and fracture analysis of quasibrittle materials. In *Fracture Mechanics of Concrete and Concrete Structures – Recent Advances in Fracture Mechanics of Concrete*. Jeju, Korea, pp. 135–143.

Bažant, Z. P., & Li, Y.-N. (1994). Penetration fracture of sea ice plate: Simplified analysis and size effect. *J. Eng. Mech. ASCE* **120**(6), 1304–1321.

Bažant, Z. P., & Li, Y.-N. (1995*a*). Penetration fracture of sea ice plate. *Int. J. Solids Struct.* **32**(3/4), 303–313.

Bažant, Z. P., & Li, Y.-N. (1997). Cohesive crack with rate-dependent opening and viscoelasticity: I. mathematical model and scaling. *Int. J. Fract.* **86**(3), 247–265.

Bažant, Z. P., & Li, Z. (1995*b*). Modulus of rupture: Size effect due to fracture initiation in boundary layer. *J. Struct. Eng. ASCE* **121**(4), 739–746.

Bažant, Z. P., & Li, Z. (1996). Zero-brittleness size-effect method for one-size fracture test of concrete. *J. Eng. Mech. ASCE* **122**(5), 458–468.

Bažant, Z. P., & Lin, F.-B. (1988*a*). Nonlocal smeared cracking model for concrete fracture. *J. Struct. Eng. ASCE* **114**(11), 2493–2510.

Bažant, Z. P., & Lin, F.-B. (1988*b*). Nonlocal yield limit degradation. *Int. J. Numer. Methods Eng.* **26**, 1805–1823.

Bažant, Z. P., Lin, F.-B., & Lippmann, H. (1993). Fracture energy release and size effect in borehole breakout. *Int. J. Num. Anal. Methods Geomech.* **17**, 1–14.

Bažant, Z. P., & Novák, D. (2000*a*). Energetic-statistical size effect in quasibrittle failure at crack initiation. *ACI Mater. J.* **97**(3), 381–392.

Bažant, Z. P., & Novák, D. (2000*b*). Probabilistic nonlocal theory for quasibrittle fracture initiation and size effect. I. theory. *J. Eng. Mech. ASCE* **126**(2), 166–174.

Bažant, Z. P., & Novák, D. (2000*c*). Probabilistic nonlocal theory for quasibrittle fracture initiation and size effect. II. application. *J. Eng. Mech. ASCE* **126**(2), 175–185.

Bažant, Z. P., & Oh, B.-H. (1983). Crack band theory for fracture of concrete. *Mater. Struct.* **16**, 155–177.

Bažant, Z. P., & Ožbolt, J. (1990). Nonlocal microplane model for fracture, damage, and size effect in structures. *J. Eng. Mech. ASCE* **116**(11), 2485–2505.

Bažant, Z. P., & Pang, S. D. (2006). Mechanics based statistics of failure risk of quasibrittle structures and size effect on safety factors. *Proc. Natl. Acad. Sci. USA* **103**, 9434–9439.

Bažant, Z. P., & Pang, S. D. (2007). Activation energy based extreme value statistics and size effect in brittle and quasibrittle fracture. *J. Mech. Phys. Solids* **55**, 91–134.

Bažant, Z. P., Pang, S.-D., Vořechovský, M., Novák, D., & Pukl, R. (2004). Statistical size effect in quasibrittle materials: Computation and extreme value theory. In V. C. Li, K. Leung, K. J. Willam, & S. Billington (eds.), *Fracture Mechanics of Concrete Structures (Proceedings of FraMCoS-5, 5th International Conference)*, pp. 153–162.

Bažant, Z. P., & Pfeiffer, P. A. (1987). Determination of fracture energy from size effect and brittleness number. *ACI Mater. J.* **84**, 463–480.

Bažant, Z. P., & Pijaudier-Cabot, G. (1988). Nonlocal continuum damage, localization instability and convergence. *J. Appl. Mech. ASME* **55**, 287–293.

Bažant, Z. P., & Pijaudier-Cabot, G. (1989). Measurement of characteristic length of nonlocal continuum. *J. Eng. Mech. ASCE* **115**(4), 755–767.

Bažant, Z. P., & Planas, J. (1998). *Fracture and Size Effect in Concrete and Other Quasibrittle Materials*. CRC Press, Boca Raton.

Bažant, Z. P., & Prat, P. C. (1988). Effect of temperature and humidity on fracture energy of concrete. *ACI Mater. J.* **85-M32**, 262–271.

Bažant, Z. P., & Schell, W. F. (1993). Fatigue fracture of high-strength concrete and size effect. *ACI Mater. J.* **90**(5), 472–478.

Bažant, Z. P., Tabbara, M. R., Kazemi, M. T., & Pijaudier-Cabot, G. (1990). Random particle model for fracture of aggregate or fiber composites. *J. Eng. Mech. ASCE* **116**(8), 1686–1705.

Bažant, Z. P., Vořechovský, M., & Novák, D. (2007). Asymptotic prediction of energetic-statistical size effect from deterministic finite element solutions. *J. Eng. Mech. ASCE* **128**, 153–162.

Bažant, Z. P., & Xi, Y. (1991). Statistical size effect in quasi-brittle structures: II. Nonlocal theory. *J. Eng. Mech. ASCE* **117**(7), 2623–2640.

Bažant, Z. P., & Xu, K. (1991). Size effect in fatigue fracture of concrete. *ACI Mater. J.* **88**(4), 390–399.

Bažant, Z. P., & Yavari, A. (2005). Is the cause of size effect on structural strength fractal or energetic-statistical? *Eng. Fract. Mech.* **72**, 1–31.

Bažant, Z. P., & Yavari, A. (2007). Response to A. Carpinteri, B. Chiaia, P. Cornetti and S. Puzzi's comments on "Is the cause of size effect on structural strength fractal or energetic-statistical?" *Eng. Fract. Mech.* **74**, 2897–2910.

Bažant, Z. P., & Yu, Q. (2004). Size effect in concrete specimens and structures: New problems and progress. In V. C. Li, K. Y. Leung, K. J. Willam, & S. L. Billington (eds.), *Fracture Mechanics of Concrete Structures, Proceedings of FraMCoS-5, 5th International Conference on Fracture Mechanics of Concrete and Concrete Structures*, pp. 153–162.

Bažant, Z. P., & Yu, Q. (2006). Reliability, brittleness and fringe formulas in concrete design codes. *J. Struct. Eng. ASCE* **132**(1), 3–12.

Bažant, Z. P., & Yu, Q. (2009). Universal size effect law and effect of crack depth on quasi-brittle structure strength. *J. Eng. Mech. ASCE* **135**(2), 78–84.

Bažant, Z. P., & Yu, Q. (2011). Size effect testing of cohesive fracture parameters and non-uniqueness of work-of-fracture method. *J. Eng. Mech. ASCE* **137**(8), 580–588.

Bažant, Z. P., Zhou, Y., Daniel, I. M., Caner, F. C., & Yu, Q. (2006). Size effect on strength of laminate-foam sandwich plates. *J. Eng. Mater. Tech. ASME.* **128**(3), 366–374.

Bažant, Z. P., Zhou, Y., Novák, D., & Daniel, I. M. (2004). Size effect on flexural strength of fiber-composite laminate. *J. Eng. Mater. Tech. ASME* **126**(1), 29–37.

Bažant, Z. P., Zhou, Y., Zi, G., & Daniel, I. M. (2003). Size effect and asymptotic matching analysis of fracture of closed-cell polymeric foam. *Int. J. Solids Struct.* **40**, 7197–7217.

Bažant, Z. P., Zi, G., & McClung, D. (2003). Size effect law and fracture mechanics of the triggering of dry snow slab avalanches. *J. Geophy. Res.* **108**(B2), 2119–2129.

Bazant, M. Z. (2000). The largest cluster in subcritical percolation. *Phys. Rev. E* **62**, 1160–1669.

Bazant, M. Z. (2002). Stochastic renormalization group in percolation. *Physica A* **316**, 451–477.

Bazant, M. Z., Choi, J., & Davidovitch, B. (2003). Dynamics of conformal maps for a class of non-Laplacian growth phenomena. *Phys. Rev. Lett.* **91**, 045503.

Beale, P. D., & Duxbury, P. M. (1988). Theory of dielectric breakdown in metal-loaded dielectrics. *Phys. Rev. B* **40**, 4641.

Becq-Giraudon, E. (2000). *Size Effect on Fracture and Ductility of Concrete and Fiber Composites*. PhD thesis, Northwestern University.

Bender, M. C., & Orszag, S. A. (1978). *Advanced Mathematical Methods for Scientists and Engineers*. McGraw-Hill, New York.

Benjamin, J. R., & Cornell, C. A. (1970). *Probability, Statistics, and Decisions for Civil Engineers*. McGraw-Hill, New York.

Beremin, F. M. (1983). A local criterion for cleavage fracture of a nuclear pressure vessel steel. *Metall. Trans.* **114A**, 2277–2287.

Bigleya, R. F., Gibelingb, J. C., Stoverc, S. M., Hazelwooda, S. J., Fyhriea, D. P., & Martin, R. B. (2007). Volume effects on fatigue life of equine cortical bone. *J. Biomech.* **40**, 3548–3554.

Bogy, D. B. (1971). Two edge-bonded elastic wedges of different materials and wedge angles under surface tractions. *J. Appl. Mech. ASME* **38**, 377–385.

Bolander, J. E., Hong, G. S., & Yoshitake, K. (2000). Structural concrete analysis using rigid-body-spring networks. *J. Comput. Aided Civil Infrastruct. Eng.* **15**, 120–133.

Bolander, J. E., & Saito, S. (1998). Fracture analysis using spring network with random geometry. *Eng. Fract. Mech.* **61**(5–6), 569–591.

Bolotin, V. V. (1969). *Statistical Methods in Structural Mechanics*. Holden-Day, San Francisco.

Borden, M. J., Hughes, T. J. R., Landis, C. M., & Verhoosel, C. V. (2014). A higher-order phase-field model for brittle fracture: Formulation and analysis within the isogeometric analysis framework. *Comp. Methods Appl. Mech. Eng.* **273**, 100–118.

Borden, M. J., Verhoosel, C. V., Scott, M. A., Hughes, T. J. R., & Landis, C. M. (2012). A phase-field description of dynamic brittle fracture. *Comp. Methods Appl. Mech. Eng.* **217-220**, 77–95.

Borino, G., Failla, B., & Parrinello, F. (2003). A symmetric nonlocal damage theory. *Int. J. Solids Struct.* **40**, 3621–3645.

Bouchaud, J.-P., & Potters, M. (2000). *Theory of Financial Risks: From Statistical Physics to Risk Management*. Cambridge University Press, Cambridge.

Bowman, D. R., & Stroud, D. (1989). Model for dielectric breakdown in metal-insulator composites. *Phys. Rev. B* **40**, 4641.

Breysse, D., & Fokwa, D. (1992). Influence of disorder of the fracture process of mortar, In *Proceedings of the International Conference, Fracture Mechanics of Concrete Structures*. London, pp. 536–541.

Broberg, K. B. (1999). *Cracks and Fracture*. Academic Press.

Broughton, J. Q., Abraham, F. F., Bernstein, N., & Kaxiras, E. (1999). Concurrent coupling of length scales: Methodology and application. *Phys. Rev. B* **60**, 2391–2403.

Buckingham, E. (1914). On physically linear systems; illustration of the use of dimensional equations. *Phys. Rev. Ser. 2* **IV**(4), 345–376.

Buckingham, E. (1915). Model experiments and the form of empirical equations. *Trans. ASME* **37**, 263–296.

Bulmer, M. G. (1979). *Principles of Statistics*. Dover, New York.

Caner, F. C., & Bažant, Z. P. (2013). Microplane model M7 for plain concrete: I. Formulation. *J. Eng. Mech., ASCE* **139**(12), 1712–1723.

Caner, F. C., Bažant, Z. P., Hoover, C. G., Waas, A. M., & Shahwan, K. (2011). Microplane model for fracturing damage of triaxially braided fiber-polymer composites. *J. Eng. Mater. Tech. ASME* **133**, 021024–1 – 021024–12.

Cannone Falchetto, A., Le, J.-L., Turos, M. I., & Marasteanu, M. O. (2014). Indirect determination of size effect on strength of asphalt mixture at low temperatures. *Mater. Struct.* **47**(1–2), 157–169.

Carmeliet, J., & de Borst, R. (1995). Stochastic approaches for damage evolution in standard and non-standard continua. *Int. J. Solids Struct.* **32**, 1149–1160.

Carpenter, W. C. (1984). Mode I and mode II stress intensities for plates with cracks of finite opening. *Int. J. Fract.* **26**, 201–214.

Carpinteri, A. (1986). *Mechanical Damage and Crack Growth in Concrete*. Martinus Nijhoff–Kluwer, Dordrecht Boston.

Carpinteri, A. (1987). Stress-singularity and generalized fracture toughness at the vertex of reentrant corners. *Eng. Fract. Mech.* **26**, 143–155.

Carpinteri, A. (1989). Decrease of apprent tensile and bending strength with specimen size: Two different explanations based on fracture mechanics. *Int. J. Solids Struct.* **25**, 407–429.

Carpinteri, A. (1994). Fractal nature of materials microstructure and size effects on apparent material properties. *Mech. Mater.* **18**, 89–101.

Carpinteri, A., Chiaia, B., & Ferro, G. (1995*a*). Multifractal scaling law: An extensive application to nominal strength size effect of concrete structures. Technical Report 50, Atti del Dipartimento di Ingegneria Strutturale, Politecnico de Torino.

Carpinteri, A., Chiaia, B., & Ferro, G. (1995*b*). Size effect on nominal tensile strength of concrete structures: Multifractality of material ligament and dimensional transition from order to disorder. *Mater. Struct.* **28**, 311–317.

Castillo, E. (1988). *Extreme Value Theory in Engineering*. Academic Press, San Diego.

Červenka, J. (1998). Applied brittle analysis of concrete structures. In H. Mihashi & K. Rokugo (eds.), *3rd International Conference on Fracture Mechanics of Concrete Structures*. Aedificatio Publishers, Freiburg, Germany, pp. 1–15.

Červenka, J., Bažant, Z. P., & Wierer, M. (2005). Equivalent localization element for crack band approach to mesh-sensitivity in microplane model. *Int. J. Numer. Methods Eng.* **62**, 700–726.

Chakraborti, B. K., & Benguigui, L. G. (1997). *Statistical Physics of Fracture and Breakdown in Disordered Solids*. Clarendon Press, Oxford.

Charles, R. J. (1958*a*). Static fatigue of glass I. *J. Appl. Phys.* **29**(11), 1549–1553.

Charles, R. J. (1958*b*). Static fatigue of glass II. *J. Appl. Phys.* **29**(11), 1554–1560.

Chatterjee, S., Kuo, Y., Lu, J., Tewg, J.-Y., & Majhi, P. (2006). Electrical reliability aspects of HfO_2 high-k gate dieletrics with TaN metal gate electrodes under constant voltage stress. *Microelectron. Reliabil.* **46**, 69–76.

Chiao, C. C., Sherry, R. J., & Hetherington, N. W. (1977). Experimental verification of an accelerated test for predicting the lifetime of organic fiber composites. *J. Comp. Mater.* **11**, 79–91.

Ciavarella, M., Paggi, M., & Carpinteri, A. (2008). One, no one, and one hundred thousand crack propagation laws: A generalized Barenblatt and Botvina dimensional analysis approach to fatigue crack growth. *J. Mech. Phys. Solids* **56**, 3461–3432.

Coffin, L. F. (1962). Low cycle fatigue–a review. *Appl. Mater. Res.* **1**(3), 129–141.

Coleman, B. D. (1958*a*). On the strength of classical fibers and fiber bundles. *J. Mech. Phys. Solids* **7**, 60–70.

Coleman, B. D. (1958*b*). The statistics and time dependent of mechanical breakdown in fibers. *J. Appl. Phys.* **29**(6), 968–983.

Coleman, T. F., & Li., Y. (1994). On the convergence of reflective Newton methods for large-scale nonlinear minimization subject to bounds. *Math. Program.* **67**(2), 189–224.

Coleman, T. F., & Li., Y. (1996). An interior, trust region approach for nonlinear minimization subject to bounds. *SIAM J. Optim.* **6**, 418–445.

Cornell, C. A. (1969). A probability-based structural code. *J. Am. Concr. Inst.* **66**(12), 974–985.

Cundall, P. A. (1971). A computer model for simulating progressive large scale movements in blocky rock systems. In *Proceedings of the International Symposium Rock Fracture*, Vol. 1, Nancy, France, paper no. II–8.

Cundall, P. A., & Strack, O. (1979). A discrete numerical model for granular assemblies. *Geotechnique* **29**(1), 47–65.

Cusatis, G., Bažant, Z. P., & Cedolin, L. (2003*a*). Confinement-shear lattice model for concrete damage in tension and compression: I. Theory. *J. Eng. Mech. ASCE* **129**(12), 1439–1448.

Cusatis, G., Bažant, Z. P., & Cedolin, L. (2003*b*). Confinement-shear lattice model for concrete damage in tension and compression: II. Computation and validation. *J. Eng. Mech. ASCE* **129**(12), 1449–1458.

Cusatis, G., Pelessone, D., & Mencarelli, A. (2011). Lattice discrete particle model (LDPM) for failure behavior of concrete. I. Theory. *Cem. Concr. Comp.* **22**, 881–890.

Cusatis, G., & Schauffert, E. A. (2009). Cohesive crack analysis of size effect. *Eng. Fract. Mech.* **76**, 2163–2173.

Daniels, H. E. (1945). The statistical theory of the strength of bundles and threads. *Proc. R. Soc. London A* **183**, 405–435.

Danzer, R., Lube, T., Supancic, P., & Damani, R. (2008). Fracture of ceramics. *Adv. Eng. Mater.* **10**(4), 275–298.

Danzer, R., Supancic, P., Pascual, J., & Lube, T. (2007). Fractue statistics of ceramics – Weibull statistics and deviations from Weibull statistics. *Eng. Fract. Mech.* **74**, 2919–2932.

da Vinci, L. (1500s). *The Notebook of Leonardo da Vinci (1945)*, Edward McCurdy, London (p. 546); and *Les Manuscrits de Léonard de Vinci*, transl. in French by C. Ravaisson-Mollien, Institut de France (1881-91), Vol. 3.

Degraeve, R., Groeseneken, G., Bellens, R., Ogier, J. L., Depas, M., Roussel, P. J., & Maes, H. E. (1998). New insights in the relation between electron trap generation and the statistical properties of oxide breakdown. *IEEE Trans. Electron Dev.* **45**(4), 904–911.

De Moivre, A. (1733). Paper reproduced in D. E. Smith: *Source Book in Mathematics*. McGraw-Hill, New York, 1929.

Dempsey, J. P., Adamson, R. M., & Mulmule, S. V. (1995). Large-scale in-situ fracture of ice. In F. H. Wittmann (ed.), *Proceedings of the 2nd International Conference Fracture Mechanics of Concrete Structures (FraMCoS-2)*, Aedificatio Publishers, Freiburg, pp. 575–684.

Dempsey, J. P., Adamson, R. M., & Mulmule, S. V. (1999). Scale effect on the insitu tensile strength and failure of first-year sea ice at resolute, NWR. *Int. J. Fract.* **95**, 347–366.

Department of National Defense of Canada (2007). *Technical Airworthiness Manual (TAM)*. Technical Report Document no. C-05-005-001/AG-001, 530 pp.

Desmorat, R., & Leckie, F. A. (1998). Singularities in bimaterials: Parametric study of an isotropic/anisotropic joint. *Eur. J. Mech. A: Solids* **17**, 33–52.

Diao, X., Mai, L. Y., & Mai, Y.-W. (1995). A statistical model of residual strength and fatigue life of composite laminates. *Comp. Sci. Technol.* **54**, 329–336.

Dimov, I. T. (2008). *Monte Carlo Methods for Applied Scientists*. World Scientific, Singapore.

Dos Santos, C., Strecker, K., Piorino Neto, F., de Macedo Silva, O. M., Baldacum, S. A., & da Silva., C. R. M. (2003). Evaluation of the reliability of Si_3N_4-Al_2O_3-CTR_2O_3 ceramics through Weibull analysis. *Mater. Res.* **6**(4), 463–467.

Duckett, K. (2005). Risk analysis and the acceptable probability of failure. *Struct. Eng.* **83**(15), 25–26.

Duffaut, P. (2013). The traps behind the failure of Malpasset arch dam, France, in 1959. *J. Rock Mech. Geotech. Eng.* **5**, 335–341.

Duffy, S. F., Janosik, L. A., Wereszczak, A. A., Suzuki, A., Lamon, J., & Thomas, D. J. (2003). Life prediction of structural components. In *Progress in Ceramic Gas Turbine Development*, Vol. 2. ASME Press, New York, pp. 553–607.

Dugdale, D. S. (1960). Yielding of steel sheets containing slits. *J. Mech. Phys. Solids* **8**, 100–108.

Dumin, D. J., Maddux, J. R., Scott, R. S., & Subramoniam, R. (1994). A model relating wearout to breakdown in thin oxides. *IEEE Trans. Electron Dev.* **41**(9), 1570–1579.

Dunn, M. L., Hui, C. Y., Labossiere, P. E. W., & Lin, Y. Y. (2001). Small scale geometric and material features at geometric discontinuities and their role in fracture analysis. *Int. J. Fract.* **110**(2), 101–121.

Dunn, M. L., Suwito, W., & Cunningham, S. J. (1996). Stress intensities at notch singularities. *Eng. Fract. Mech.* **57**(4), 417–430.

Dunn, M. L., Suwito, W., & Cunningham, S. J. (1997). Fracture initiation at sharp notches: Correlation using critical stress intensities. *Int. J. Solids Struct.* **34**(29), 3873–3883.

Eliáš, J., Vořechovský, M., Skoček, J., & Bažant, Z. P. (2015). Stochastic discrete meso-scale simulations of concrete fracture: Comparison to experimental data. *Eng. Fract. Mech.* **135**(1), 1–16.

Elishakoff, I. (1983). *Probabilistic Methods in the Theory of Structures*. John Wiley & Sons, Toronto, Canada.

Ellingwood, B. R. (2006). Mitigating risk from abnormal loads and progressive collapse. *J. Perfor. Constr. Fact.* **20**(4), 315–323.

Eringen, A. C. (1966). A unified theory of thermomechanical materials. *Int. J. Eng. Sci.* **4**, 179–202.

Eringen, A. C. (1972). Linear theory of nonlocal elasticity and dispersion of plane waves. *Int. J. Eng. Sci.* **10**, 425–425.

Eringen, A. C., & Edelen, D. G. B. (1972). On nonlocal elasticity. *Int. J. Eng. Sci.* **10**, 233–248.

Evans, A. G. (1972). A method for evaluating the time-dependent failure characteristics of brittle materials – and its application to polycrystalline alumina. *J. Mater. Sci.* **7**, 1173–1146.

Evans, A. G. (1974). Analysis of strength degradation after sustained loading. *J. Am. Ceram. Soc.* **57**(9), 410–411.

Evans, A. G., & Fu, Y. (1984). The mechanical behavior of alumina. In *Fracture in Ceramic Materials*. Elsevier, Amsterdam, pp. 56–88.

Evans, R. H., & Marathe, M. S. (1968). Microcracking and stress-strain curves in concrete for tension. *Mater. Struct.* **1**, 61–64.

Eyring, H. (1936). Viscosity, plasticity, and diffusion as examples of absolute reaction rates. *J. Chem. Phys.* **4**, 283–291.

Feller, W. (1957). *Introduction to Probability Theory and Its Applications*, 2nd edition, John Wiley & Sons, New York.

Fett, T. (1991). A fracture-mechanical theory of subcritical crack growth in ceramics. *Int. J. Fract.* **54**, 117–130.

Fett, T., & Munz, D. (1991). Static and cyclic fatigue of ceramic materials. In *Ceramics Today – Tomorrow's Ceramics*. Elsevier, Amsterdam, pp. 1827–1835.

Fett, T., & Munz, D. (1993). Difference between static and cyclic fatigue effects in alumina. *J. Mater. Sci. Lett.* **12**, 352–354.

Fisher, R. A., & Tippett, L. H. C. (1928). Limiting form of the frequency distribution the largest and smallest number of a sample. *Proc. Cambridge Philos. Soc.* **24**, 180–190.

Fréchet, M. (1927). Sur la loi de probailité de l'écart maximum. *Ann. Soc. Polon. Math. (Cracow)* **6**, 93.

Freed, Y., & Banks-Sills, K. (2008). A new cohesive zone model for mixed mode interface fracture in bimaterials. *Eng. Fract. Mech.* **75**, 4583–4593.

Freudenthal, A. M. (1956). Safety and probability of structural failure. *ASCE Trans.* **121**, 1337–1397.

Freudenthal, A. M. (1968). Statistical approach to brittle fracture. In *Fracture: An Advanced Treatise*, Vol. 2. Academic Press. New York, pp. 591–619.

Freudenthal, A. M. (1981). *Selected Papers by Alfred M. Freudenthal: Civil Engineering Classics*. American Society of Civil Engineers, New York.

Freudenthal, A. M., Garrelts, J. M., & Shinozuka, M. (1966). The analysis of structural safety. *J. Struct. Eng. Div. ASCE* **92**(ST1), 267–324.

Galilei, G. (1638). *Discorsi e Dimostrazioni Matematiche Intorno a Due Nuove Scienze*. Elsevirii, Leiden.

Gauss, C. F. (1809). *The Heavenly Bodies Moving about the Sun in Conic Sections*. Reprint, Dover, New York, 1963.

Geers, M. G. D., Peerlings, R. H. J., Brekelmans, W. A. M., & de Borst, R. (2000). Phenomenological nonlocal approaches based on implicit gradient-enhanced damage. *Acta Mechan.* **144**(1–2), 1–15.

Gettu, R., Bažant, Z. P., & Karr, M. E. (1990). Fracture properties and brittleness of high-strength concrete. *ACI Mater. J.* **87**, 608–618.

Ghidini, G., Brazzelli, D., Clementi, C., & Pellizzer, F. (1999). Charge trapping mechanism under dynamic stress and its effect on failure time. In *Proceedings of the 37th Annual Reliability Physics Symposium*. IEEE, San Diego.

Ghoniem, N. M., Busso, E. P., Kioussis, N., & Huang, H. (2003). Multiscale modeling of nanomechanics and micromechanics: An overview. *Philos. Mag.* **83**(31–34), 3475–3528.

Glasstone, S., Laidler, K. J., & Eyring, H. (1941). *The Theory of Rate Processes*. McGraw-Hill, New York.

Gnedenko, B. V. (1943). Sur la distribution limite du terme maximum d'une série aléatoire. *Ann. Math.* **44**, 423–453.

Gomez, F. J., & Elices, M. (2003). A fracture criterion for sharp V-notched samples. *Int. J. Fract.* **123**(3–4), 163–175.

Graham-Brady, L. L., Arwade, S. R., Corr, D. J., Gutiérrez, M. A., Breysse, D., Grigoriu, M., & Zabara, N. (2006). Probability and materials: From nano- to macro-scale: A summary. *Prob. Eng. Mech.* **21**, 193–199.

Grassl, P., & Bažant, Z. P. (2009). Random lattice-particle simulation of statistical size effect in quasi-brittle structures failing at crack initiation. *J. Eng. Mech. ASCE* **135**(2), 85–92.

Grenestedt, J. L., & Hallstrom, S. (1997). Crack initiation from homogeneous and bimaterial corners. *J. Appl. Mech. ASME* **64**, 811–818.

Griffith, A. A. (1921). The phenomenon of rupture in solids. *Philos. Trans.* **221A**, 582–593.

Gross, B. (1996). Least squares best fit method for the three parameter Weibull distribution: Analysis of tensile and bend specimens with volume or surface flaw failure. *NASA Technical Report* **TM-4721**, 1–21.

Gumbel, E. J. (1958). *Statistics of Extremes*. Columbia University Press, New York.

Györgyi, G., Moloney, N. R., Ozogány, K., & Rácz, Z. (2008). Finite-size scaling in extreme statistics. *Phys. Rev. Lett.* **100**, 210601.

Hahn, H. T., & Kim, R. Y. (1975). Proof testing of composite materials. *J. Comp. Mater.* **9**, 297–311.

Haldar, A., & Mahadevan, S. (2000*a*). *Probability, Reliability, and Statistical Methods in Engineering Design*. John Wiley & Sons, New York.

Haldar, A., & Mahadevan, S. (2000*b*). *Reliability Assessment Using Stochastic Finite Element Analysis*, John Wiley & Sons, New York.

Halpin, J. C., Johnson, T. A., & Waddoups, M. E. (1972). Kinetic fracture models and structural reliability. *Int. J. Fract. Mech.* **8**, 465–466.

Harlow, D. G., & Phoenix, S. L. (1978*a*). The chain-of-bundles probability model for the strength of fibrous materials I: Analysis and conjectures. *J. Comp. Mater.* **12**, 195–214.

Harlow, D. G., & Phoenix, S. L. (1978*b*). The chain-of-bundles probability model for the strength of fibrous materials II: A numerical study of convergence. *J. Comp. Mater.* **12**, 314–334.

Harlow, D. G., Smith, R. L., & Taylor, H. M. (1983). Lower tail analysis of the distribution of the strength of load-sharing systems. *J. Appl. Prob.* **20**, 358–367.

Hasofer, A. M., & Lind, N. C. (1974). Exact and invariant second moment code format. *J. Eng. Mech. Div. ASCE* **100**(EM1), 111–121.

Hastings, M. B. (2001). Fractal to nonfractal phase transition in the dielectric breakdown model. *Phys. Rev. Lett.* **87**, 175502.

Heilmann, H., Hilsdorf, H., & Finsterwalder, K. (1969). Festigkeit und Verformung von Beton unter Zugspannungen. Heft 203, Deutscher Ausschuss für Stahlabeton.

Hill, R. (1963). Elastic properties of reinforced solids: Some theoretical principles. *J. Mech. Phys. Solids* **11**, 357–362.

Hillerborg, A., Modéer, M., & Petersson, P. E. (1976). Analysis of crack formation and crack growth in concrete by means of fracture mechanics and finite elements. *Cem. Concr. Res.* **6**, 773–782.

Hillig, W. B., & Charles, R. J. (1964). Surfaces, stress-dependent surface reaction, and strength. In V. F. Zackay (ed.) *Proceedings of the 2nd Berkeley International Materials Conference*. John Wiley and Sons, New York.

Hoover, C. G., & Bažant, Z. P. (2013). Comprehensive concrete fracture tests: Size effects of types 1 & 2, crack length effect and postpeak. *Eng. Fract. Mech.* **110**, 281–289.

Hoover, C. G., & Bažant, Z. P. (2014*a*). Cohesive crack, size effect, crack band and work-of-fracture models compared to comprehensive concrete fracture tests. *Int. J. Fract.* **187**(1), 133–143.

Hoover, C. G., & Bažant, Z. P. (2014*b*). Comparison of the Hu-Duan boundary effect model to size-shape effect law for quasi-brittle fracture based on new comprehensive fracture tests. *J. Eng. Mech. ASCE* **140**(3), 480–486.

Hoover, C. G., & Bažant, Z. P. (2014*c*). Universal size-shape effect law based on comprehensive concrete fracture tests. *J. Eng. Mech. ASCE* **140**(3), 473–479.

Hoover, C. G., Bažant, Z. P., Vorel, J., Wendner, R., & Hubler, M. H. (2013). Comprehensive concrete fracture tests: Description and results. *Eng. Fract. Mech.* **114**, 92–103.

Hoshide, T. (1995). Proof testing and subsequent fatigue characteristics in ceramics. In H. Kishimoto, T. Hoshide, & N. Okabe (eds.), *Cyclic Fatigue in Ceramics*. Elsevier and The Society of Materials Science, Japan, pp. 207–232.

Hrennikoff, A. (1941). Solution of problems of elasticity by the framework method. *J. Appl. Mech.* **12**, 169–175.

Hudson, J. A., Brown, E. T., & Fairhurst, C. (1971). Optimizing the control of rock failure in servo-controlled laboratory tests. *Rock Mech.* **3**, 217–224.

Hughes, B. P., & Chapman, G. P. (1966). The complete stress-strain curve for concrete in direct tension. *RILEM Bull. (Paris)* **30**, 95–97.

Hutchinson, J. W., & Suo, Z. (1992). Mixed-mode cracking in layered materials. *Adv. Appl. Mech.* **29**, 63–191.

Ibnabdeljalil, M., & Phoenix, S. L. (1995). Creep rupture of brittle matrix composite reinforced with time dependent fibers: scalings and Monte Carlo simulations. *J. Mech. Phys. Solids* **43**(6), 897–931.

Iguro, M., Shioya, T., Nojiri, Y., & Akiyama, H. (1985). Experimental studies on shear strength of large reinforced concrete beams under uniformly distributed load. *Conc. Lib. Int.* **5**, 137–154.

Ironside, J. G., & Swain, M. V. (1998). Ceramics in dental restorations – A review and critical issues. *J. Austral. Ceram. Soc.* **34**(2), 78–91.

Irwin, G. R. (1958). Fracture. In W. Flügge (ed.), *Handbuch der Physik*, Vol. VI. Springer-Verlag, Berlin, pp. 551–590.

Jirásek, M., & Bažant, Z. P. (1995*a*). Macroscopic fracture characteristics of random particle systems. *Int. J. Fract.* **69**, 201–228.

Jirásek, M., & Bažant, Z. P. (1995*b*). Particle model for quasibrittle fracture and application to sea ice. *J. Eng. Mech. ASCE* **121**, 1016–1025.

Kanninen, M. F., & Popelar, C. H. (1985). Advanced fracture mechanics. Oxford University Press, New York.

Kaplan, M. F. (1961). Crack propagation and the fracture concrete. *ACI J.* **58**(11), 591–610.

Karihaloo, B. L., Abdalla, H. M., & Xiao, Q. Z. (2003). Size effect in concrete beams. *Eng. Fract. Mech.* **70**(7–8), 979–993.

Kausch, H. H. (1978). *Polymer Fracture*, Springer-Verlag, Berlin Heidelberg, New York.

Kawai, T. (1978). New discrete element models and their application to seismic response analysis of structures. *Nucl. Eng. Des.* **48**, 207–229.

Kawakubo, T. (1995). Fatigue crack growth mechanics in ceramics. In H. Kishimoto, T. Hoshide, & N. Okabe (eds.) *Cyclic Fatigue in Ceramics*. Elsevier and The Society of Materials Science, Japan, pp. 123–137.

Kaxiras, E. (2003). *Atomic and Electronic Structure of Solids*. Cambridge University Press, Cambridge.

Kesler, C. E., Naus, D. J., & Lott, J. L. (1972). Fracture mechanics – Its applicability to concrete. In *Proceedings of International Conference on the Mechanical Behavior of Materials*, Kyoto, Japan, pp. 113–124.

Kfouri, A. P., & Rice, J. R. (1977). Elastic-plastic separation energy rate for crack advance in finite growth steps. In D. M. R. Taplin (ed.), *Fracture 1977 (Proceedings of the 4th International Conference on Fracture, ICF4)*, Vol. 1, University of Waterloo, Canada, pp. 43–59.

Khare, R., Mielke, S. L., Paci, J. T., Zhang, S. L., Ballarini, R., Schatz, G. C., & Belytschko, T. (2007). Coupled quantum mechanical/molecular mechanical modeling of the fracture of defective carbon nanotubes and graphene sheets. *Phys. Rev. B* **75**(7), 075412.

Khare, R., Mielke, S. L., Schatz, G. C., & Belytschko, T. (2008). Multiscale coupling schemes spanning the quantum mechanical, atomistic forcefield, and continuum regimes. *Comp. Methods Appl. Mech. Eng.* **197**, 3190–3202.

Kim, Y.-H., & Lee, J. C. (2004). Reliability characteristics of high-*k* dielectrics. *Microelectron. Reliabil.* **44**, 183–193.

Kirane, K., & Bažant, Z. P. (2015*a*). Microplane damage model for fatigue of quasibrittle materials: Subcritical crack growth, lifetime and residual strength. *Int. J. Fatigue* **70**, 93–105.

Kirane, K., & Bažant, Z. P. (2015*b*). Size effect in Paris law for quasibrittle materials analyzed by the microplane constitutive model M7. *Mech. Res. Commune.* **68**, 627–631.

Kirchner, H. P., & Walker, R. E. (1971). Delayed fracture of alumina ceramics with compressive surface layers. *Mater. Sci. Eng.* **8**, 301–309.

Kittl, P., & Diaz, G. (1988). Weibull's fracure statistics, or probabilistic strength of materials: State of the art. *Res. Mechan.* **24**, 99–207.

Kittl, P., & Diaz, G. (1989). Some engineering applications of the probabilistic strength of materials. *Appl. Mech. Rev.* **42**, 108–112.

Kittl, P., & Diaz, G. (1990). Size effect on fracture strength in the probabilistic strength of materials. *Reliabil. Eng. Syst. Saf.* **28**, 9–21.

Knauss, W. C. (1973). On the steady propagation of a crack in a viscoelastc sheet; Experiment and analysis. In H. H. Kausch (ed.), *The Deformation in Fracture of High Polymers*. Plenum Press, New York, pp. 501–541.

Knauss, W. C. (1974). On the steady propagation of a crack in a viscoelastc plastic sold. *J. Appl. Mechan. ASME* **41**, 234–248.

Knott, J. F. (1973). *Fundamentals of Fracture Mechanics*. Butterworth, London.

Kramers, H. A. (1941). Brownian motion in a field of force and the diffusion model of chemical reaction. *Physica* **7**, 284–304.

Krausz, A. S., & Krausz, K. (1988). *Fracture Kinetics of Crack Growth*. Kluwer Academic, Dordrecht, the Netherlands

Krayani, A., Pijaudier-Cabot, G., & Dufour, F. (2009). Boundary effect on weight function in nonlocal damage model. *Eng. Fract. Mech.* **76**(14), 2217–2231.

Kröner, E. (1966). Continuum mechanics and range of atomic cohesion forces. In T. Yokobori, T. Kawasaki, & J. Swedlow (eds.), *Proceedings of the 1st International Conference on Fracture*. Japanese Society for Strength and Fracture of Materials. Sendai, Japan, p. 27.

Kröner, E. (1967). Elasticity theory of materials with long range cohesive forces. *Int. J. Solids Struct.* **3**, 731–742.

Labossiere, P. E. W., Dunn, M. L., & Cunningham, S. J. (2002). Application of bimaterial interface corner failure mechanics to silicon/glass anodic bonds. *J. Mech. Phys. Solids* **50**, 405–433.

Le, J.-L. (2011). General size effect on strength of bi-material quasibrittle structures. *Int. J. Fract.* **172**, 151–160.

Le, J.-L. (2012). A finite weakest link model of lifetime distribution of high-*k* gate dielectrics under unipolar AC voltage stress. *Microelectron. Reliabil.* **52**, 100–106.

Le, J.-L. (2015). Size effect on reliability indices and safety factors of quasibrittle structures. *Struct. Saf.* **52**, 20–28.

Le, J.-L., Ballarini, R., & Zhu, Z. (2015). Modeling of probabilistic failure of polycrystalline silicon MEMS structures. *J. Am. Ceram. Soc.* **98**(6), 1685–1697.

Le, J.-L., & Bažant, Z. P. (2009). Strength distribution of dental restorative ceramics: Finite weakest link model with zero threshold. *Dent. Mater.* **25**(5), 641–648.

Le, J.-L., & Bažant, Z. P. (2011). Unified nano-mechanics based probabilistic theory of quasibrittle and brittle structures: II. Fatigue crack growth, lifetime and scaling. *J. Mech. Phys. Solids* **59**, 1322–1337.

Le, J.-L., & Bažant, Z. P. (2012). Scaling of static fracture of quasibrittle structures: Strength, lifetime and fracture kinetics. *J. Appl. Mech. ASME* **79**, 031006.

Le, J.-L., & Bažant, Z. P. (2014). A finite weakest-link model of lifetime distribution of quasibrittle structures under fatigue loading. *J. Math. Mech. Solids* **19**, 56–70.

Le, J.-L., Bažant, Z. P., & Bazant, M. Z. (2009). Crack growth law and its consequences on lifetime distributions of quasibrittle structures. *J. Phys. D: Appl. Phys.* **42**, 214008.

Le, J.-L., Bažant, Z. P., & Bazant, M. Z. (2011). Unified nano-mechanics based probabilistic theory of quasibrittle and brittle structures: I. Strength, crack growth, lifetime and scaling. *J. Mech. Phys. Solids.* **59**, 1291–1321.

Le, J.-L., Bažant, Z. P., & Yu, Q. (2010). Scaling of strength of metal-composite joints: II. Interface fracture analysis. *J. Appl. Mech. ASME* **77**, 011012.

Le, J.-L., Cannone Falchetto, A., & Marasteanu, M. O. (2013). Determination of strength distribution of quasibrittle structures from mean size effect analysis. *Mech. Mater.* **66**, 79–87.

Le, J.-L., & Eliáš, J. (2016). A probabilistic crack band model for quasibrittle fracture. *J. Appl. Mech. ASME* **83**(5), 051005.

Le, J.-L., Eliáš, J., & Bažant, Z. P. (2012). Computation of probability distribution of strength of quasibrittle structures failing at macro-crack initiation. *J. Eng. Mech. ASCE* **138**(7), 888–899.

Le, J.-L., Manning, J., & Labuz, J. F. (2014). Scaling of fatigue crack growth in a rock. *Int. J. Rock Mech. Miner Sci.* **72**, 71–79.

Le, J.-L., Pieuchot, M., & Ballarini, R. (2014). Effect of stress singularity magnitude on scaling of strength of quasibrittle structures. *J. Eng. Mech. ASCE* **140**(5), 04014011.

Le, J.-L., & Xue, B. (2013). Energetic-statistical size effect in fracture of bimaterial hybrid structures. *Eng. Fract. Mech.* **111**, 106–115.

Lee, H. L., Park, S. E., & Hahn, B. S. (1995). Modeling of cyclic fatigue stress for life prediction of structural ceramics. *J. Mater. Sci.* **30**, 2521–2525.

Lee, J., & Fenves, G. L. (1998). Plastic-damage model for cyclic loading of concrete structures. *J. Eng. Mech. ASCE* **124**(8), 892–900.

Lee, S. G., Ma, Y., Thimm, G. L., & Verstraeten, J. (2008). Product lifecycle management in aviation maintenance, repair and overhaul. *Comput. Indust.* **59**, 296–303.

Leguillon, D. (2002). Strength or toughness? A criterion for crack onset at a notch. *Eur. J. Mech. A. Solids* **21**, 61–72.

Leicester, R. H. (1969). The size effect of notches. In *Proceedings of the 2nd Australasian Conference on Mechanics of Structural Materials*. Melbourne, pp. 4.1–4.20.

Lekhnitskii, S. G. (1963). *Theory of Elasticity of an Anisotropic Body*. Holden-Day, San Francisco.

Lemaire, M. (2009). *Structural Reliability*. ISTE and John Wiley & Sons, London and Hokoken.

Lennard-Jones, J. E. (1924). On the determination of molecular fields II. From the equation of state of a gas. *Proc. R. Soc. London A* **106**(738), 463–477.

Levy, M., & Salvadori, M. (1992). *Why Buildings Fall Down*. W. W. Norton, New York.

Li, X., & Marasteanu, M. O. (2006). Investigation of low temperature cracking in asphalt mixtures by acoustic emission. *Road Mater. Pavement Des.* **7/4**, 491–512.

Li, Y.-N., & Bažant, Z. P. (1994). Penetration fracture of ice plate: 2D analysis and size effect. *J. Eng. Mech. ASCE* **120**(7), 1481–1498.

Li, Y.-N., & Bažant, Z. P. (1997). Cohesive crack with rate-dependent opening and viscoelasticity: II. Numerical algorithm, behavior and size effect. *Int. J. Fract.* **86**(3), 267–288.

Lin, Q., & Labuz, J. F. (2013). Fracture of sandstone characterized by digital image correlation. *Int. J. Rock Mech. Miner Sci.* **60**, 235–245.

Liu, D., & Fleck, N. A. (1999). Scale effect in the initiation of cracking of a scarf joint. *Int. J. Fract.* **95**, 66–88.

Liu, W. K., Karpov, E. G., & Park, H. S. (2005). *Nano Mechanics and Materials: Theory, Multiscale Methods and Applications*. John Wiley & Sons, West Sussex, UK.

Liu, X. H., Suo, Z., & Ma, Q. (1999). Split singularities: Stress field near the edge of silicon die on polymer substrate. *Acta Mater.* **47**, 67–76.

Lohbauer, U., Petchelt, A., & Greil, P. (2002). Lifetime prediction of CAD/CAM dental ceramics. *J. Biomed. Mater. Res.* **63**(6), 780–785.

Mahesh, S., & Phoenix, S. L. (2004). Lifetime distributions for unidirectional fibrous composites under creep-rupture loading. *Int. J. Fract.* **127**, 303–360.

Malvar, J. L., & Warren, G. E. (1988). Fracture energy for three-point-bend tests on single-edge-notched beams. *Exp. Mech.* **28**(3), 266–272.

Marder, M. (2004). Effect of atoms on brittle fracture. *Int. J. Fract.* **130**, 517–555.

Mariotte, E. (1686). *Traité du mouvement des eaux*, posthumously edited by M. de la Hire; Eng. transl. by J. T. Desvaguliers, London (1718), p. 249; also *Marriotte's Collected Works*, 2nd ed., The Hague (1740).

Marti, P. (1989). Size effect in double-punch tests on concrete cylinders. *ACI Mater. J.* **86**, 597–601.

McKinney, K. R., & Rice, R. W. (1981). Specimen size effects in fracture toughness testing of heterogeous ceramics by the notch beam test. In S. W. Freiman, & S. R. Fuller (eds.), *Fracture Mechanics Methods for Ceramics, Rocks and Concrete*. ASTM Special Technical Publication No. 745. American Society for Testing Materials, Philadelphia, pp. 118–126.

McPherson, J. W., Khamankar, R. B., & Shanware, A. (2000). Complementary model for intrinsic time-dependent dielectric breakdown in SiO_2 dielectrics. *J. Appl. Phys.* **88**, 5351–5359.

Melchers, R. E. (1987). *Structural Reliability, Analysis & Prediction*. John Wiley & Sons, New York.

Mihashi, H. (1983). Stochastic theory for fracture of concrete. In F. H. Wittmann (ed.), *Fracture Mechanics of Concrete*. Elsevier, Amsterdam, pp. 301–339.

Mihashi, H., & Izumi, M. (1977). Stochastic theory for concrete fracture. *Cem. Concr. Res.* **7**, 411–422.

Mihashi, H., & Zaitsev, J. W. (1981). Statistical nature of crack propagation. In F. H. Wittmann (ed.), Report to RILEM TC50-FMC, Vol. Section 4-2.

Miyakawa, T., Ichiki, T., Mitsuhashi, J., Miyamoto, K., Tada, T., & Koyama, T. (2007). Study of time dependent dielectric breakdown distribution in ultrathin gate oxide. *Jpn. J. Appl. Phys.* **46**(28), 691–692.

Morse, P. M. (1929). Diatomic molecules according to the wave mechanics. II. Vibrational levels. *Phys. Rev.* **34**, 57–64.

Mulmule, S. V., Dempsey, J. P., & Adamson, R. M. (1995). Large-scale in-situ ice fracture experiments–part II: Modeling efforts. In Ice Mechanics – 1995 (ASME *Joint Applied Mechanics and Materials Summer Conference*). ASME, New York.

Munz, D., & Fett, T. (1999). *Ceramics: Mechanical Properties, Failure Behavior, Materials Selection*. Springer-Verlag, Berlin.

Nallathambi, P. (1986). *Fracture Behaviour of Plain Concretes*. PhD thesis, University of New Castle, Callaghan, Australia.

Nielsen, L. F. (1996). Lifetime and residual strength of wood subjected to static and variable load. Technical Report R-6, Department of Structural Engineering and Materials, Technical University of Denmark.

Niemeyer, L., Pietronero, L., & Wiesmann, H. J. (1984). Fractal dimension of dielectric breakdown. *Phys. Rev. Lett.* **52**, 1033–1036.

Nordic Committee for Building Structures (NKB) (1978). Recommendation for loading and safety regulations for structural design. Technical Report 36.

Ogawa, T. (1995). Fatigue crack growth of monolithic and composite ceramics. In H. Kishimoto, T. Hoshide, & N. Okabe (eds.), *Cyclic Fatigue in Ceramics*. Elsevier and The Society of Materials Science, Japan, pp. 167–188.

Okabe, N., & Hirata, H. (1995). High temperature fatigue properties for some types of SiC and Si_3N_4 and the unified strength estimation method. In H. Kishimoto, T. Hoshide, & N. Okabe (eds.), *Cyclic Fatigue in Ceramics*. Elsevier and The Society of Materials Science, Japan, pp. 245–276.

Omeltchenko, A., Yu, J., Kalia, R. K., & Vashishta, P. (1997). Crack front propagation and fracture in a graphite sheet: A molecular dynamics study on parallel computers. *Phys. Rev. Lett.* **78**(11), 2148–2151.

Ortiz, M., & Pandolfi, A. (1999). Finite-deformation irreversible cohesive elements for three-dimensional crack-propagation analysis. *Int. J. Numer. Methods Eng.* **44**(9), 1267–1282.

Palmer, A. C., & Rice, J. R. (1973). The growth of slip surfaces on the progressive failure of over-consolidates clay. *Proc. R. Soc. London A.* **332**(527–548).

Pang, S.-D., Bažant, Z. P., & Le, J.-L. (2008). Statistics of strength of ceramics: Finite weakest link model and necessity of zero threshold. *Int. J. Fract.* **154**, 131–145.

Paris, P. C., & Erdogan, F. (1963). A critical analysis of crack propagation law. *J. Basic Eng.* **85**, 528–534.

Park, K., Paulino, G. H., & Roesler, J. R. (2009). A unified potential-based cohesive crack model for mixed-mode fracture. *J. Mech. Phys. Solids* **57**(6), 891–908.

Park, S. E., & Lee, H. L. (1997). Prediction of static fatigue life of ceramics. *J. Mater. Sci. Lett.* **16**, 1352–1353.

Peerlings, R. H. J., de Borst, R., Brekelmans, W. A. M., & de Vree, J. H. P. (1996). Gradient enhanced damage for quasi-brittle materials. *Int. J. Numer. Methods Eng.* **39**(19), 3391–3403.

Peerlings, R. H. J., Geers, M. G. D., de Borst, R., & Brekelmans, W. A. M. (2001). A critical comparison of nonlocal and gradient-enhanced softening continua. *Int. J. Solids Struct.* **38**(44–45), 7723–7746.

Peirce, F. T. (1926). Tensile tests of cotton yarns – V. The weakest link. *J. Textile Inst.* **17**, 355–368.

Petch, N. J. (1954). The cleavage strength of polycrystals. *J. Iron Steel Inst.* **174**, 25–28.

Petersson, P. E. (1981). Crack growth and development of fracture zones in plain concrete and similar materials. Report TVBM-1006, Division of Building Materials, Lund Institute of Technology, Lund, Sweden.

Phillips, R. (2001). *Crystals, Defects and Microstructures: Modeling Across Scales.* Cambridge University Press, Cambridge.

Phoenix, S. L. (1978*a*). The asymptotic time to failure of a mechanical system of parallel members. *SIAM J. Appl. Math.* **34**(2), 227–246.

Phoenix, S. L. (1978*b*). Stochastic strength and fatigue of fiber bundles. *Int. J. Fract.* **14**(3), 327–344.

Phoenix, S. L., Ibnabdeljalil, M., & Hui, C.-Y. (1997). Size effects in the distribution for strength of brittle matrix fibrous composites. *Int. J. Solids Struct.* **34**(5), 545–568.

Phoenix, S. L., & Tierney, L.-J. (1983). A statistical model for the time dependent failure of unidirectional composite materials under local elastic load-sharing among fibers. *Eng. Fract. Mech.* **18**(1), 193–215.

Pietronero, L., & Wiesmann, H. J. (1988). From physical dielectric breakdown to the stochastic fractal model. *Z. Phys. B: Condensed Matter* **70**, 87–93.

Pietruszczak, S., & Mróz, Z. (1981). Finite element analysis of deformation of strain-softening materials. *Int. J. Numer. Methods Eng.* **17**, 327–334.

Pijaudier-Cabot, G., & Bažant, Z. P. (1987). Nonlocal damage theory. *J. Eng. Mech. ASCE* **113**(10), 1512–1533.

Planas, J., & Elices, M. (1988). Conceptual and experimental problems in the determination of the fracture energy of concrete. In *Proceedings of International Workshop on "Fracture Toughness and Fracture Energy, Test Methods for Concrete and Rock."* Tohoku University, Sendai, Japan, pp. 203–212.

Planas, J., & Elices, M. (1989). Size effect in concrete structures: Mathematical approximation and experimental validation, In J. Mazars & Z. P. Bažant (eds.), *Cracking and Damage, Strain Localization and Size Effect (Proceedings of France–U.S. Workshop).* Elsevier, Amsterdam, pp. 462–476.

Qian, Z., & Akisanya, A. R. (1998). An experimental investigation of failure initiation in bonded joints. *Acta Mater.* **46**, 4895–4904.

Rackwitz, R. (1976). Practical probabilistic approach to design. Bulletin No. 112, Comité European du Béton.

Rackwitz, R., & Fiessler, B. (1976). Note on discrete safety checking when using nonnormal stochastic models for basic variables. In *Load Project Working Session*, MIT, Cambridge, MA.

Rackwitz, R., & Fiessler, B. (1978). Structural reliability under combined random load sequences. *Comp. Struct.* **9**(5), 484–494.

Redner, S. (2001). *A Guide to First-Passage Processes*. Cambridge University Press, Cambridge.

Reedy E. D., Jr. (2000). Comparison between interface corner and interfacial fracture analysis of an adhesively-bonded butt joint. *Int. J. Solids Struct.* **37**, 2429–2442.

Rezakhani, R., & Cusatis, G. (2016). Asymptotic expansion homogenization of discrete fine-scale models with rotational degrees of freedom for the simulation of quasi-brittle materials. *J. Mech. Phys. Solids* **88**, 320–345.

Rice, J. R. (1967). Mechanics of crack tip deformation and extension by fatigue. In *Fatigue Crack Propagation*. Special Technical Publication 415. ASTM, Philadelphia, pp. 247–309.

Rice, J. R. (1968a). In *Fracture, Mathematical analysis of the mechanics of fracture* H. Liebowitz, ed., Vol. 2, Academic Press, 192–308.

Rice, J. R. (1968b). Path independent integral and approximate analysis of strain concentrations by notches and cracks. *J. Appl. Mech. ASME* **35**, 379–386.

Rice, J. R. (1988). Elastic fracture mechanics concepts for interface cracks. *J. Appl. Mech. ASME* **55**, 98–103.

RILEM TC 89-FMT (1990). Size effect method for determining fracture energy and process zone of concrete. *Mater. Struct.* **23**, 461–465.

RILEM TC QFS (2005). Quasibrittle fracture scaling and size effect – Final report. *Mater. Struct.* **37**, 547–568.

Rinne, H. (2009). *The Weibull Distribution. A Handbook.* CRC Press, Boca Raton, London/New York.

Risken, H. (1989). *The Fokker-Planck Equation*. Springer-Verlag, Berlin.

Ritchie, R. O. (2005). Incomplete self-similarity and fatigue crack growth. *Int. J. Fract.* **132**, 197–203.

Ritchie, R. O., Knott, J. F., & Rice, J. R. (1973). On the relation between critical tensile stress and fracture toughness in mild steel. *J. Mech. Phys. Solids* **21**, 395–410.

Rocco, C. G. (1995). *Size Dependence and Fracture Mechanisms in the Diagonal Compression Splitting Test.* PhD thesis, Universidad Politecnica de Madrid.

Rüsch, H., & Hilsdorf, H. (1963). Deformation characteristics of concrete under axial tension. Voruntersuchungen bericht (preliminary report), Munich.

Sabnis, G. M., & Mirza, S. M. (1979). Size effect in model concretes. *J. Struct. Eng. Div. ASCE* **105**(6), 1007–1020.

Sakai, T., & Fujitani, K. (1989). A statistical aspect on fatigue behavior of alumina ceramics in rotating bending. *Eng. Fract. Mech.* **32**(4), 653–664.

Sakai, T., & Hoshide, T. (1995). Statistical aspect on fatigue fracture of structural ceramics under cyclic loads. In H. Kishimoto, T. Hoshide, & N. Okabe (eds.), *Cyclic Fatigue in Ceramics*. Elsevier and The Society of Materials Science, Japan, pp. 189–206.

Salem, J. A., Nemeth, N. N., Powers, L. P., & Choi, S. R. (1996). Reliability analysis of uniaxially ground brittle materials. *J. Eng. Gas Turbines & Power* **118**, 863–871.

Salviato, M., Chau, V. T., Li, W., Bažant, Z. P., & Cusatis, G. (2016). Direct testing of gradual postpeak softening of fracture specimens of fiber composites stabilized by enhanced grip stiffness and mass. *J. Appl. Mech. ASME*, **83**, 11103 (16 pp.).

Salviato, M., Kirane, K., & Bažant, Z. P. (2014). Statistical distribution and size effect of residual strength after a period of constant load. *J. Mech. Phys. Solids* **64**, 440–454.

Schauffert, E. A., & Cusatis, G. (2011). Lattice discrete particle model for fiber-reinforced concrete. I: Theory. *J. Eng. Mech. ASCE* **138**(7), 826–833.

Schauffert, E. A., Cusatis, G., Pelessone, D., O'Daniel, J. L., & Baylot, J. T. (2011). Lattice discrete particle model for fiber-reinforced concrete. II: Tensile fracture and multiaxial loading behavior. *J. Eng. Mech. ASCE* **138**(7), 834–841.

Schlangen, E., & van Mier, J. G. M. (1992). Experimental and numerical analysis of micromechanisms of fracture of cement-based composites. *Cem. Concr. Res.* **14**, 105–118.

Sedov, L. I. (1959). *Similarity and Dimensional Methods in Mechanics*. Academic Press, New York.

Seweryn, A. (1994). Brittle fracture criterion for structures with sharp notches. *Eng. Fract. Mech.* **47**, 673–681.

Sglavo, V. M., & Green, D. J. (1995). Threshold stress intensity factor in soda-lime silicate glass by interrupted static fatigue test. *J. Eur. Ceram. Soc.* **16**, 645–651.

Sglavo, V. M., & Renzi, S. (1999). Fatigue limit in borosilicate glasses by interrupted static fatigue test. *Phys. Chem. Glasses* **40**(2), 79–84.

Shioya, T., & Akiyama, H. (1994). Application to design of size effect in reinforced concrete structures. In H. Mihashi, H. Okamura, & Z. P. Bažant (eds.), *Size Effect in Concrete Structures*. E & FN Spon, London, pp. 409–416.

Sinclair, G. B., Okajima, M., & Griffin, J. H. (1984). Path independent integrals for computing stress intensity factors at sharp notches in elastic plates. *Int. J. Numer. Methods Eng.* **20**(6), 999–1008.

Smith, J., Cusatis, G., Pelessone, D., Landis, E., O'Daniel, J. L., & Baylot, J. (2014). Discrete modeling of ultra-high-performance concrete with application to projectile penetration. *Int. J. Impact Eng.* **65**, 13–32.

Smith, R. L. (1982). The asymptotic distribution of the strength of a series-parallel system with equal load sharing. *Ann. Probab.* **10**(1), 137–171.

Soong, T. T. (2004). *Fundamentals of Probability and Statistics for Engineers*. John Wiley & Sons, Hoboken, NJ.

Stanley, P., & Inanc, E. Y. (1985). Assessment of surface strength and bulk strength of a typical brittle material. In S. Eggwertz & N. Lind (eds.), *Probabilistic Methods. I. The Mechanics of Solids and Structures*. Springer-Verlag, Berlin, pp. 231–251.

Stathis, J. H. (1999). Percolation models for gate oxide breakdown. *J. Appl. Phys.* **86**(10), 5757–5766.

Stillinger, F. H., & Weber, T. A. (1985). Computer simulation of local order in condensed phases of silicon. *Phys. Rev. B* **31**, 5256–5271.

Stroh, A. N. (1958). Dislocations and cracks in anisotropic elasticity. *Philos. Mag.* **3**, 625–646.

Studarta, A. R., Filser, F., Kochera, P., & Gauckler, L. J. (2007). Fatigue of zirconia under cyclic loading in water and its implications for the design of dental bridges. *Dent. Mater.* **23**, 106–114.

Studarta, A. R., Filser, F., Kochera, P., Lüthy, H., & Gauckler, L. J. (2007). Cyclic fatigue in water of veneer framework composites for all-ceramic dental bridges. *Dent. Mater.* **23**, 177–185.

Suné, J. (2001). New physics-based analytic approach to the thin-oxide breakdown statistics. *IEEE Trans. Electron Dev. Lett.* **22**(6), 296–298.

Suné, J., Placencia, I., Barniol, N., Farrés, E., Martín, F., & Aymerich, X. (1990). On the breakdown statistics of very thin sio$_2$ films. *Thin Solid Films* **185**, 347–362.

Suné, J., Tous, S., & Wu, E. Y. (2009). Analytical cell-based model for the breakdown statistics of multilayer insulator stacks. *IEEE Trans. Electron Dev. Lett.* **30**(12), 1359–1361.

Suresh, S. (1998). *Fatigue of Materials*. Cambridge University Press, Cambridge.

Syroka-Korol, E., & Tejchman, J. (2014). Experimental investigation of size effect in reinforced concrete beams failing by shear. *Eng. Struct.* **58**, 63–78.

Tadmor, E. B., & Miller, R. E. (2011). *Modeling Materials: Continuum, Atomistic and Multiscale Techniques*. Cambridge University Press, Cambridge.

Tadmor, E. B., Ortiz, M., & Phillips, R. (1996). Quasicontinuum analysis of defects in solids. *Philos. Mag.* **73**, 1529–1563.

Tang, T., Bažant, Z. P., Yang, S., & Zollinger, D. (1996). Variable-notch one-size test method for fracture energy and process zone length. *Eng. Fract. Mech.* **55**(3), 383–404.

Tersoff, J. (1988). New empirical approach for the structure and energy of covalent systems. *Phys. Rev. B* **37**, 6991.

Thomas, D. J., Verrilli, M., & Calomino, A. (2002). Stress-life behavior of a C/SiC composite in a low partial pressure of oxygen environment part III: Life prediction using probabilistic residual strength rupture model. In H.-T. Lin and M. Singh (eds.), *Ceramic Engineering and Science Proceedings*. The American Ceramic Society, Westerville, Vol. 23(3), pp. 453–460.

Thouless, M. D., Hsueh, C. H., & Evans, A. G. (1983). A damage model of creep crack growth in polycrystals. *Acta Metall.* **31**(10), 1675–1687.

Timoshenko, S. P., & Goodier, J. N. (1951). *Theory of Elasticity*. McGraw-Hill, New York Toronto London.

Tinschert, J., Zwez, D., Marx, R., & Ausavice, K. J. (2000). Structural reliability of alumina-, feldspar-, leucite-, mica- and zirconia-based ceramics. *J. Dent.* **28**, 529–535.

Tippett, L. H. C. (1925). On the extreme individuals and the range of samples. *Biometrika* p. 364.

Tobolsky, A., & Erying, H. (1943). Mechanical properties of polymeric materials. *J. Chem. Phys.* **11**, 125–134.

Turner, C. H., Wang, T., & Burr, D. B. (2001). Shear strength and fatigue properties of human cortical bone determined from pure shear tests. *Calcif. Tissue Int.* **69**, 373–378.

Turos, M. I., Cannone Falchetto, A., Tebaldi, G., & Marasteanu, M. O. (2012). Determining the flexural strength of asphalt mixtures using the bending beam rheometer. In A. Scarpas, N. Kringos, I. Al-Qadi, & A. Loizos (eds.), *Proceedings of the 7th RILEM Conference on Cracking in Pavements*. Springer, Netherlands, pp. 11–20.

van der Hofstad, R., & Redig, F. (2006). Maximal clusters in non-critical percolation and related models. *J. Stat. Physics* **122**(4), 671–703.

Vanmarcke, E. (2010). *Random Fields Analysis and Synthesis*. World Scientific, Singapore.

var der Hofstad, R., & Redig, F. (2006). Maximal clusters in non-critical percolation and related models. *J. Stat. Phys.* **122**, 671–703.

van Vliet, M. R., & van Mier, J. G. (2000). Experimental investigation of size effect in concrete and sandstone under uniaxial tension. *Engineering Fracture Mechanics*, **65**(2–3): 165–188.

Vashy, A. (1892). Sur les lois de similitude en physique. *Annales Telégraphiques.* **19**, 25–28.

von Mises, R. (1936). La distribution de la plus grande de n valeurs. *Rev. Math. Union Interbalcanique* **1**, 1.

Vořechovský, M., & Sadílek, V. (2008). Computational modeling of size effects in concrete specimens under uniaxial tension. *Int. J. Fract.* **154**, 27–49.

Wagner, H. D. (1989). Stochastic concepts in the study of size effects in the mechanical strength of highly oriented polymeric materials. *J. Polym. Sci.* **27**, 115–149.

Wagner, H. D., Schwartz, P., & Phoenix, S. L. (1986). Lifetime statistics for single Kevlar 49 filaments in creep-rupture. *J. Polym. Sci.* **21**, 1868–1878.

Walraven, J. (1995). Size effects: Their nature and their recognition in building codes. *Studi Ricerche (Politecnico di Milano)* **16**, 113–134.

Walsh, P. F. (1972). Fracture of plain concrete. *Indian Concrete J.* **46**(11), 469–470, 476.

Walsh, P. F. (1976). Crack initiation in plain concrete. *Mag. Concr. Res.* **28**, 37–41.

Wanger, H. D., Phoenix, S. L., & Schwartz, P. (1984). A study of statistical variability in the strength of single aramid filaments. *J. Comp. Mater.* **18**, 312–338.

Wawersik, W., & Fairhurst, C. (1970). A study of brittle rock fracture in laboratory compression experiments. *Int. J. Rock Mech. Miner. Sci.* **7**(5), 561–575.

Weertman, J. (1966). Rate of growth of fatigue cracks calculated from the theory of infinitesimal dislocation distributed on a crack plane. *Int. J. Fract.* **2**, 460–467.

Weibull, W. (1939). The phenomenon of rupture in solids. *Proc. R. Sweden Inst. Engrg. Res.* **153**, 1–55.

Weibull, W. (1951). A statistical distribution function of wide applicability. *J. Appl. Mech. ASME* **153**(18), 293–297.

Wiederhorn, S. M. (1967). Influence of water vapor on crack propagation in soda-lime glass. *J. Am. Ceram. Soc.* **50**(8), 407–414.

Wiederhorn, S. M., & Bolz, L. H. (1970). Stress corrosion and static fatigue of glass. *J. Am. Ceram. Soc.* **53**(10), 543–548.

Wilk, G. D., Wallace, R. M., & Anthony, J. M. (2001). High-*k* gate dielectrics: Current status and materials properties considerations. *J. Appl. Phys.* **89**(10), 5243–5275.

Williams, M. L. (1952). Stress singularities resulting from various boundary conditions in angular corners of plates in extension. *J. Appl. Mech.* **74**, 526–528.

Williams, T., & Baxer, S. C. (2006). A framework for stochastic mechanics. *Prob. Eng. Mech.* **21**(3), 247–255.

Wisnom, M. (1992). The relationship between tensile and flexural strength of unidirectional composite. *J. Comp. Mater.* **26**, 1173–1180.

Witten, T. A., & Sander, L. M. (1981). Diffusion-limited aggregation, a kinetic critical phenomenon. *Phys. Rev. Lett.* **47**, 1400–1403.

Wittmann, F. H., ed. (1995). *Fracture Mechanics of Concrete Structures, Proceedings of Second International Conference*. Aedificatio Publishers, Freiburg.

Wnuk, M. P. (1974). Quasi-static extension of a tensile crack contained in viscoelasto plastic solid. *J. Appl. Mech. ASME* **41**, 234–248.

Xu, M., Tabarraei, A., Paci, J. T., Oswald, J., & Belytschko, T. (2012). A coupled quantum/continuum mechanics study of graphene fracture. *Int. J. Fract.* **173**(2), 163–173.

Xu, X. F. (2007). A multiscale stochastic finite element method on elliptic problems involving uncertainties. *Comp. Methods Appl. Mech. Eng.* **196**, 2723–2736.

Yang, J. N. (1978). Fatigue and residual strength degradation for graphite/epoxy composites under tension-compression cyclic loadings. *J. Comp. Mater.* **12**, 19–39.

Yang, J. N., & Liu, M. D. (1977). Residual strength degradation model and theory of periodic proof tests for graphite/epoxy laminates. *J. Comp. Mater.* **11**, 176–202.

Yavuz, B. O., & Tessler, R. E. (1993). Threshold stress intensity for crack growth in silicon carbide ceramics. *J. Am. Ceram. Soc.* **76**(4), 1017–1024.

Young, T. (1807). *A Course of Lectures on Natural Philosophy and the Mechanical Arts*, Vol. I. Joseph Johnson, St Paul's Church Yard, London, p. 144.

Yu, Q., Bažant, Z. P., Bayldon, J. M., Le, J.-L., Caner, F. C., Ng, W. H., & Waas, A. M. (2010). Scaling of strength of metal-composite joints: I. Experimental investigation. *J. Appl. Mech. ASME* **77**, 011011.

Yu, Q., Bažant, Z. P., & Le, J.-L. (2013). Scaling of strength of metal-composite joints: III. Numerical simulation. *J. Appl. Mech. ASME* **80**, 054593.

Yu, Q., Le, J.-L., Hubler, M. H., Wendner, R., Cusatis, G., Bažant, Z. P. (2016). Comparison of main models for size effect on shear strength of reinforced and prestressed concrete beams. *Structural Concrete (fib)*, **17**(5), 778–789.

Zaitsev, J. W., & Wittmann, F. H. (1974). A statistical approach to study the mechanical behavior of porous materials under multiaxial state of stress. In *Proceedings of the 1973 Sympoisum on Mechanical Behavior of Materials*. Kyoto, Japan, p. 705.

Zech, B., & Wittmann, F. H. (1977). A complex study on the reliability assessment of the containment of a pwr, Part II. probabilitic approach to describe the behavior of materials. In T. A. Jaeger & B. A. Boley (eds.), *Transactions of the 4th International Conference on Structural Mechanics in Reactor Technology*. Vol. H, J1/11, European Communities, Brussels, Belgium, pp. 1–14.

Zhurkov, S. N. (1965). Kinetic concept of the strength of solids. *Int. J. Fract. Mech.* **1**(4), 311–323.

Zhurkov, S. N., & Korsukov, V. E. (1974). Atomic mechanism of fracture of solid polymer. *J. Polym. Sci.* **12**(2), 385–398.

Zubelewicz, A., & Bažant, Z. P. (1987). Interface modeling of fracture in aggregate composites. *J. Eng. Mech. ASCE* **113**(11), 1619–1630.

Author Index

Subject Index